T0318776

Case Studies in Food Retailing and Distribution

Related titles

Case Studies in the Traditional Food Sector
(ISBN: 9780081010075)

Case Studies in Food Retailing and Distribution
(ISBN: 9780081020371)

Case Studies in the Wine Industry
(ISBN: 9780081009444)

Woodhead Publishing Series in Consumer Science and Strategic Marketing

Case Studies in Food Retailing and Distribution

Series Editors

Alessio Cavicchi

Cristina Santini

Edited by

John Byrom

Dominic Medway

WOODHEAD
PUBLISHING
An imprint of Elsevier

Woodhead Publishing is an imprint of Elsevier
The Officers' Mess Business Centre, Royston Road, Duxford, CB22 4QH, United Kingdom
50 Hampshire Street, 5th Floor, Cambridge, MA 02139, United States
The Boulevard, Langford Lane, Kidlington, OX5 1GB, United Kingdom

British Library Cataloguing-in-Publication Data
A catalogue record for this book is available from the British Library

Library of Congress Cataloging-in-Publication Data
A catalog record for this book is available from the Library of Congress

ISBN: 978-0-08-102037-1 (print)

ISBN: 978-0-08-102038-8 (online)

For information on all Woodhead Publishing publications
visit our website at https://www.elsevier.com/books-and-journals

Working together
to grow libraries in
developing countries

www.elsevier.com • www.bookaid.org

Publisher: Charlotte Cockle
Acquisition Editor: Megan R. Ball
Editorial Project Manager: Karen R. Miller
Production Project Manager: Joy Christel Neumarin Honest Thangiah
Cover Designer: Greg Harris

Typeset by MPS Limited, Chennai, India

Contents

Contributor biographies

Abhishek (abhishek@iima.ac.in) is an Associate Professor in Marketing at IMT Ghaziabad, India. His research interests are in the fields of marketing communications, digital and mobile marketing, and consumer behavior in retail contexts. His research has been accepted for publication in refereed journals like *Journal of Retailing and Consumer Services, Marketing Intelligence and Planning, Asian Case Research Journal, South Asian Journal of Management, Journal of Indian Business Research, International Journal of Indian Culture and Business Management, Vikalpa, Vision*, and *Decision*. He has also written more than twenty cases and prepared six simulation games which have been used in different institutions in India.

Rebecca Abushena (r.abushena@mmu.ac.uk) is a Senior Lecturer in Retail Operations at Manchester Metropolitan University, United Kingdom. She has previously worked as a researcher at Manchester Business School, where she completed her PhD, and in management consultancy. Her research interests include qualitative methodology techniques, multinational retailing firm internationalization, and interfirm linkages.

Paul W. Ballantine (paul.ballantine@canterbury.ac.nz) is a Professor of Marketing and Head of the Business School at the University of Canterbury in Christchurch, New Zealand. His research interests include retailing, consumption behavior (particularly the negative aspects of consumption), and social and ethical issues in marketing. His recent publications have appeared in outlets including the *Journal of Retailing and Consumer Services, Journal of Marketing Management, The International Review of Retail, Distribution and Consumer Research, Journal of Brand Management* and the *Journal of Consumer Behaviour.*

Andrés Barrios (andr-bar@uniandes.edu.co) is an Associate Professor of Marketing at the Universidad de Los Andes, Bogota, Colombia. His research analyses marketing strategies at the intersection between development, poverty, and social conflict. Dr Barrios's work has been published in several outlets including *Journal of Service Research, Journal of Business Research*, and *Journal of Public Policy and Marketing*, among others. He is also a board member of the International Society of Markets and Development.

Pelin Bicen (pbicen@suffolk.edu) is an Associate Professor of Marketing in the Sawyer Business School, Suffolk University, USA. In her research, she explores market systems as a dynamic ensemble of institutions, and how they contribute to the evolution of systems of governance. Some of the questions she focuses on are how do organizations manage the competitive dynamism of markets and what is the role of governance mechanisms in the institutionalization of markets? Her research has been published in a variety of leading marketing and management journals including *Journal of Business and Industrial Marketing, Journal of Business Research, Journal of Marketing Theory and Practice, Journal of Marketing Development and Competitiveness, Communications of the Association for Information Systems, Creativity and Innovation Management,* and *Innovation: Management, Policy, & Practice* among many others. She has also presented her research at many national and international conferences.

Daniel Boller (daniel.boller@unisg.ch) is a Doctoral Student in Marketing at the Institute for Customer Insight, University of St. Gallen, Switzerland, and Visiting Student Researcher at Stanford Graduate School of Business, USA. His research is primarily concerned with consumer decision making in computer-mediated environments.

Kathryn Boys (kaboys@ncsu.edu) is an Assistant Professor in the Department of Agricultural and Resource Economics at North Carolina State University, USA. Her research is in the fields of agribusiness marketing and management, and international trade. In particular, she is interested in issues related to the economics of food safety and quality, linkages between the food system and economic development, and the international trade of agri-food products.

John Byrom (john.byrom@manchester.ac.uk) is a Lecturer in Marketing at the University of Manchester, United Kingdom. His research interests include retail marketing, marketing management and consumer behavior. He research has been published in *European Journal of Marketing, Journal of Marketing Management, Journal of Business Research, Marketing Theory, Cities,* and *The Service Industries Journal,* amongst others, and he has authored chapters in edited books in various areas of marketing.

Schmidt H. Dadzie (dadzieschmidt@yahoo.com) is a graduate of the MSc in International Business Economics from Aalborg University, Denmark, and the Bachelor of Business Administration from KNUST School of Business, Ghana. He is a guest lecturer in research philosophy and quantitative methods at Niels Brock Copenhagen Business College, Denmark. His research interests lie in international retailing and consumer behavior – specifically in the evolution of retail business models and consumer interactions with retail brand touchpoints in sub-Saharan Africa.

Francesca De Canio (francesca.dek@gmail.com) is a Researcher in Marketing at the University of Modena and Reggio Emilia, Italy. Francesca's research interests lie within the areas of consumer behavior, multichannel retailing, digital marketing, and collaborative consumption. Her PhD research focused on the benefits of shopping across multiple retailing channels and on the effect of the "need for touch" on channel choice. She has published her research in a number of international journals, including *Journal of Retailing and Consumer Services.*

Susanne Doppler (susanne.doppler@hs-fresenius.de) is currently Professor of Event Management and Subject Coordinator for Tourism and Event Management at the Hochschule Fresenius Heidelberg, Germany. She has broad research interests at the interface between sales and marketing, with a special focus on experiences and event marketing within the sales circle. Specific interests include the role of involvement in attendees' perceived experience quality at trade fairs, congresses and exhibitions, with a special focus on B2B, effects of strategic dramaturgy at the trade fair boost, and building attendees' trust and satisfaction through sustainable business practices and events. Furthermore, she has undertaken research in the field of mobile apps, games and virtual reality applications as marketing tools, e.g., for tourism destinations.

Johanna F. Gollnhofer (jogo@sam.sdu.dk) is an Assistant Professor of Marketing at the University of Southern Denmark, and a Research Associate at the University of St. Gallen, Switzerland. Johanna's work is primarily concerned with consumer behavior in the context of food. Her research has been published in international journals such as the *Journal of Public Policy & Marketing*, *Journal of Marketing Management* and the *Journal of Macromarketing*.

Louise Grimmer (louise.grimmer@utas.edu.au) is a Lecturer in Marketing in the Tasmanian School of Business and Economics at the University of Tasmania, Australia. Louise completed a PhD in retail marketing and strategy with a focus on local and independent retailers. She teaches retailing and marketing at postgraduate and undergraduate level and her current research interests include retail precinct marketing, crisis communications in the retail industry, supermarket and department store retailing, and how the digital economy and technology are transforming traditional modes of shopping. She is currently leading the only-known longitudinal study of small retail firms which examines the resources and strategies used by higher performing retail firms. Louise has published in a number of journals including *Journal of Small Business Management* and *Journal of Consumer Behaviour*. She is regularly called on to comment on retail matters in the local, national, and international media.

C. Michael Hall (michael.hall@canterbury.ac.nz) is a Professor of Marketing at the University of Canterbury, New Zealand. He is also a Docent in the Department of Geography, University of Oulu, Finland; and a Visiting Professor in the School of Business and Economics, Linnaeus University, Kalmar, Sweden. He has published widely on tourism, regional development, environment change, and food, wine and gastronomy.

Beverley Hill (beverley.hill@swansea.ac.uk) is a Senior Lecturer in Marketing in the School of Management, Swansea University, United Kingdom. She is a critical marketer, specializing in language and discourse, who researches communication, culture and consumption. She analyses text and talk in promotional, corporate, and consumer contexts to understand how institutional practices shape social life. She has published in the *Journal of Marketing Management*, the *British Food Journal*, and the *Journal of Media and Culture*.

Neal H. Hooker (hooker.27@osu.edu) is Professor of Food Policy in the John Glenn College of Public Affairs, The Ohio State University, USA. His research explores public policy, marketing and management issues within global food supply chains. He is particularly interested in how food safety, nutrition, and sustainability attributes are communicated, controlled, and (where appropriate) certified. He has published more than 70 journal articles and book chapters on the economics and marketing of food quality, product recalls, international food trade and the demand for and supply of sustainable and functional food and drink.

Muhammad Azman Ibrahim (muhazman@gmail.com) is a Lecturer in Marketing at the Center for University of Wollongong Programs at INTI International College in Subang Jaya, Malaysia. Recently, he graduated with a PhD in Marketing from the University of Canterbury, New Zealand under the supervision of Profs C. Michael Hall and Paul Ballantine. His research focuses on food marketing/food retailing and halal marketing. As a young researcher, he is dedicated to publishing more research articles and working with other researchers in future.

Sonja Lahtinen (sonja.lahtinen@staff.uta.fi) is a doctoral candidate in marketing in the Faculty of Management at the University of Tampere, Finland. Sonja is building her research interests within the field of corporate social responsibility and corporate sustainability. In her dissertation, she is scrutinizing connections between transformative businesses and environmental and social well-being by applying a multilevel approach, recognizing the individual, organizational, and institutional levels.

Morven G. McEachern (m.mceachern@hud.ac.uk) is Professor of Sustainability and Ethics at the University of Huddersfield, United Kingdom. Her research interests primarily lie in the area of consumer behavior within a variety of sustainable production and consumption contexts. This research has been presented at a number of international conferences and in a range of academic journals such as *Consumption Markets & Culture*, *Sociology*, *Journal of Marketing Management* and the *Journal of Business Ethics*. She has co-edited various journal special issues around her related research interests; contributed to edited books (e.g. *The Ethical Consumer*, Sage, 2005) and is co-author of *Contemporary Issues in Green and Ethical Marketing* (Routledge, 2014).

Sarah Maddock (sarah.maddock@rau.ac.uk) is a Principal Lecturer in Marketing in the Centre for Business and Enterprise at the Royal Agricultural University, United Kingdom. Sarah has a specific interest in the food and agricultural sectors and currently lectures on marketing issues in agri-food businesses, including consumer behavior and food choices. She has published papers on food marketing topics in the *British Food Journal*, *Nutrition & Food Science* and *Appetite*.

Alan J. Malter (amalter@uic.edu; PhD in Marketing, University of Wisconsin-Madison; MSc in Agricultural Economics, University of Illinois at Urbana-Champaign) is an Associate Professor of Marketing and Associate Head of the Department of Managerial Studies at the University of Illinois at Chicago, USA. Alan's current research examines geographic branding, embodied and process knowledge in managerial and consumer decision making, and consumer healthcare decisions. His research has been published in the *Journal of Marketing Research*, *Journal of Marketing*, *Journal of Consumer Psychology*, *Journal of International Marketing*, *International Journal of Research in Marketing*, *Journal of Product Innovation Management*, *MIT Sloan Management Review*, and *World Bank Technical Papers*, among others. He has received many grants and awards, including the 2005 Buzzell Best Paper Award from the Marketing Science Institute, and was a finalist for the 2012 *JMR* O'Dell Award. He is a former agricultural trade analyst and has consulted on export development for The World Bank.

Gianluca Marchi (gianluca.marchi@unimore.it) is Full Professor of Management in the Marco Biagi Department of Economics at the University of Modena and Reggio Emilia, Italy. His studies are concerned with firms' internationalization processes, governance models in crossborder alliances, knowledge transfer mechanisms, and innovation strategies in industrial and retail sectors. He has published in various leading academic journals, including *Research Policy, Journal of Business Research, Technovation, European Journal of Marketing, European Management Review*, and *Business History*.

Elisa Martinelli (elisa.martinelli@unimore.it) is an Associate Professor of Management in the Marco Biagi Department of Economics at the University of Modena and Reggio Emilia, Italy. Elisa's work is primarily concerned with retailing and consumer behavior, with particular reference to private labels, retail brand extension and customer loyalty. She has published in a variety of leading academic journals, including *The Service Industries Journal, The International Review of Retail, Distribution and Consumer Research, International Journal of Entrepreneurial Behavior & Research, Journal of International Consumer Marketing*, and the *British Food Journal*.

Dominic Medway (d.medway@mmu.ac.uk) is Professor of Marketing in the Institute of Place Management at Manchester Metropolitan University, United Kingdom. Dominic's work is primarily concerned with the complex interactions between places, spaces, and those who manage and consume them, reflecting his academic training as a geographer. He has published extensively in a variety of leading academic journals, including *Environment and Planning A, Tourism Management, Journal of Environmental Psychology, Cities, European Journal of Marketing*, and *Marketing Theory*.

Felix A. Nandonde (nandonde@sua.ac.tz) is a Lecturer in Marketing at Sokoine University of Agriculture, Morogoro, Tanzania. Felix has graduated with a PhD in Business Economics from Aalborg University, Denmark, an MSc from the University of Newcastle upon Tyne, United Kingdom, and a Bachelor of Business Administration from Mzumbe University, Tanzania. He teaches undergraduate courses such as Business Communication and Business Strategy, and Management Strategy on the MBA. He also teaches International Marketing of Agribusiness Products for the MBA in Agribusiness. Felix's work is primarily concerned with market linkages, retail and consumer behavior. His research has appeared in *Journal of African Business, African Management Review, Ethiopian Journal of Business and Economics, Journal of Business Research, British Food Journal, Journal of Language, Technology & Entrepreneurship in Africa*, and *International Journal of Retail & Distribution Management*.

Rosemarie Neuninger (rosemarie.neuninger@otago.ac.nz) is a Research Fellow in the Department of Marketing at the University of Otago, New Zealand. Rosemarie's work is primarily concerned with consumer food choice, consumer behavior and decision-making in different product categories. Her research interests also include food and beverage marketing, food safety and entomophagy.

Selcen Öztürkcan (selcen.ozturkcan@ju.se) is an Assistant Professor of Business Administration in the Jönköping International Business School (JIBS) at Jönköping University, Sweden. Besides JIBS, she is also affiliated with Bahcesehir

University, Turkey, as Professor of New Media (on leave of absence), Helsinki School of Business, Finland, as Visiting Professor, and Sabanci University, Turkey, as Network Professor of Marketing in the Brand Practice Forum. Selcen's work is primarily concerned with digital consumer behavior, reflecting her academic training as an engineer. More about her, including her updated research portfolio, is accessible at http://www.selcenozturkcan.com.

Michelle Phillipov (michelle.phillipov@adelaide.edu.au) is a Lecturer in Media at the University of Adelaide, Australia. Her research examines the ways that food media is impacting on public debates about food production and consumption, influencing the production methods and marketing strategies of agricultural businesses, and shaping new relationships between media and food industries. She is author of three books, including *Media and Food Industries: The New Politics of Food* (Palgrave, 2017) and *Fats: A Global History* (Reaktion, 2016). Research for her chapter was funded by an Australian Research Council Discovery Early Career Researcher Award (2014-2016, DE140101412).

Donatella Privitera (donatella.privitera@unict.it) is an Associate Professor of Tourism Geography in the Department of Educational Sciences at the University of Catania, Italy. Prior to becoming a fulltime academic, she had work experience in a multinational company in the marketing area. Donatella's research is primarily concerned with the interactions between places, tourism, cities and technologies. She has presented papers at several conferences and published in international books and academic journals. She is a reviewer for several national and international journals and a member of the editorial boards of the *International Journal of Sustainable Economies Management* and *Ottoman: Journal of Tourism and Management Research.*

Hiran Roy (hiranroy6@gmail.com) holds a PhD in Management from University of Canterbury, Christchurch, New Zealand. Before joining academia he worked extensively in the hospitality industry. He received his MBA from the University of Guelph, Ontario, Canada. He is an independent researcher and a hospitality educator. Hiran is also a recipient of a New Zealand Commonwealth Scholarship and Fellowship Plan Award. His research interests are in sustainable local food systems and local food marketing in hospitality and tourism industry contexts. Hiran's work has been published in a variety of leading academic journals including *Journal of Sustainable Tourism* and *Journal of Destination Marketing and Management.*

Hannu Saarijärvi (hannu.saarijarvi@uta.fi) is Professor of Service and Retailing in the Faculty of Management at the University of Tampere, Finland. Hannu's scholarly work is focused on emerging retailing phenomena such as transformational food retailing, C2C e-commerce and the reverse use of customer data. In addition, he is interested in topics such as customer value, customer relationships and service logic. He has published in a variety of academic journals, including *Journal of Retailing and Consumer Services, European Business Review, Journal of Strategic Marketing, British Food Journal, AMS Review,* and *The International Review of Retail, Distribution and Consumer Research.*

Marcos Santos (marcosfsantos@gmail.com) is an Associated Researcher for Observatorio SCALA in Bogotá, Colombia, and a Professor in the Multivix Faculty in Vitoria, Brazil. Teaching in universities and faculties for more than 15 years, Marcos's work reflects his multidisciplinary formation focusing on quantitative methods of researching consumer behavior, subsistence markets, transformative research consumption, innovation and entrepreneurship. He has published more than 28 articles nationally and internationally.

Leigh Sparks (leigh.sparks@stir.ac.uk) is Professor of Retail Studies and Deputy Principal at the University of Stirling, United Kingdom. His research concerns spatial-structural change in retailing, an interest which began with his undergraduate geography dissertation in Cambridge. He has published extensively in leading academic journals, has produced several books, and has been an advisor to the Scottish Government on retail and town matters. Leigh is Chair of Scotland's Towns Partnership and runs a blog on retail matters at www.stirlingretail.com.

Adrienne Steffen (Adrienne.Steffen@hs-fresenius.de) has been a Professor of Consumer Behaviour and Market Research since September 2012 at the Hochschule Fresenius Heidelberg, Germany, and its predecessor institution. Adrienne Steffen received her BBA in International Management at the International University in Germany and studied at the University of Michigan, USA, and the ESC Rennes, France. She completed her PhD (entitled "Affective Reaction to Critical Shopping Experiences in Shopping Centres-a Mixed Method Approach") and Post Graduate Diploma in Research in Business Management at the University of Strathclyde in Glasgow, United Kingdom. Adrienne has recently published in the area of consumer marketing where she focuses on retail experience management, consumer—salesperson interactions and sustainable consumption.

Sujo Thomas (sujo.thomas@ahduni.edu.in) is a member of faculty in the Marketing Department of Amrut Mody School of Management at Ahmedabad University, India. Sujo's research interests are in the area of consumer behavior; specifically, he likes to work on topics related to cause related marketing, online grocery retailing and social media marketing. He is an avid case writer and has published cases in a variety of reputed outlets, including IIM Ahmedabad Case Unit, The Case Centre-ECCH, and ET-Times Publishing Group India.

Deniz Tunçalp (tuncalp@itu.edu.tr) is Associate Professor of Management in the Department of Management Engineering at Istanbul Technical University, Turkey. Dr Tunçalp is primarily concerned with technological changes and organizational adaptation, working at the intersection of organization studies, information systems, and entrepreneurship. He has published in a variety of leading academic journals, including *Organizational Research Methods*, *Management Decision*, *Journal of Organizational Ethnography*, *Operations Research*, and *Journal of Global Information Technology Management*.

Sanket Vatavwala (f17sanketv@iimidr.ac.in) is a Doctoral Student at the Indian Institute of Management Indore in the area of marketing management. His research interests are in the fields of business-to-business marketing, destination branding and social media marketing.

Xiaojin Wang (xiaojin.wang2010@uky.edu) is a Postdoctoral Researcher in the John Glenn College of Public Affairs, The Ohio State University, USA. His research focuses on international agricultural trade, food marketing, and industrial organization in the agricultural and food sector.

Gary Warnaby (g.warnaby@mmu.ac.uk) is Professor of Retailing and Marketing at Manchester Metropolitan University, United Kingdom. His research interests focus on the marketing of places and retailing. Results of this research have been published in various academic journals in both the management and geography disciplines, including *Environment and Planning A, Journal of Business Research, Journal of Marketing Management, Marketing Theory, Consumption Markets & Culture, European Journal of Marketing, International Journal of Management Reviews, Area, Cities* and *Local Economy.* He is co-author of *Relationship Marketing: A Consumer Experience Approach* (Sage, 2010), co-editor of *Rethinking Place Branding: Comprehensive Brand Development for Cities and Regions* (Springer, 2015) and has contributed to numerous edited books.

Series Preface

The recent background document published by the European Commission for the high-level event on "FOOD 2030: Research & Innovation for Tomorrow's Nutrition & Food Systems," held in 2016, underlines how "the European food sector unites centuries of know-how with innovation in areas such as packaging, storage, transportation, and marketing. Thanks to its size and importance, the European food sector acts as a global benchmark. It is diversified yet standardized, traditional yet highly innovative, local but integrated, and consumer driven."

This description explains well the complexity of a sector in which many challenges for all stakeholders stand on a tradition-innovation continuum. According to several market research companies, consumer trends in the years ahead will be affected by changes driven by different motivations, such as health awareness and preference for niche products, and by a growing influence of new technologies for both food processing and communication.

Thus, today's food and beverage companies have to face multifaceted consumer demand. Recently, Forbes, in its "State of the Wine Industry 2018," stated that companies should be vigilant, market savvy, and adaptable. In fact, a renewed attention to consumers' emerging needs and trends is motivating profound industry transformations at various levels. Innovation can be a leverage for meeting customer demand; the market can also create the chance for a rejuvenation of mature products. Companies should understand how to optimize the flow of information and the inputs they gain from the market.

Thus, a synchronized combination of different disciplines like economics, psychology, sociology, marketing, management, anthropology, neuroscience, and statistics (just to cite the most relevant) as well as their relations to sensory analysis is necessary to increase the accuracy of forecasting and detect the probability of consumers' food choice by adopting a huge variety of qualitative and quantitative research methods. According to the DG-RTD of the European Commission, this combination of different fields, addressed to understand how people think, perceive and behave about food, and its production; is part of a multidisciplinary area of research of increasing importance, that crosses both social and natural sciences: consumer science.

This series, focused on consumer science and strategic marketing, provides practical information, through real cases and field-based research, to support practitioners in understanding how research in the field of consumer science is relevant for marketing strategies.

The "Woodhead Publishing Series in Consumer Science and Strategic Marketing" presents the tangible economic and financial outcomes obtained by the joint work of sensory scientists, marketing researchers, and agribusiness managers and outlines communication methods and practices that support research and development in the food and beverage sectors. Volumes in this series present successful examples and provide the foundation for further theoretical investigation.

Volumes in this series address several research questions, including the following:

- Which market trends and challenges can be observed at the international level?
- How did research in the field of consumer science become relevant for the marketing strategies of SMEs?
- Which tangible economic outcomes have been obtained by the joint work of consumer scientists, researchers, and consultants in marketing field, agribusiness managers, and SMEs owners?
- Which challenges are faced in order to make the most of R&D?

Practitioners in the food industry, including marketing, communication, and R&D managers, entrepreneurs, and subject matter experts in food and beverage sectors; undergraduate and postgraduate students studying business, agriculture, food engineering and technology; and academics teaching courses in the fields of agribusiness, applied marketing, or business strategy; are sure to find the "Woodhead Publishing Series in Consumer Science and Strategic Marketing" a useful source of reference.

Alessio Cavicchi[1] and Cristina Santini[2]

Series editors

[1]University of Macerata, Italy,

[2]University San Raffaele Roma, Rome, Italy

The changing nature of food retailing and distribution: Using one case to understand many

John Byrom[1] **and Dominic Medway**[2]

[1]Alliance Manchester Business School, University of Manchester, United Kingdom, [2]Institute of Place Management, Manchester Metropolitan University, United Kingdom

The distribution and retailing of food has undergone remarkable levels of transition and change over the last century or so. There are few better examples of this than the case of the UK-based multinational retailer Tesco. The company's origins hark back to 1919 when Jack (Jacob) Cohen, the son of Jewish immigrants, founded a market stall in the East End of London, following demobilization from the Air Corps. Initially, Cohen started selling unmarked tins of war groceries; one of his skills was apparently being able to identify the contents of a tin through merely shaking it. By 1924, the brand name Tesco had emerged—a combined acronym bringing together the initials of Cohen's tea supplier (T E Stockwell), coupled with the first two letters of his surname (Ryle, 2013).

Over the next 40 years, Cohen gradually developed the Tesco business through organic growth and acquisition of competitors (Clark and Chan, 2014), such that by the 1960s, the company was trading from over 800 stores—typically in "high street" locations within town centers. In this first half of the company's existence, one of the most radical developments was the introduction of self-service approaches to retailing. Cohen had visited America after World War II and seen the innovations being made there in this aspect of retail business. This was very different to the traditional behind-the-counter service which had dominated food retailing in the UK and Europe up to that point in time. Inspired by what Cohen had seen in the United States, the first self-service Tesco store opened in 1948 in St Albans, Hertfordshire; and by 1954, this vision had developed further with the opening of Tesco's first supermarket in Maldon, Essex. Such changes reflected an increasing consumer desire for both choice and speed of service through one-stop grocery shopping.

The next significant development in Tesco's retail story was the abolition of resale price maintenance (RPM) in 1964. Up to this point, manufacturers had set the price at which their products could be sold, but Cohen lobbied parliament for a change in the law on this issue. The parliamentary vote to abolish RPM was won narrowly. It was a watershed. The decision meant that price competition for retailing was now a reality in the UK and retailers were not confined to simply competing on their service offer. In turn, this had very significant implications for the size of retail organizations. Most notably, bigger retailers with more stores had greater bargaining power with their suppliers. This allowed them to push down purchase prices and then pass these discounts on to the retail consumer. Over the next 20 years, one result of this was that many small independent retailers went out of business across the UK, as they could no longer compete effectively with bigger operators such as Tesco. By the same token, with every small retailer that went to the wall, the slice of UK consumers' grocery spends with large operators such as Tesco grew ever greater.

In the early 1980s, however, Tesco was beginning to face problems. The portfolio of stores built up by Cohen over his leadership of the company had been somewhat haphazard and lacking in strategic focus. The next significant phase of development for the organization came with a more *laissez faire* approach to retail planning around the period of the Thatcher government. This allowed Tesco, under the leadership of Ian (later Lord) MacLaurin and David Malpas, to begin an ambitious program of building edge-of-town and out-of-town stores, drawing the focus for grocery retailing away from traditional high streets and town centers (Ryle, 2013). Yet by the mid-1990s, there was concern that such developments had gone too far, and were beginning to damage the vitality and viability of the UK's urban areas — a message also embodied in revisions of the UK Government's Planning Policy Guidance Note 6 (PPG6) in 1993 and 1996. Tesco responded quickly to this change in planning focus (as did other major UK grocery retailers such as J Sainsbury), developing a new series of in-town retail formats such as Tesco Metro and Tesco Express, whilst also acquiring competitors in the convenience store sector. This put the company's presence firmly back on the UK high street and coincided with an increase in residential living within town and city centers, which provided a re-emergent urban customer base.

By 1999, Tesco's portfolio in the UK covered over 600 stores and accounted for 15% of UK grocery market share (Pal et al., 2001). Following an "Ansoffian" strategic growth logic, the company had by this time already started looking at international horizons as a means of expanding their market and customer base. The first successful international foray, following earlier problematic attempts, was into Hungary in 1994 (Ryle, 2013). Growth continued, in both the UK and overseas, under Terry Leahy's tenure as CEO (1997−2011). This was seemingly relentless, to the point where Tesco was reported to take one pound in every seven spent in British shops (Wallop, 2007).

By 2018, Tesco could claim existing or past operations in at least 10 other countries outside the UK adopting various market entry strategies from partnering to straightforward acquisition. This accounted for over 6800 stores globally (Tesco plc, 2018a). Another significant development from 2000 onwards was the introduction of Tesco's online retail offer, including the home delivery of groceries. The company was one of the first movers in this area, and by 2018 was serving 500,000 UK customers per week through this platform (Tesco plc, 2018b). In such instances, the traditional side of a retail business, encompassing the idea of a physical outlet with tangible goods on shelves, which is visited by customers, looks increasingly anachronistic.

In recent years, Tesco's success and growth has not been without its difficulties, and it would seem that being a significant multinational operator acts as a magnet for both media and consumer criticism. Concerns have been raised over the company's supply chain practices, and in particular the alleged bullying of producers to extract optimum buying terms for the company, which might not necessarily be passed on to the consumer, but instead realized as company profit (Simms, 2007). Others have argued that a lack of concern by retailers such as Tesco over their supply chain safety and practices lay at the heart of the horsemeat scandal of 2013, where beef products were found to also include equine ingredients (Nelson, 2013). It has also been suggested that under intense pressure by retailers to drop prices, suppliers will be forced into cutting corners in this manner (Ruddick, 2013).

Irrespective of the legitimacy of, and reasons for, these kinds of concerns, it would seem that they are largely indicative of an era in which retailers' integration of the supply chain has become so complete that they can be held responsible for everything from product design and manufacturing, through to the distribution and promotion of a variety of goods and services, ranging from everyday groceries to comparison goods and financial service products (McGoldrick, 2002). Whilst this level of corporate hegemony and oligopoly is of concern to some, it does mean that consumers can collectively act to exert pressure on supply chain practices very quickly through mechanisms such as "buycotting" (Neilson, 2010), which impacts on retailers directly. From a more positive perspective, and perhaps as a happier outcome of episodes like the horsemeat scandal, many consumers are becoming increasingly concerned about the authenticity and provenance of the products they consume, whether this be where meat has come from, or whether a particular item was produced using ethically sourced ingredients—for example, Fairtrade cotton in a t-shirt or dolphin-friendly tuna in a tin (Baker, 2015).

At this point, any reader might be asking where the story of Tesco's development and growth as a business actually takes us. We would argue that its discussion above is because it is an illustrative case and a microcosm of the many macro and microenvironmental forces that have affected food retailing and its supply chain over the last century; particularly in terms of technological and social change over the last two to three decades. Indeed, in this respect, the Tesco case only serves to support the long-standing aphorism that the only constant in retailing is change. More significantly for this text, those same environmental forces that have impacted Tesco and its food retailing and distribution business are also clearly evident in the rich variety of cases drawn together into this edited volume on the topic. The rest of this chapter pulls together some of the common themes emerging from the book in this respect.

We saw with Tesco's story that the company has been central to waves of retail consolidation and concentration in the UK through, amongst other activities, the acquisition of competitors. This is a direct consequence of the perceived economic imperative within macro and micro environmental contexts for many retail businesses to maximize their profit, and a sober reminder that one operator's success in the food retail supply chain can be reflected in another's failure. However, the cases in this volume also show that alongside this competitive battleground of capitalist retail endeavor, there may still be room for smaller operators to survive and indeed prosper in food retailing, provided their offer has unique qualities that cannot be easily replicated by larger scale operators.

Further, it would appear that smaller retail players can be effective under a variety of organizational and tenure structures. For example, in Chapter 1, McEachern and Warnaby demonstrate the ability of independent cooperative retailers in Greater Manchester, UK (close to the birthplace of the cooperative movement) to carve out a distinctive niche in the food sector by placing an importance on community values at all levels within the business. And Grimmer (in Chapter 2) reveals how two Tasmanian independent food retailers are able to compete successfully through differentiation in a heavily consolidated Australian grocery retail market, dominated by two major corporate players. In a similar manner, Neuninger in Chapter 3 indicates how, specifically in New Zealand wine retailing, a concentration of retail power may result in a loss of specialist product knowledge and staff. This situation leaves a service gap in wine retailing, allowing the specialist operator to compete effectively by sharing valuable product knowledge and insights on wine with customers. On a less positive note, the rise and demise of Turkish online retailer TazeDirekt, outlined by Ozturkcan and Tuncalp in Chapter 4, demonstrates that irrespective of how uniquely different a food retailing businesses is, if it expands too quickly and outstrips its resource capabilities, then it is likely to succumb to the harsh economic realities that any other business might suffer. Put otherwise, a unique food retailing offer is of little consequence if its positioning is wrong and/or when the books cannot be balanced.

Macro environmental social forces clearly altered the nature of Tesco's food retailing business. For example, it was noted how consumer desire for choice and speed of service through one-stop food shopping influenced the growth of Tesco's supermarket offer. In Chapter 5, Dadzie and Nandonde recount how similar consumer motivations are beginning to change the nature of the retail landscape in developing economic contexts such as Ghana, where there is a move away from traditional markets to supermarket operations in which consumers can obtain all their grocery needs under one roof. However, social forces are manifest in other more subtle ways, reflecting the changing priorities of consumers in respect of food supply and retailing. Thus, increasing focus is being placed on issues of food provenance and authenticity (matters brought into sharper focus by incidents such as the horsemeat scandal referred to above). In this volume, we see such issues reflected in Chapter 6, where Privitera and Abushena identify how the BonAppetour food sharing platform for home dining experiences delivers consumers greater transparency and authenticity in food preparation, as well as engendering social connections through communal food experiences. Equally, Roy, Hall, and Ballantine, in their case study of the supply chain between farmers and restaurants in Canada and New Zealand (see Chapter 7), signal the perceived benefits of local food provenance for key stakeholders involved in the supply chain from field to fork. Chapter 8 by Bicen and Malter details the institutional arrangements governing the supply of table olives from the Gemlik region of Turkey, highlighting further how linkages to place and locality feature increasingly in the supply and distribution of food products.

There is also an increasing societal imperative placed on personal and family wellbeing through food consumption, certainly in the contexts of developed, or rapidly developing, economies. We see such concerns captured in Thomas, Abhishek and Vatavwala's account of the growth of Ayurvedic products (based on notions of wellbeing through ancient medicine) in the Indian food supply chain (see Chapter 9). Two cases in the book covering the growth of organic food retailing, in the USA by Wang, Boys, and Hooker (see Chapter 10), and in Malaysia by Ibrahim, Hall, and Ballantine (see Chapter 11), also demonstrate how society's concerns over what we eat are affecting consumption patterns in both Western and Asian contexts.

We also see evidence in this book that a focus on wellbeing can move beyond the domain of the individual and the family, to encompass wider societal change for collective good. For example, in Chapter 12, Santos and Barrios recount how Danone in Brazil have adopted a social partnering business model with NGOs to stimulate a more effective direct distribution and sales approach for their products. This provides welcome employment for many women from disadvantaged neighborhoods. Similarly, in Chapter 13, Saarijärvi, Sparks, and Lahtinen outline a future for food retailing which draws on data to transform lives with the retail customer base and wider society in critical areas such as health and wellbeing. Steffen and Doppler's contribution (see Chapter 14) identifies how the German organic supermarket Alnatura has employed sustainable business practices for the wider benefit of society, but also with the clear understanding that such actions can act as levers to develop consumer trust and satisfaction in their brand.

This volume has very much focused on developments that have already occurred, and in many cases are still occurring, in the field of food retailing and distribution. But what is the future for the sector? In many ways, like our above reflections on what has gone before, the practice of futurology is also usefully understood through the lens of macro and micro environmental forces. The impacts of technology are likely to be particularly significant. Driverless vehicles and the Internet of Things, embodied in devices like smart fridges, will inevitably change the speed and efficiency of food distribution. Further advances in mobile technology and apps are likely to facilitate this process further. In Chapter 15, Boller and Gollnhofer also reveal how technological and theoretical advances in analysis techniques could be used to examine patent data, thereby identifying future food retail trends; and the projected implications of these for food supply chain management and food distribution.

But the future is not just about the impact(s) of technology. Paradoxical trends of urbanization and counter urbanization to rural areas in different parts of the world are likely to produce very different outcomes for the distribution of food. In the case of the former, this might involve the challenges of providing a stable food supply chain to high-rise and congested mega cities, where the last mile, or even the last 100 meters, of fulfillment can be the most logistically challenging − the lift up a tower block or a tortuous urban one-way system. Urban grocery fulfillment also requires picking centers that take up huge amounts of space, which is likely to bring new challenges for the urban property market. By way of illustration, only recently it was reported that organizations in London are struggling to find enough land for retail distribution centers (Evans, 2017). By contrast, rural contexts provide a very different set of problems, in which the dispersal and lack of density of the population results in a rapid reduction in the economies of scale that make Internet-based means of delivery and fulfillment viable for retail organizations. This presents concerns as to how rural areas may be served by food retailing and distribution in the future.

Finally, if food retailing is increasingly reduced to a process dependent on automated and technologically driven interaction efficiencies, then it is likely that society will seek other ways to interact with food and obtain the social and experiential benefits that may have previously characterized traditional food shopping. This book demonstrates how this can occur in many ways for consumers, including: the use of online-ordered home meal kits in the UK as recounted by Hill and Maddock in Chapter 16, the authenticity sought through the traditional farmer's market, as detailed by Roy, Hall, and Ballantine for New Zealand and Canada, mentioned above; and individuals building an experiential relationship with food through television cooking shows with integrated food advertising, as outlined by Phillipov in Chapter 17, in an Australian context. If anything, this demonstrates that food retailing and distribution is set to be differentiated in the future, not just by convenience of purchase for the consumer or the efficiencies and speed of delivery and fulfillment, and not just by the quality of the food products themselves, but also by a whole range of uniquely perceived factors from the consumer's perspective relating to food performance, spectacle, entertainment and theater. As shown in Chapter 18 by Martinelli, De Canio, and Marchi for an Italian food retailer, some operators are already incorporating these qualities into their retail offer, through mechanisms such as private label products and innovative concept stores.

Similarly, and returning to the Tesco case with which we began this introduction, it is instructive to note the company's renewed emphasis on improving the customer experience (Rigby, 2016), which has included trialing innovative in-store concessions that reflect "on-trend" developments in food retailing and consumption and introduce a sense of retail theater into more conventional store environments (Hughes and Flavián, 2015).

References

Baker, J. (2015). The rise of the conscious consumer: Why businesses need to open up. *The Guardian,* 2 April. 〈https://www.theguardian.com/women-in-leadership/2015/apr/02/the-rise-of-the-conscious-consumer-why-businesses-need-to-open-up⟩ Accessed 29 January 2018.

Clark, T., & Chan, S. P. (2014). A history of Tesco: The rise of Britain's biggest supermarket. *The Telegraph,* 4 October. ⟨http://www.telegraph.co.uk/finance/newsbysector/retailandconsumer/2788089/A-history-of-Tesco-The-rise-of-Britains-biggest-supermarket.html⟩ Accessed 29 January 2018.

Evans, J. (2017). Alarm raised on dearth of London warehouse space. *Financial Times,* 2 February. ⟨https://www.ft.com/content/26036984-e7cd-11e6-893c-082c54a7f539⟩ Accessed 1 February 2018.

Hughes, D., & Flavián, M. (2015). *Tesco bringing theatre & food service to big box retailing?* ⟨https://supermarketsinyourpocket.com/2015/04/08/tesco-bringing-theatre-food-service-to-big-box-retailing/⟩ Accessed 1 February 2018.

McGoldrick, P. (2002). *Retail marketing* (2nd ed.). Maidenhead: McGraw Hill.

Neilson, L. A. (2010). Boycott or buycott? Understanding political consumerism. *Journal of Consumer Behaviour, 9*(3), 214−227.

Nelson, F. (2013). Slavery, not horse meat, is the real scandal on our doorstep. *The Telegraph,* 14 February. ⟨http://www.telegraph.co.uk/foodanddrink/foodanddrinknews/9870692/Slavery-not-horse-meat-is-the-real-scandal-on-our-doorstep.html⟩ Accessed 29 January 2018.

Pal, J., Bennison, D., Clarke, I., & Byrom, J. (2001). Power, policy networks and planning: The involvement of major grocery retailers in the formulation of Planning Policy Guidance Note 6 since 1988. *The International Review of Retail, Distribution and Consumer Research, 11*(3), 225−246.

Rigby, C. (2016). *Tesco sets out improvements to customer experience as it unveils first-half figures.* ⟨http://internetretailing.net/2016/10/tesco-sets-improvements-customer-experience-unveils-first-half-figures/⟩ Accessed 1 February 2018.

Ruddick, G. (2013). Bullying culture' blamed for horse meat crisis. *The Telegraph,* 19 February. ⟨http://www.telegraph.co.uk/finance/newsbysector/retailandconsumer/9880861/Bullying-culture-blamed-for-horse-meat-crisis.html⟩ Accessed 29 January 2018.

Ryle, S. (2013). *The making of Tesco: A story of British shopping.* London: Transworld.

Simms, A. (2007). *Tescopoly: How one shop came out on top and why it matters.* London: Constable.

Tesco plc. *Our businesses.* (2018a). ⟨https://www.tescoplc.com/about-us/our-businesses/⟩ Accessed 29 January 2018.

Tesco plc. *History.* (2018b). ⟨https://www.tescoplc.com/about-us/history/⟩ Accessed 16 January 2018.

Wallop, H. (2007). £1 in every seven now spent in Tesco. *The Telegraph,* 16 April. ⟨http://www.telegraph.co.uk/news/uknews/1548742/1-in-every-seven-now-spent-in-Tesco.html⟩ Accessed 16 January 2018.

Community building strategies of independent cooperative food retailers

1

Morven G. McEachern[1] and Gary Warnaby[2]
[1]Huddersfield Business School, University of Huddersfield, United Kingdom, [2]Institute of Place Management, Manchester Metropolitan University, United Kingdom

1.1 Introduction

The UK retail marketplace has faced continual change and disruption (Fernie, Fernie, & Moore, 2015). However, academic literature discussing these developments in the context of food retailing has tended to focus on the larger retail multiples (e.g., Burt & Sparks, 2003; Clarke, 2000; Wrigley, 1994), ignoring the fact that many small independent retailers outperform larger companies in terms of sales growth (Goodfellow, 2014). Where attention has fallen on independent food retailers, it has often been in a rural context (e.g., Byrom, Medway, & Warnaby, 2001; Byrom, Medway, & Warnaby, 2003), typically highlighting their social role(s). This has left a significant gap in academic inquiry relating to the community-based retail aspects of *urban* food retailing, which we explore from the perspective of independent cooperative retailers. Thus, a unique insight into the urban independent cooperative food retailer and their complex links between community, place, and social relations are advanced here. This chapter first considers the empirical literature surrounding independent food retailers and their respective communities, before illustrating our key findings and conclusions.

1.2 The independent retailer: Problems and prospects

The major focus of the limited literature on the small independent retailer, to date, relates to the perceived disadvantages of this retail form *vis-à-vis* its multiple retailer counterparts (see Clarke & Banga, 2011). Smith and Sparks (2000) summarize the problems and difficulties faced by small retailers in terms of:

- *Inadequacies in the trading environment*—i.e., competition from multiple retailers, economic and social change to the detriment of the small retailer, and locational difficulties in terms of spatial marginalization.

Case Studies in Food Retailing and Distribution. DOI: https://doi.org/10.1016/B978-0-08-102037-1.00001-3

- *Inadequacies in the retail form*—i.e., a less efficient and effective operating cost base than larger formats, lack of availability of investment capital, supply problems (e.g., lack of economies of scale etc.); all of which compound the above changes in the trading environment.
- *Inadequacies in management*—i.e., limited expertise in, and knowledge of, management techniques.

Consequently, a major focus of existing research has been on reasons for their decline (see Coca-Stefaniak, Hallsworth, Parker, Bainbridge, & Yuste, 2005), and how it might be arrested, either through policy intervention (see Clarke & Banga, 2011; Kirby, 1981) or strategic action on the part of retailers themselves (see Megicks, 2001; Megicks & Warnaby, 2008). However, much of this research has a rural (and isolated), rather than urban, context; for example, in terms of investigating how the drawbacks of operating in peripheral areas and regions can be overcome (see Byrom et al., 2001, 2003; Jussila, Lotvonen, & Tykkyläinen, 1992). Later work on their community role (see Calderwood & Davies, 2012, 2013) reflects the fact that small retailers are more likely to be located in more rural areas.

In light of such difficulties, the continuance of small independent retailers may depend on their ability to perform various roles. Smith and Sparks (2000) posit that these include:

- *Consumer supply of products and services*—which may occur in a variety of contexts, ranging from isolated areas, where small independents may be the only shops available, and thus used for all purchases; to circumstances where the small shop is a "destination" shop, arising from a particular product/service specialism.
- *Diversity, "color," and choice*—arising from such specialism, and the fact that small retailers might offer an alternative, nonstandard format and customer offer in contrast to the homogeneity of a retail landscape dominated by multiple retailers.
- *Dynamism and local adaption*—arising from the fact that small independents are often a source of retail innovation. Smith and Sparks note that the ease of entry/exit in this sector can create volatility and dynamism; the latter aspect manifest in better understanding of local markets and appreciation of customer requirements.
- *Economic linkages with other businesses*—via the supply chains that provide the products sold by small retailers. Smith and Sparks stress that such linkages also arise from the fact that small retailers also consume a range of other products/services (including public services such as refuse collection etc.), which are more likely to be locally oriented.
- *Employment generation and maintenance*—especially self-employment, as the small shop may be a seed-bed for entrepreneurship.

A key factor is the extent to which small independent retailers can develop distinctive competence, which might enable them to achieve some degree of competitive advantage (or merely survive). Small independent retailer strategies have been investigated in detail by Megicks (2001), who identifies five generic competitive strategy types:

- *Buying group merchants*—i.e., members of larger buying groups who act as traditional merchants, assembling and merchandising stock lines and delivering them with a service level tailored to the needs of their identified customer base.

- *Full-service strategists*—i.e., demonstrating a strong customer focus and growing through diversifying activities into new products/markets, as well as service improvements. They are more proactive in marketing activities.
- *Specialist vendors*—i.e., conventional retailers of specialist goods, with a strong emphasis on merchandising and providing high levels of service and unique, quality products.
- *Indistinct traders*—i.e., "*distinguished by a lack of distinction*" (p. 323), lacking real initiative, and not particularly active in pursuing growth opportunities.
- *Free-standing merchants*—i.e., similar to buying group merchants in strategic orientation and modus operandi, but not part of buying groups, and therefore more autonomous.

This literature emphasizes the importance of market orientation (see Megicks & Warnaby, 2008), which resonates with some issues identified above, in that effectively responding to a more detailed understanding of the needs of a locally oriented customer base may be a source of competitive advantage. Byrom et al. (2003) imply that such local embeddedness and the consequent knowledge gained could be a source of both market-and product-led strategic expansion strategies. However, resonating with Megicks' grouping of "indistinct traders," Byrom et al., also identify a strategy of "strategic stasis," whereby the primary aim is to maintain the status quo in terms of customer base, turnover, and profit (perhaps reflecting the extent of the difficulties facing these retailers, as outlined above).

1.3 The independent retailer and their role in the local community

Given the importance of local knowledge and adaptation to the small independent retailer, and their widely acknowledged "social" role (Calderwood & Davies, 2012; Clarke & Banga, 2011; Smith & Sparks, 2000), could the community-run or small cooperative shop be an independent retail form more able to mitigate the negative impact of trends that have led to the decline of the small independent retailer more generally? Arguably, the retail cooperative movement provides a sustainable retail format for small food retailers, and one which can be pursued either as a retailers' buying group, retailers' cooperative, and/or a retailers' cooperative retail chain (Kennedy, 2016). However, much research around retail cooperation as a strategy for independent retailers ignores the importance of the community-led, "social" role advocated by Clarke and Banga (2011) and Smith and Sparks (2000). This is despite the fact that the retail sector cannot be fully understood without reflecting upon the "interrelated systems of which it is a part" (McArthur, Weaven, & Dant, 2016, p. 281). That is to say, the place providing a spatial context, and the community relations inherent within, is paramount to the retailer's success.

Smith and Sparks (2000, p. 208) note, "an independent small shop may also provide a sense of community or identity both for a place and for its inhabitants". Aside from the exceptions listed above (albeit their social focus revolves around a rural location), few retail studies acknowledge this observation and consequently, community aspects have generally been treated as an "exogenous part of the environment" where in fact, the community should be recognized as being "completely endogenous to the enterprise" (Peredo and Chrisman, 2006, p. 310). Clarke and Banga (2011) explore this social role of the small retailer further, identifying four key aspects:

- *A "hub" for communities*—by providing an arena for social interaction, thereby helping to meet a variety of social, sustainability, and ethical needs (Megicks, 2007), through the facilitation of relationship-building, and creating emotional connections in a friendly environment.
- *Vital for the disadvantaged and socially excluded*—especially in terms of helping to meet the food shopping requirements of, for example, the elderly, financially deprived, socially excluded, and less mobile; groups which may be concentrated in geographically isolated communities (including in deprived urban areas).
- *Enhancing consumer choice and access*—to help consumers, such as those above who suffer poor economic and physical access to retail facilities (especially in relation to multiple retail provision), make more "informed" consumption choices, by increasing the "repertoire" of retail venues they frequent.
- *Creating consumer value*—potentially accomplished through three main ways: (1) the *generic features of small stores*, such as the ability to adapt to local situations in terms of providing a product range and level of service tailored to local needs; (2) the development of *specialist store formats* (e.g., offering better ease of access through location); and (3) by small retailers *targeting* their activities on specific consumer groups.

Also taking a more socially led view, Majee and Hoyt (2011) argue that the cooperative model in particular helps to build social capital between members and other stakeholders, which in turn strengthens both business and local community. In addition to calling for further research in this area, Moufahim, Wells, and Canniford (2017) urge marketing scholars to "dig deeper" when it comes to conceptualizing "community" as a unit of analysis. Therefore, to help advance our understanding of the role of independent cooperative food retailers and their respective community building activities, this chapter explores the complex links between retailing, community, place, and social relations.

Semi-structured, in-depth qualitative interviews were carried out with owners, store managers, and/or members from three food cooperatives located across the United Kingdom's Greater Manchester conurbation between late 2015 and 2016. Although a relatively small sample size, a contrasting case-type approach allowed for an in-depth understanding of these independent retailers and their community value to emerge (Mason, 2010). Each interview was recorded and transcribed verbatim and subsequent analysis involved coding and the development of initial themes (Strauss & Corbin, 1990). Intercoder cross-checking was conducted in parallel by the authors to facilitate the identification, development and refinement of themes (Miles & Huberman, 1994). Independent cooperative retailers featured in this research include WC#1, a worker's cooperative that has been operational since the 1970s; WC#2, another worker's cooperative, established in the mid-1990s; and a community cooperative (i.e., CC), which has been operational for three years.

1.4 Findings and discussion

The main themes identified from our data extend the concept of *community retailing* from a focus of just concentrating on the local *immediate community* to include also a *community of values* and a *supply chain community*, thereby suggesting a broader, more diffuse spatiality relating to the concept of community, beyond that of the immediate locale. Contrary to previous research (see Coca-Stefaniak et al., 2005), these independent retail cooperatives saw themselves as a viable alternative to the retail multiples. They enjoyed strong loyalty from customers who lived in the local area, as well as from those who shared a values-driven ethos, and displayed resilient financial growth despite the tough economic climate. Our thematic analysis now illustrates these additional dimensions of community-led retailing: *community of values, immediate community,* and the *supply chain community*.

1.4.1 Community of values

Contrary to Megicks's (2001) strategy of "indistinct traders" and Byrom et al.'s (2003) "strategic stasis," none of the independent food retail cooperatives could be accused of merely maintaining the status quo. Instead, there appeared to be a dynamic, values-led market orientation around providing safe, healthy, quality, and ethical food/nonfood products for their customers, acting sustainably, and creating opportunities for (and protecting) the workforce:

> *Our cooperative exists to encourage the optimum health of its customers and staff by providing quality vegetarian food and advice, whilst maintaining a caring, sustainable, democratic and ethical business environment for its workers (WC#1).*
> *Our aim is to trade honestly and ethically with our growers, suppliers and producers; promote local produce; offer a choice of organic foods free of artificial additives, preservatives and colours; and operate a 'No Junk Policy' (CC).*

Similarly, WC#2's statement of purpose echoes a strong focus on secure employment, equal opportunities, healthy consumption, fair trade, and a cooperative mindset:

> *The fundamental reasoning was to have a business that people wanted to shop in. It was basic ingredients. They were affordable. They were sourced with a little bit more care. Not too dissimilar to the Rochdale pioneers' basic history of the Co-op (WC#2).*

Importantly for all cooperatives, an *internal* community of shared values was ensured by having the right people to help develop and grow the business. For example:

> *They have got to build up the values of a cooperative and know what they are working towards; i.e. something where the food is sourced in a particular way, as well as a particular kind of food. So they have got to be on board with that element of it, as well as be willing or wanting to work cooperatively (CC).*

While independent food retail cooperatives had no problems in attracting new members with similar social and ethical values, many discussed the management struggles around balancing and securing the right skillsets:

> *I've got sixty-five member directors and then we've got five probationers. We have a quarterly members meeting. So strategy and policy are agreed upon. They can only be agreed by everybody. It's difficult, because we do require a consensus when decision making ... There is less resentment if everyone is involved in that consensual process. We are trying to spend this year looking at ways that we can keep this kind of structure with that engagement and keep it dynamic and get the new members to feel as closely, as much ownership of people who have been there longer. There are challenges (WC#2).*

> *It's a community co-op where people from the community put their money into it and the board reflects that ... The people at the shop are paid, but the board membership and participation other than that is all done on the fringes of everyone's time (CC).*

Although Smith and Sparks (2000) highlight "inadequacies in management" as a common problem amongst small retailers, the flexible management structure adopted by independent food retail cooperatives appeared to facilitate increased levels of internal trust and, more importantly, business innovation for the "social good":

> *There is a natural hierarchy that kind of comes and changes over time. It's worked very well in the sense it's helped us be quite an innovative and creative hard working co-op (WC#1).*
> *There is a built in mechanism for people to resign after three years. I think it's quite a positive thing as long as you've got a fundamental stability to be bringing new and innovative ideas to fresh people (CC).*

In addition, this sense of social innovation extended to cover spin-off, entrepreneurial ventures. WC#2, in particular, talked at length about a fund that it operated which aimed to "*try and encourage people to set up similar businesses ... it's a great thing to be able to do ... As individuals we are free to go and support any projects we choose to.*" In so doing, these independent food retail cooperatives managed to create a sustainable business model that successfully publicized their *community* values to local, national, and international members/stakeholders, as well as differentiating their values from larger food retail competitors.

1.4.2 Immediate community

Contrary to Peredo and Chrisman's (2006) notion that *community* is generally treated as an afterthought, distinct competences around the cooperative business model were evident in terms of a variety of values-led activities, benefiting communities, society, and the planet (see also Clarke & Banga, 2011; Megicks, 2007). This involved providing a strong sense of community for their customers and residents both within the local and global area, as well as attempts to protect the environment:

> *We have the 1%, 4% fund, which are attached to our wages and so they accurately reflect growth and prosperity ... We made a donation to a local community centre that was installing energy at St John's [a local school] and that's revenue going back to local people ... the 4% kind of accepts that we are part of quite an equal global trading system and much of the harm is in the Global South ... securing their future with their community (WC#2).*

> *We pay a reasonable amount of money for all our recycling with a social enterprise and all our cardboard, tins, veg waste and all sorts are taken away. One of our members did a massive, brilliant campaign on plastic bags. He did a big installation on all the railings, you know, all along by the path was covered in plastic bags and got lots of children involved and really raised awareness of the problem in the seas and everything. We haven't had plastic bags in here for about eight years now. All the plastic that the animals were choking on it, sort of thing. Plastic doesn't disappear. It's horrible stuff (WC#1).*

> *The general line is that we keep waste to a minimum and I know that we often cook it. Say, for example, we've made soups or smoothies by reusing vegetables that would otherwise have been wasted. But I think, we always seem to be on a constant quest for improvement in terms of waste (CC).*

Difficulties relating to "inadequacies in the trading environment," and/or "inadequacies in the retail form" (see Smith & Sparks, 2000), appeared to be an issue for recently established cooperatives only. For example, CC spoke of limitations surrounding their economies of scale in comparison to other more established cocoperatives:

> *There is a catch 22 isn't there? Because [WC#2] is large now - it can price well. Whereas, because we are not, it's hard, it's difficult. Because of our resources, it's difficult to be price competitive. Obviously, the more you sell the cheaper it could get (CC).*

In contrast, the more established cooperatives (i.e., WC#1, WC#2) experienced much less marginalization in these areas, preferring instead to adopt a differentiation focus (i.e., akin to a "specialist vendor"—see Megicks, 2001) in an attempt to remove themselves from direct price competition against other food retailers:

We sell a lot of things other people don't sell, and that's why we've survived. We are very proactive in doing different things ... There is a big call for the nutritional area, you know and I'm really, really busy up there, it's like a GP surgery up there some mornings ... I do blood testing and blood pressure days and all that sort of thing (WC#1).

We always saw our competition as multiples and position ourselves - so our organic food, fresh produce is often competitive with non-organic and supermarkets. For normal shopping, we very much wanted to be complementary within our local shopping area and we are part of this wonderful trading centre of independents and there is a symbiotic relationship there. We try and encourage people to set up similar businesses... We were helped by other co-ops etc. and we are happy to help and support (WC#2).

Clear evidence of the independent cooperatives' specialist supply of products and services impacted on the range of customer profiles who shopped there. WC#1 and CC spoke of the advantages of being located near a transient university/school population and WC#2 discussed their appeal to ethnic communities:

Our demographic is very well educated with people who work at the university as well as students. Actually that's not our main customer, it is a big part, but it's not what we survive on, really. We know a lot of our customers and we've a lot of regulars and we have a kind of community feel, even though it can be transient - a lot of people work in the area over the years. I think as we offer all the niche products it brings people in (WC#1);

We are surrounded by Asian communities, lots of communities that have got a great heritage of cooking and we've got a big Ethiopian customer base and it's the kind of people who like to cook and like ingredients ... it's great. There is a real wide range of ages and people and essentially more younger people as well (CC).

As opposed to the rural community retailer (see Byrom et al., 2001, 2003; Calderwood & Davis, 2012, 2013; Jussila et al., 1992), none of the cooperatives spoke of any locational difficulties as business was not always directed toward local customers. Here, WC#2 estimated "*the percentage [of customers] that are within two to five miles is really high.*" Subsequently, there was much discussion around how urban independent food retail cooperatives had to work hard to change out-dated perceptions of their business model:

In the olden days it was vegetarian and it was one of the first no smoking vegetarian businesses - 46 years ago that was very radical ... Vegetarian isn't a big thing now. You can get vegetarian food anywhere (WC#1);

The perception of wholefood shops were they were exclusive - they were pious ... There is only so much you can control. It's like, let's try to make it welcoming and fun and that people come in and it's just a shop, a community shop and not to feel that it's not a shop for them (WC#2).

Consequently, in a bid to change such perceptions, a lot of effort went into communications/PR activities. With most independent food retail cooperatives being involved in activities with local school children, some also implemented social media strategies and participated in local festivals:

> We've done a bit of work in schools and had children come in and see what was going on and that kind of outreach and providing things like apprenticeships and that way of giving back to the community (CC).

> Primary schools come into us and we'll do little tours and talk about, depending on what they are studying, whether it's nutrition or fairtrade or growing [your own food]. We cook with the local community allotment and do talks about growing . . . we feel it's part of our responsibility as a co-op just to spread the word and show what we are doing (WC#2).

> Our social media is good. I think we've got about 5,000 Twitter followers, and 6,000 on Facebook. The café do a lot of events . . . They do some festivals, but you know, it's more fairs and stuff they do (WC#1).

1.4.3   Supply chain community

Positive supply chain relations were paramount to creating a sustainable business which delivered on ethical, social, and environmental values. Linking together with a *community of values* and the *immediate community*, supply chain activities often extended to global suppliers as well as local ones:

> We've built up relationships over the years with UK growers . . . who have got an incredibly vulnerable business. One of the best things that [WC#2] does for growers is trade as we do, plan, clear them [i.e. stock] quickly and not try and depress the prices. While accepting that we do trade within what is our eco-global trading system and we are part of that . . . We can't say that every supply line is perfect; but we kind of are constantly trying (WC#2).

Another notable difference from independent small rural retailers which permits urban independent food retail cooperatives to maximize scales of economies is the reciprocal relationships with other larger cooperatives:

> A lot of what we sell is very high-end in the shop. But we also have a special relationship with other co-ops, so our wall of beans beats any Tescos hands down, because it's all organic. Our buying power isn't as big as theirs, but our organic eggs are much cheaper than the Co-op[1]. We have some mutual benefits all together and that's why some of them are knocking spots off supermarkets. As we work together to support each other (WC#1).

[1] The 'Co-op' referred to in this context is the Co-operative Retail Group.

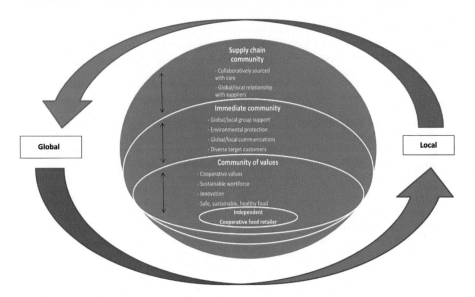

Figure 1.1 Community and the independent cooperative food retailer.

1.5   Conclusions

Through a contrasting case-type approach, our study sheds light upon the urban independent cooperative food retailer and its complex links between community, place, and social relations. This involves extending the concept of *community retailing* from just a focus on the local *immediate community*, to include also a *community of values* and a *supply chain community*. It recognizes a much more fluid and dynamic conceptualization of *community*, with a broader, more diffuse spatiality and ethicality beyond that of the immediate locale (see Fig. 1.1).

Therefore, in addition to looking at *community* in terms of spatial proximity, these independent retail cooperatives were strategically positioned to capture loyalty from their *immediate community* as well as another layer of ethically value-driven consumers and suppliers (i.e., community of values; supply chain community) which reside outside the "local." That is to say, these cooperative retailers actively sought and engaged local customers and residents around seasonal events (e.g., pumpkin growing workshops), cookery workshops (e.g., reducing food waste), and school competitions (e.g., designing a hessian shopping bag), to name but a few. Moreover, another layer of *community* engagement activities for these co-ops revolved around organizing workshops and events for national/global suppliers and customers looking to trade with and purchase ethical product alternatives, respectively. In so doing, each of the independent cooperative food retailers used a number of online and more traditional communication practices and tactics to mobilize local/global support, expand networks, and in some cases bring about socially responsible actions.

Overall, this case study offers significant learning points for both academicians and practitioners. From an academic perspective, we extend the concept of *community* retailing beyond that of spatial boundaries and instead advocate that community is embedded within all strategic operations concerning customers, suppliers, and other local/global stakeholders, and not just as an afterthought to cater for customers within the geographic locale of the store. Such an approach may prove advantageous for retail practitioners, whereby instead of trying to compete directly on aspects such as product pricing and/or opening hours, a more cooperatively led engagement with customers and other stakeholders in the food supply chain can lead to the creation of greater social value across local, national, and global communities. Thus, helping to bring about a greater "common good" for consumers and suppliers, as well as a collective, *community* drive for social change.

Acknowledgment

This study was funded by The Academy of Marketing Research Initiative Fund (2015−16).

References

Burt, S. L., & Sparks, L. (2003). Power and competition in the UK retail grocery market. *British Journal of Management, 14*(September), 237−254.

Byrom, J., Medway, D., & Warnaby, G. (2001). Issues of provision and 'remoteness' in rural food retailing: a case study of the Southern Western Isles of Scotland. *British Food Journal, 103*(6), 400−413.

Byrom, J., Medway, D., & Warnaby, G. (2003). Strategic alternatives for small retail businesses in rural areas. *Management Research News, 26*(7), 33−49.

Calderwood, E., & Davies, K. (2012). The trading profiles of community retail enterprises. *International Journal of Retail & Distribution Management, 40*(8), 592−606.

Calderwood, E., & Davies, K. (2013). Localism and the community shop. *Local Economy, 28*(3), 339−349.

Clarke, I. (2000). Retail power, competition and local consumer choice in the UK grocery sector. *European Journal of Marketing, 34*(8), 975−1002.

Clarke, I., & Banga, S. (2011). The economic and social role of small stores: A review of UK evidence. *International Review of Retail, Distribution and Consumer Research, 20* (2), 209−237.

Coca-Stefaniak, A., Hallsworth, A. G., Parker, C., Bainbridge, S., & Yuste, R. (2005). Decline in the British small shop independent retail sector: Exploring European parallels. *Journal of Retailing and Consumer Services, 12*, 357−371.

Fernie, J., Fernie, S., & Moore, C. (2015). *Principles of retailing* (2nd ed.). Abingdon: Routledge.

Goodfellow, C. (2014). *SME retailers account for largest growth in sales*. Available at www. realbusiness.co.uk (accessed 26th May 2016).

Jussila, H., Lotvonen, E., & Tykkyläinen, M. (1992). Business strategies of rural shops in a peripheral region. *Journal of Rural Studies, 8*(2), 185−192.

Kennedy, A. M. (2016). Re-imagining retailers' co-operatives. *The International Review of Retail, Distribution and Consumer Research, 26*(3), 304−322.

Kirby, D.A. (1981). Aid for village shops: A consideration of the forms of aid and scope for aid in Britain. In: *Workshop paper URPI S3, unit for retail planning information*, Reading.

Majee, W., & Hoyt, A. (2011). Cooperatives and community development: A perspective on the use of cooperatives in development. *Journal of Community Practice, 19*(1), 48−61.

McArthur, E., Weaven, S., & Dant, R. (2016). The evolution of retailing: A meta review of the literature. *Journal of Macromarketing, 36*(3), 272−286.

Mason, M. (2010). Sample size and saturation in PhD studies using qualitative interviews. *Forum Qualitative Sozialforschung/Forum: Qualitative Social Research, 11*(3).

Megicks, P. (2001). Competitive strategy types in the UK independent retail sector. *Journal of Strategic Marketing, 9*, 315−328.

Megicks, P. (2007). Levels of strategy and performance in UK small retail businesses. *Management Decision, 45*(3), 484−502.

Megicks, P., & Warnaby, G. (2008). Market orientation and performance in small independent retailers in the UK. *International Review of Retail, Distribution and Consumer Research, 18*(1), 105−119.

Miles, M. B., & Huberman, A. M. (1994). *An expanded sourcebook of qualitative data analysis* (2nd ed.). Thousand Oaks, CA: Sage.

Moufahim, M., Wells, V., & Canniford, R. (2017). *JMM special issue call for papers, special issue on the consumption, politics and transformation of community.* Available at http://www.jmmnews.com/the-consumption-politics-and-transformation-of-community/ (accessed 6 February, 2017).

Peredo, A. M., & Chrisman, J. J. (2006). Toward a theory of community-based enterprise. *Academy of Management Review, 31*(2), 309−328.

Smith, A., & Sparks, L. (2000). The role and function of the independent small shop: The situation in Scotland. *International Review of Retail, Distribution and Consumer Research, 10*(2), 205−226.

Strauss, A. L., & Corbin, A. C. (1990). *Basics of qualitative research: Grounded theory procedures and techniques.* Newbury Park, CA: Sage.

Wrigley, N. (1994). After the store wars: Towards a new area of competition in UK food retailing? *Journal of Retailing and Consumer Services, 1*(1), 5−20.

Disrupting the giants: How independent grocers respond to the supermarket duopoly in Tasmania, Australia

2

Louise Grimmer
Tasmanian School of Business and Economics, University of Tasmania, Hobart, Australia

> *"Together, [Coles and Woolworths] take in more than 70 cents of every dollar spent in Australian supermarkets"*
>
> *(Knox, 2015)*

> *"Small is beautiful"*
>
> *(Schumacher, 1973)*

2.1 Introduction

Food retailing is an essential economic, social, and cultural activity and a successful retail sector is a critical part of local communities. Small, independent grocers contribute to their local area through the provision of high-quality, locally sourced products, by supporting local farmers and producers, and by providing employment opportunities. Importantly, they also provide an alternative to the larger supermarkets. While small grocers also make a significant contribution to local, state, and national economies, in a global market, small firms are facing increasing challenges.

Australia has one of the most concentrated grocery markets in the world, with the sector dominated by two supermarket giants—Coles and Woolworths. Together, Coles and Woolworths employ more Australians than any other entity, other than state governments, and are amongst the 20 largest retailers globally (Knox, 2015). The dominance of large retailers in the Australian market makes it extremely difficult for small, independent stores to operate successfully. In Tasmania, the island state to the south of mainland Australia, two family-owned food retailing businesses—Hill Street Grocer and Salamanca Fresh—are exceptions that prove the rule (Knox, 2015).

2.2 The Australian retail landscape

From an industry perspective, retailing is Australia's second largest employer (the largest being the health care and social assistance industry), followed by construction and manufacturing (ABS, 2012). The Australian retail industry employs around 11% of the working population and makes a significant contribution to Australia's economic output.

Coles (owned by Wesfarmers) and Woolworths (Woolworths Limited) have interests that extend beyond supermarket retailing, including the provision of petrol, liquor, finance and insurance products, hotels and gambling, general merchandise and hardware. Together they operate nearly 1000 supermarkets, 700 petrol stations, more than 1000 bottle shops and hotels, as well as more than 500 variety and hardware stores. In the context of Australian supermarket retailing, these giant corporations now operate effectively as a powerful duopoly.

In 2010, Kantar Retail (the global retailing consultancy) released the 2010−15 Global Retail Opportunity Index Ranking, and Australia ranked number 7 out of the top 34 countries (Badillo, 2010). The index is based on three factors: the size of the retail market, market risk, and inflation-adjusted growth forecasts. Of note, Australia's sustained growth resulted in a higher ranking (at the time) than the United Kingdom, France, Japan, and Germany. Retailing is more global than ever and large retailers continue to trade across borders. While the opportunities vary by region (and within regions), the large global retailers continue to compete with one another in the search for international growth (Evans, 2011). In this rather "fertile" retail landscape, Coles and Woolworths have grown to become Australia's top two retailers and both corporations are in the top 20 (in terms of revenue) in the list of global retail firms.

There are around 140,000 retail businesses in Australia (Productivity Commission, 2014) and the food retailing sector is diverse with considerable differences between retailers in terms of business size, format, and offerings; these differences are further highlighted across different regions throughout the country.

2.3 Retailing in Tasmania

In the southern Australian island state of Tasmania, with a population of just over half a million (ABS, 2013), two family-owned food retailing businesses have challenged the Coles/Woolworths duopoly. The state's population is concentrated into three regions: Hobart in the south, Launceston in the north, and Burnie and Devonport in the northwest. Compared with the majority of the Australian mainland states, Tasmania has a small population (with little population growth), high rates of unemployment, low socioeconomic status indicators, an ageing population, and low rates of private and government investment (DIRD, 2012). The environment for small food retailers is challenging: there is a shortage of retail property available for rent/purchase, and retail rents are relatively high (Productivity Commission, 2014). In addition, there is a distinct lack of shopping "clusters" in local neighborhoods and a growing trend toward "big box" or high-rent and anchored shopping centers (housing a Coles or Woolworths supermarket).

Although in mainland Australia, German-owned Aldi and the US wholesaler Costco are making significant gains in market share (and it is rumored German-owned Lidl is also poised to enter the Australian market), neither Aldi nor Costco currently operate in Tasmania. Therefore, in the context of Tasmanian supermarket retailing, Coles and Woolworths are the "big two" throughout the state, with Woolworths operating 31 supermarkets, and Coles 16. It should also be noted that the IGA (Independent Grocers Australia) group also operates 88 small to medium stores in Tasmania. However, although the IGA stores outnumber Coles and Woolworths, these two corporations still represent almost 90% of market share in Tasmania and the state has been characterized as home to a "hyper-concentration of Coles and Woolworths" (Glaetzer, 2015).

Over the past decade, two family-owned and run independent grocers—Hill Street Grocer and Salamanca Fresh—have increased their market share in Tasmania (and accordingly their buying power) through the provision of high-quality products, an enjoyable customer experience, expansion of stores (mostly in southern Tasmania), and a range of creative offerings that have set them apart from their smaller competitors as well as from Coles and Woolworths. From original single stores, they have systematically taken over existing grocers (many of them IGA stores) or started from scratch in empty premises. Both firms have a strong brand identity and are known for stocking local products, the provision of attractive stores, and for superior customer service.

Salamanca Fresh was originally conceived as a single store in 1982 situated on Hobart's historic Salamanca Place, and then known as the Salamanca Fresh Fruit Market. The Salamanca area is now a thriving hub for tourists, business visitors, and locals who live and work in the area. Since its inception, this waterfront area has seen increases in the number of people living in Salamanca and nearby Battery Point, and the University of Tasmania has opened a large and modern building nearby to house its Institute for Marine and Antarctic Science. The Commonwealth Scientific and Industrial Research Organization has offices nearby, and many of the old wharf sheds and associated buildings have been renovated and put to use housing special events, festivals, and conferences. The area has dramatically changed since Salamanca Fresh opened as a result of the influx of visitors, local residents, and city workers. The location of this Salamanca Fresh flagship store is therefore ideal to serve the local market. Nestled alongside art galleries, restaurants, bars, and specialty shops, the Salamanca Fresh store is a landmark grocer and a popular shopping destination for local travelers, Airbnb guests, and those on their way home from work. The business is owned and operated by the Behrakis family and there are now seven stores operating in the south of the state employing over 100 staff.

In 2001, the three Nikitaras brothers and their families renovated and rebranded an original family-owned corner store in the suburb of West Hobart in southern Tasmania to create the original Hill Street store (now known as Hill Street Grocer). In addition to the bricks and mortar stores, Hill Street Grocer also offers an extensive online shopping option. Hill Street Grocer operates nine stores throughout Tasmania (six in southern Tasmania, two in the north, and one in the northwest), and there are plans for at least two new stores to open in the south in the next year.

Hill Street Grocer and Salamanca Fresh stores are strategically located either in high socioeconomic status neighborhoods or in lower socioeconomic status suburbs with no other local grocery offering. The expansion imperative means both firms are able to better negotiate with local suppliers and create and sustain growth through economies of scale. Both firms recognize the growing demand from consumers who want to shop locally, want to know where (and how) their food is sourced or produced, and are increasingly shunning the impersonal corporatization of the large supermarket chains.

2.4   Differentiation strategy

One of the most frequently applied approaches to retailing from the field of strategic management has been the work of Porter (1980, 1985) on competitive advantage (e.g., Alexander & Veliyath, 1993; Ellis & Calantone, 1994; Ellis & Kelley, 1992; Morschett, Swoboda, & Schramm-Klein, 2006). The term "competitive advantage," first used in the 1970s (South, 1980), is the process through which a firm can identify, develop, and then take advantage of a sustainable business edge over its competitors and refers to the ability of a firm to create value for its customers that exceeds the cost of creating that value (Ellis & Kelley, 1992; Ellis & Calantone, 1994). Although a firm may have a number of specific strengths and weaknesses compared with its competitors, Porter (1980) proposed two basic types of competitive strategy: cost leadership and differentiation. Traditionally, the corner store, or the local grocer, could compete on convenience, customer service, and product offering (differentiation) but usually not on price (cost leadership). Independent grocers usually have to make a choice between the two strategies and select only one, as most firms cannot be "all things to all people."

In Australia, Coles and Woolworths clearly pursue a cost leadership strategy. That is, they are concerned with ensuring cost reductions across all departments (e.g., labor, distribution, control systems) (Ellis & Kelley, 1992; Porter, 1980), as well as ensuring that they still maintain a level of service and quality. Achieving and maintaining a cost leadership strategy usually depends on the firm possessing advantages in the industry such as market share, access to financial capital, purchase of specific equipment, supply chains, and logistics. As a result of their cost leadership strategy, Coles and Woolworths are able to command high margins and reinvest in equipment and processes which allow them to maintain and sustain their low cost proposition.

On the other hand, differentiation sets apart a firm's product or service from its competitors and allows them to offer something unique within their industry, for example, offering high quality products as well as a reputation for excellent customer service (Day & Wensley, 1988; Wortzel, 1987), as in the case of Hill Street Grocer and Salamanca Fresh. The differentiation strategy targeted by Hill Street Grocer and Salamanca Fresh does not mean that they ignore cost; rather, cost is not the primary concern. Through effective differentiation, both firms are able to focus on a narrow segment of the market, and as a result they enjoy brand loyalty from customers, higher margins, and increased market share.

2.5 Taking on the giants

How do Hill Street Grocer and Salamanca Fresh enact their differentiation strategy? This can be examined by considering how they both approach the 4Ps of marketing.

2.5.1 Product

Walking into a Hill Street Grocer or Salamanca Fresh store, it is apparent that the offering is of particularly high quality, with attention to detail and an efficient yet friendly atmosphere. Products are attractively presented with a clear focus on Tasmanian produce (across a wide range of product lines) and this is one of the major drawcards for local shoppers. The Hill Street Grocer experience could be described as more "up-market" than that offered by Salamanca Fresh. In addition to the in-store experience, Hill Street Grocer are also more sophisticated in terms of their online presence, their product offerings, and their marketing, and in 2016 Hill Street Grocer won the title of National IGA Retailer of the Year.

Both stores offer a one-stop shopping experience with a limited range of general merchandise. They offer a plentiful array of fresh and locally grown fruit and vegetables, as well as an extensive range of gourmet delicatessen products, cheese, wine, and international food products. Many smaller producers are being dropped from the larger supermarket shelves as the major players move to stock only global or national brands or increasingly their own private label products. The small grocery industry, therefore, plays an important role in supporting and promoting smaller producers and innovative products and providing an outlet for consumers to discover new and different products.

In addition to food and grocery products, Hill Street Grocer recently launched a signature "Hill Street Grocer Home" range which is housed in the new flagship store (around the corner from the original Hill Street Grocer corner store) as well as the Devonport store. The range offers homewares, kitchenware, a hamper collection, a cheese room, local handmade chocolates, and an in-store florist. Customers can also purchase Hill Street Grocer Home items online and have flowers, hampers, and gifts delivered directly to friends and family. Hill Street Grocer also offers customers a wine and cheese loyalty club, and a range of catering options and ordering systems, including office fruit box delivery, Christmas and Easter delivery, click and collect, gift hampers, gift cards, and even a fuel discount voucher program in conjunction with their own single petrol station and nine other fuel outlets around the state. During the Christmas shopping season, both grocers are also known locally for stocking much higher quality local, seasonal holiday food products than Coles and Woolworths, and Hill Street Grocer also offers a preorder service for gourmet Christmas products.

Wahl (1992) describes the most important elements that shoppers look for in a grocery/food store, and 25 years later those criteria are still relevant for the contemporary consumer. Wahl's elements were cleanliness, all products prices labeled, good produce department, pleasant and knowledgeable staff, low prices, use-by dates on products, a good meat department, well-stocked shelves, unit pricing on shelves, and convenient store locations. As is detailed below, Hill Street Grocer and Salamanca Fresh exceed all of Wahl's critical elements.

2.5.2 Price

The target market for both Hill Street Grocer and Salamanca Fresh are middle-aged, well-educated, higher earning professionals and families who are interested in healthy eating, ethical consumption, and who are looking for a pleasant shopping experience.

On many products, particularly fruits and vegetables, Hill Street Grocer and Salamanca Fresh are often cheaper than Coles and Woolworths, and both firms offer competitive weekly specials on a range of products, including food products and general merchandise. Notably, the home delivery charge for Hill Street Grocer is the same as that charged by Coles and Woolworths. However, in general, the prices are higher than those offered by the larger chains. In addition, a number of the exclusively stocked gourmet food products are higher in price and the Tasmanian wine products are usually more expensive than local liquor retailers. In Tasmania (and Australia), alcoholic products are not allowed to be sold in supermarkets but must be purchased from stand-alone liquor stores. However, independent grocery stores can sell wine (and beer or cider) but only if it is Tasmanian.

2.5.3 Place

Hill Street Grocer and Salamanca Fresh stores are visually appealing and pleasant places to shop. Giant tubs and rows of attractively presented fruit and vegetables greet customers as they enter a Hill Street Grocer or Salamanca Fresh store. Customer service is professional, friendly, and prompt. Each store has a large number of checkouts and staff are regularly called to open additional stations when it is busy in store. Personalized service is emphasized—there are no self-service checkouts—and management and staff know many of the customers by name. Both Hill Street Grocer and Salamanca Fresh have focused on retaining existing staff (whenever possible) from businesses that have been taken over by their brands, and in many of the suburban or outer suburban areas they hire staff who live locally.

The Hill Street Grocer in West Hobart is so popular that the store employs staff solely to direct traffic in the store's rather modest car park, and the store in Devonport also houses a popular and bustling Hill Street Café.

While Hill Street Grocer has an expansion strategy across the state, Salamanca Fresh have limited their growth just in southern Tasmania, where they continue to rebadge existing stores in various Hobart suburbs. Salamanca Fresh sites are strategically located in suburbs populated by their target market and all seven stores including their original flagship store in Hobart's historic Salamanca Place are performing strongly.

We are experiencing exceptional consumer demand and growth in our business and we recognize the need to make sure we have a strong presence in growing areas in the South (Beniuk, 2015).

2.5.4 Promotion

Hill Street Grocer promotes itself as "your local store" and employs a sophisticated marketing strategy. The firm operates a comprehensive website (www.hillstreetgrocer.com) which includes online shopping (click and collect as well as home delivery in many locations). Customers can also shop online from the Hill Street Grocer Home range.

There is a blog, an online recipe collection and an email newsletter, which provides all the latest specials, exclusive promotions, and news. Weekly specials are emailed to their member customer list. There is a quarterly newsletter, *Provisions*, which is available online and in store. As well as providing seasonal recipes and cooking tips, the newsletter profiles local producers and growers includes book reviews, interviews with local restaurateurs, celebrity chefs, sommeliers, gardeners, and authors. At Christmas, Hill Street Grocer is the "go to" store for gourmet holiday supplies with customers able to preorder basic and deluxe Christmas food and trimmings—locally made plum puddings and mince pies, Tasmanian seafood, game, ham, and turkey. Goods are packed and ready for customer pick up as late as Christmas Eve.

Salamanca Fresh's tagline is "famous for freshness since 1982" and their main focus has always been on the fruit and vegetable offering; with their branding and marketing reflecting this. As stated earlier, the original name was the "Salamanca Fresh Fruit Market." The grocer underwent an extensive rebrand in 2002 to become Salamanca Fresh with expanded stores, staff uniforms, environment-friendly shopping bags, and more modern signage and marketing collateral. The website is fairly basic (www.salamancafresh.com.au), but it is updated with weekly specials and recipe ideas. Salamanca Fresh's strategy is to drive customers into the stores and it does not offer online shopping. Salamanca Fresh's emphasis has always been on guaranteeing fresh products for their customers through well-established supply chains and relationships with growers and producers.

2.6 Foes or allies?

Both Hill Street Grocer and Salamanca Fresh have purposefully and strategically adopted a differentiation strategy. Specifically, they distinguish their offerings from the two major players through stocking high-quality (local and seasonal) fresh produce, local artisanal, and international gourmet food products as well as dedicated meat, delicatessen, and cheese counters and a range of preprepared, "ready to heat" gourmet meals.

With both Hill Street Grocer and Salamanca Fresh pursuing a rapid expansion model, they have faced criticism from some in the industry that they are limiting the success of other small grocers as well as impacting retail diversity in the Tasmanian market. However, growth may be the only option for the survival of the independent grocery sector, as operating more stores equals greater buying power and market share.

Although on the face of it Hill Street Grocer and Salamanca Fresh are competitors, in many ways they are actually allies. Although they are not pursuing a traditional "horizontal marketing" strategy (e.g., Lewis, Byrom, & Grimmer, 2015; Morris, Kocak, & Ozer, 2007), both firms acknowledge that if the independent sector flourishes, it is beneficial for all the smaller players in the industry. A strong independent grocery sector is an effective way to compete against Coles and Woolworths.

> It's good if they [Salamanca Fresh] do well and it's good if we [Hill Street Grocer] do well. It's important the smaller independents do better. We have to give people a reason to shop locally and independently (Glaetzer, 2015).

2.7 Shopping trends and the impact on consumers

There have been a number of social, cultural, and economic trends that have impacted on the success of Tasmania's two leading independent grocers. Consumers are increasingly continuing to embrace local shopping and support small businesses and they are increasingly aware of ethical and sustainable issues with regard to the sourcing, purchasing, and consumption of food products.

2.7.1 Shop local movement

Both Hill Street Grocer and Salamanca Fresh have enjoyed the resurgence in the "shop local" movement, which in tandem with the popularity of the "eat local" trend has put the focus on fresh, sustainable, and local produce (e.g., Aucoin & Fry, 2015). Consumers are increasingly concerned with the provenance of food, the sustainability of the processes by which the food has been produced and has traveled, and the importance of the ethical nature of the relationship between retailer, supplier, and producer (Nonini, 2013).

> Customers have certainly become far more discerning in how and where they shop. Shopping locally is the preference of the majority of our customers (Beniuk, 2015).

Indeed, the target market for Hill Street Grocer and Salamanca Fresh are educated, time-poor, higher earning professionals who are concerned with the provenance of their food. Independent grocers are more readily able to provide these consumers with such information, as they usually stock a variety of products from smaller, local producers. This delivers their customers high levels of local provenance for the various product lines stocked; something which larger supermarkets Coles and Woolworths find it harder to do.

2.7.2   Ethical consumption

Another trend, related to the shop local and shop small movements, is the increasing concern amongst particular groups of consumers about the provenance of the food and general grocery items they purchase. More and more consumers want to know that the products they buy have made their way to the retailer through an ethical and sustainable supply chain. They are interested in organic food, local food systems, and sustainable agricultural practices, including the ethical treatment of animals (Beagan, Ristovski-Slijepcevic, & Chapman, 2010). "Ethical" consumption or "green" consumption promotes the notion that in making everyday food choices, consumers can "think critically, buy more selectively, and seek out information on the environmental and social costs involved in their daily meals" (Johnston, 2008: 239).

With the focus on providing fresh Tasmanian local produce, as well as other local products including bread, cheese, chocolates, and more, both stores maintain strong relationships with farmers, producers, and local suppliers. This has benefits for the state's farmers and producers of artisan products.

> If they're [Hill Street Grocer and Salamanca Fresh] concentrating on Tasmanian produce then of course that's better for the agricultural producers (Beniuk, 2015).

2.7.3   Time-poor lifestyle and the luxury of choice

Echoing the global trend across developed nations, Tasmanians are becoming increasingly time-poor, and shoppers with higher disposable incomes are seeking ways to minimize the time they spend on shopping and other nonleisure activities such as housework and home maintenance. These consumers have the "luxury of choice" in choosing shopping options that may involve paying slightly higher prices as a trade-off for convenience. In this regard, both Hill Street Grocer and Salamanca Fresh offer their customers a value for money proposition. The beauty of a local grocer is that the stores are smaller, often more conveniently located with easy parking facilities, and offer an increasing range of high-quality products including ready-to-eat gourmet meals. The interior of the store is welcoming, the staff are friendly and the entire shopping trip takes less time on average than a trip to the larger supermarket chain. For Hill Street customers, shopping can be ordered online and then delivered to the door or picked up from the store on the way home.

2.7.4 Tourism and the "MONA effect" in Tasmania

Another factor which has affected the level of sophistication of product offerings and the expansion of stores into particular regions or suburbs in Tasmania is the incredible growth in tourism experienced by Tasmania over the past 5 years. Once considered by many on the "mainland" as a "cultural backwater," Tasmania has recently "come of age." Thanks to the stellar popularity of multimillionaire David Walsh's Museum of Old and New Art (MONA) in Hobart, as well as its associated mid-winter and summer festivals ("Dark Mofo" and "Mona Foma"), the numbers of culturally sophisticated and discerning travelers to Tasmania has been increasing at such a rate that in 2015 the global travel authority Lonely Planet voted Tasmania as one of the world's top visitor regions (Tourism Tasmania, 2015). This new cohort of tourists are interested in food and wine experiences and they are keen to sample the high-quality fresh fruit and vegetables, artisan cheese, breads and small goods, and world-class wine that Tasmania is producing.

The state's now solidly cemented reputation as a "must-visit" destination for "foodies," coupled with the rise of the sharing accommodation (including many Airbnb travelers who want to "live like the locals"), means that shops selling high-end, good quality, and locally produced products are in strong demand. Airbnb visitors, in particular, are seeking a more authentic experience; many consider themselves to be travelers as opposed to tourists, and this self-perception underscores their passion for more genuine, and therefore, local travel experiences. They want to discover more and make a deeper connection with their destination; they want to shop, eat, and drink where the locals do.

In addition to the provision of quality products, both Hill Street Grocer and Salamanca Fresh offer pleasant retail premises. They have carefully designed their stores to enhance the shopping experience and appeal to consumers through the provision of attractive displays and merchandising highlighting local and artisanal products. Increasingly, more locals and visitors to Tasmania are shopping at independent grocery stores, which are becoming the hub of local neighborhoods, as well as playing a vital role in supporting local growers and suppliers.

2.8 The future for the Tasmanian grocery sector

Clearly, physical retail continues to play a very important role for both consumers and retailers. Independent "local" grocers such as Hill Street Grocer and Salamanca Fresh are successfully meeting consumer demand for locally produced and sourced products presented in a pleasant shopping environment. People still want to see and touch products before they purchase them, they want to enjoy the social role that "bricks and mortar" shopping provides, and they want to be part of an exchange that benefits others—farmers, artisans, niche food producers—and provides employment for local people.

Given that the Australian supermarket and grocery sector is one of the most competitive in the world, the future of the industry can perhaps be described as "polarizing." The sector may well be moving toward a choice between either very large and impersonal supermarkets selling national brands and private supermarket label products, and smaller, independent stores supporting local producers and suppliers. Indeed, as the Tasmanian industry shows, the growing popularity of small, independent, and local shops highlights the clear distinction between, on one hand, the differentiated offering provided by grocery stores such as Hill Street Grocer and Salamanca Fresh and, on the other hand, the more cost-driven proposition from the large national supermarkets such as Coles and Woolworths.

What then are the implications for those grocery stores and supermarkets that fall somewhere in-between? As Porter (1980) notes, a firm that is not able to develop a differentiation or cost-leadership strategy is unlikely to enjoy sustainable competitive advantage. These firms will effectively be "stuck" in the middle and suffer from low profitability. These retailers will either be small grocery stores and individual supermarkets which are too small to achieve competitive economies of scale required for cost leadership or they will be too large (but not large enough to compete with the "duopoly") to build reliable supply networks with local producers and suppliers and therefore unable to offer product differentiation. As the retail industry, in Australia, and internationally, continues the trend toward consolidation, these are the types of firms that will be unable to compete and will fall by the wayside.

It is in this competitive milieu that Hill Street Grocer and Salamanca Fresh have grown their businesses from original single-site stores to their current chains. They successfully exploit a niche market that is not served by the very large supermarkets in Tasmania. Both family-run stores recognize what their customers want, and it is different from that offered by the large supermarket chains. Hill Street Grocer and Salamanca Fresh are succeeding in the game of differentiation and are excelling at "disrupting the giants."

References

Alexander, L. D., & Veliyath, R. (1993). Matching competitive strategy with grocery store format: an investigation of the performance implications. *Journal of Strategic Management, 1*(1), 3−19.

Aucoin, M., & Fry, M. (2015). Growing local food movements: farmers' markets as nodes for products and community. *The Geographical Bulletin, 56*(2), 61−78.

Australian Bureau of Statistics (ABS). 2012. 1301.0 Year Book Australia 2012. http://www.abs.gov.au/ausstats/abc@.nsf/Lookup/by%20Subject/1301.0~2012~Main%20Featues~Employment%20in%20Australian%20Industry~241.

Australian Bureau of Statistics (ABS). (2013). National regional profile: Tasmania. http://www.abs.gov.au/AUSSTATS/abs@nrp.nsf/Latestproducts/6Population/People12007-2011?opendocument&tabname=Summary&prodno=6&issue=2007-2011.

Badillo, F. (2010). *Global retail outlook*. Columbus, OH: Kantar Retail.

Beagan, B. L., Ristovski-Slijepcevic, S., & Chapman, G. E. (2010). 'People are just becoming more conscious of how everything's connected': 'ethical' food consumption in two regions of Canada. *Sociology, 44*(4), 751−769.

Beniuk, D. (2015). Much more waiting in store as independents take on Big Two in major expansion. *The Mercury*. Available from http://www.themercury.com.au/lifestyle/much-more-waiting-in-store-as-independents-take-on-big-two-in-major-expansion/news-story/7a8846dadecfe3ef87e8ded870553a55.

Day, G. S., & Wensley, R. (1988). Marketing theory with a strategic orientation. *Journal of Marketing, 47*, 79−89.

Department of Infrastructure and Regional Development (DIRD). (2012). *Diversifying Tasmania's economy: analysis and options—Final report*. Canberra, ACT: Commonwealth of Australia.

Ellis, B., & Calantone, R. (1994). Understanding competitive advantage through a strategic retail typology. *Journal of Applied Business Research, 10*(2), 23−33.

Ellis, B., & Kelley, S. W. (1992). Competitive advantage in retailing. *International Review of Retail, Distribution and Consumer Research, 2*(4), 381−396.

Evans, J. R. (2011). Retailing in perspective: the past is a prologue to the future. *The International Review of Retail, Distribution and Consumer Research, 21*(1), 1−31.

Glaetzer, S. (2015). Independents deliver price check on supermarket giants. *The Mercury*. Available from http://www.themercury.com.au/lifestyle/tasweekend-independents-deliver-price-check-on-supermarket-giants/news-story/18e298806b66b26b832ae92afc27b6ad.

Johnston, J. (2008). The citizen-consumer hybrid: Ideological tensions and the case of whole food market. *Theory and Society, 37*, 229−279.

Knox, M. (2015). *Supermarket monsters: the price of Coles and Woolworths' dominance*. Collingwood: RedBack.

Lewis, G. K., Byrom, J., & Grimmer, M. (2015). Collaborative marketing in a premium wine region: The role of horizontal networks. *International Journal of Wine Business Research, 27*(3), 203−219.

Morris, M. H., Kocak, A., & Ozer, A. (2007). Coopetition as a small business strategy: implications for performance. *Journal of Small Business Strategy, 18*(1), 35−55.

Morschett, D., Swoboda, B., & Schramm-Klein, H. (2006). Competitive strategies in retailing: an investigation of the applicability of Porter's framework for food retailers. *Journal of Retailing and Consumer Services, 13*(4), 275−287.

Nonini, D. M. (2013). The local-food movement and the anthropology of global systems. *American Ethnologist, 40*(2), 267−275.

Porter, M. (1980). *Competitive strategy: techniques for analyzing industries and competitors*. New York, NY: Free Press.

Porter, M. (1985). *Competitive advantage: creating and sustaining superior performance*. New York, NY: Free Press.

Productivity Commission. (2014). *Relative costs of doing business in Australia: Retail trade, research report*. Canberra, ACT: Australian Government Productivity Commission.

Schumacher, E. F. (1973). *Small is beautiful: economics as if people mattered*. New York, NY: Harper & Row.

South, S. E. (1980). Competitive advantage: the cornerstone of strategic thinking. *The Journal of Business Strategy, 1*, 15−25.

Tourism Tasmania. (2015). Public relations guide for tourism operators: How to take advantage of Lonely Planet's announcement that Tasmania is one of the world's Top 10 Regions to visit in 2015. http://www.tourismtasmania.com.au/__data/assets/pdf_file/0017/25235/Public-Relations-Guide-LP.pdf.

Wahl, M. (1992). *In-store marketing: a new dimension in the share wars*. Winston-Salem, NC: Wake Forrest University Press.

Wortzel, L. H. (1987). Retailing strategies for today's mature marketplace. *The Journal of Business Strategy, 7*(4), 45−56.

The effect of concentration of retail power on the specialist knowledge of retail staff in the food and beverage sector: A case study of wine retail

3

Rosemarie Neuninger
Department of Marketing, University of Otago, Dunedin, New Zealand

3.1 Introduction

Prior to World War II, the grocery market in Western countries consisted of small independent grocery stores, which were owned and run by local families. These families made an effort to get to know their customers and to maintain a personal connection with them. They "knew where their customers lived, how many children they had, even how they liked their meat cut" (Koch, 2013, p. 6). The customer would ask at the counter for the food items, and grocery shopping had a large focus on customer service. Customers would regularly visit the same store to purchase their products because they were able to obtain store credits. In 1916, Piggly Wiggly, an American supermarket chain, introduced the self-service grocery store where customers were able to walk through the store and select their own products (Koch, 2013).

In contemporary society, most individuals purchase their groceries from convenience stores, supermarkets or hypermarkets, and increasingly online. Convenience stores are usually small (average selling area of 2,800 square feet), closer to consumers, and charge higher prices than supermarkets (Hovhannisyan & Bozic, 2016). These stores are often run by small independent retailers and consumers use them to do their top-up shopping. In contrast, supermarkets are relatively large (average selling area of 4,000–27,000 square feet), on city outskirts, and offer more convenient shopping hours. Here, consumers are able to buy food and household products at more affordable prices. Recently, the market has been dominated by a limited number of large-format, multiple-store retailers where consumers do their bulk shopping (Dobson, Waterson, & Davies, 2003). The food superstore or hypermarket/supercenter has become the dominant retail format in developed and developing countries (Gustafsson, Jönson, Smith, & Sparks, 2006). Here the traditional supermarket has been complemented by warehouse stores, supercenters, and combination stores (Binkley & Connor, 1998). These changes have led to the decline of

Case Studies in Food Retailing and Distribution. DOI: https://doi.org/10.1016/B978-0-08-102037-1.00003-7

independent smaller stores (Hawkes, 2008) and to growing consolidation and concentration amongst retailers.

Recently, governments and organizations have become concerned about the concentration of retail power in a small number of businesses in the food sector (OECD, 2013). Concerns have been expressed in North America and European countries (Dobson et al., 2003; Ivey, 2004), the Asia-Pacific region (Dixon, Hattersley, & Isaacs, 2014), and South America (Hawkes, 2008). Some governments are placing restrictions on the development of large supermarkets to protect small stores (Hawkes, 2008). Although the main concern about this concentration of power has been its effects on prices for consumers and farmers, the concentration of retail power might also have other effects. For example, it may lead to a loss of expertise amongst retail staff with respect to the food products they are selling, particularly where large supermarket chains out-compete specialty retailers with specialist knowledge. This reduction in knowledge may affect the stores' inventory, as retailers with expert knowledge can use this to purchase high-quality and specialty food products. The reduction in knowledge may also affect the ability of staff to communicate with consumers about food products.

3.2   Wine retailers as a case study

For some product categories, expert advice is more important than for others. Aqueveque (2006) argues that expert opinion is particularly important for "experience goods" (products whose quality is difficult to judge in advance), such as wine, because it reflects a measure of objective quality and reduces the consumer's perceived risk associated with purchase decisions. When purchasing "experience goods," less knowledgeable consumers might desire guidance from an expert. In such cases, consumers might consider wine retail managers, either in a supermarket or in a wine specialty store, as experts and seek their advice.

This chapter will investigate the effect of current retail trends on staff expertise using a case study of supermarkets and wine retailers in New Zealand. Knowledge of wines amongst staff in different retailers will be explored, using wine competition results as a proxy. The chapter will investigate their relationships and communication of their expertise with customers. The chapter will also consider the experience of New Zealand consumers when shopping for wine, in particular regarding their interactions with retail staff at different retailers. This study has possible implications for the theoretical framework of service-centered marketing logic (Vargo & Lusch, 2004).

3.3   The concentration of food retail power in New Zealand

Grocery shopping in New Zealand has followed worldwide trends. The concentration of retail power has developed more recently than in, for example, the USA, but is now arguably more extensive. Today, in New Zealand, most grocery retailing is concentrated into two large companies: Foodstuffs and Progressive Enterprises. These companies own all the major supermarkets in the country. Foodstuffs are New Zealand's biggest grocery distributor. Foodstuffs North Island Ltd. owns the supermarkets: New World, Pak'n Save, Four Square, Shoprite and Write Price, as well as the off-license liquor chain Liquorland, and two wholesalers: Gilmours and Trends Wholesale (Foodstuffs North Island Ltd., 2017). Foodstuffs South Island Ltd. operates the supermarkets: New World, Pak'n Save, Four Square, On the Spot Convenience and Raeward Fresh, the off-license liquor chain Henry's, and the wholesaler Trends Wholesale Ltd. (Foodstuffs New Zealand, 2017). Progressive Enterprises is the second largest grocery company in New Zealand and is part of Woolworths Limited (the Australian parent company). It operates the Countdown supermarkets and is franchisor of the Super Value and Fresh Choice supermarkets (Progressive Enterprises Limited, 2017). These companies operate the majority of New Zealand's grocery retailing. The share of wine sales in New Zealand super-markets has increased from 30% in the early 2000s to almost 50% in 2016 (ANZ Bank New Zealand Limited, 2016). This rise may be due to the supermarkets' abil-ity to bulk sell alcohol very cheaply, and also to sell wine within one-stop-shopping, allowing them to out-compete standalone wine retailers.

3.4 Method

This chapter uses a case study of supermarkets and wine retailers in Dunedin, New Zealand. As the study was exploratory, two qualitative research techniques—focus groups and in-depth interviews—were chosen for their ability to capture rich per-spectives; advantages that might be less easily achieved with quantitative methods, such as surveys. The two techniques allowed for a deeper exploration of partici-pants' words and emotions (Neuninger, 2017) regarding the topic of investigation (Krueger & Casey, 2015). Focus groups have the unique property of allowing unre-stricted interaction amongst participants, who are able to express their ideas sponta-neously and to share their thoughts and feelings. One-on-one interviews, on the other hand, are useful when investigating topics that are too complex for focus groups, and that require a deep understanding of the experiences of different indivi-duals and the meanings they make of those experiences (Neuninger, 2017).

3.4.1 Data collection and analysis

Following the method of Krueger and Casey (2015), four focus groups were con-ducted. A total of 44 wine drinkers, 22 females and 22 males, participated. Participants were allocated to a focus group according to their level of involvement with wine (Neuninger, Mather, & Duncan, 2017). The discussions explored their attitudes toward different wine retailers, and details of their wine purchase decision-making. Then, seven in-depth interviews were conducted (Gubrium,

Holstein, Marvasti, & Mckinney, 2012) with three wine specialist store managers, two managers of general off-license chains, and two department managers of grocery retailers (one from Foodstuffs and the other one from Progressive Enterprises). A set of standardized questions was prepared and the interviews were semi-structured in style. The participants were interviewed to explore their perceptions and opinions on their own levels of expertise in the wine retail environment and to understand the types of questions that consumers ask them. Each session lasted from 30 minutes to over an hour. All focus groups and in-depth interviews were audiorecorded and transcribed verbatim (Johnson & Rowlands, 2012), and thematic analysis was used (Braun & Clarke, 2006; Neuninger et al., 2017). The transcripts were analyzed for emerging themes concerning: (1) the retail environment where consumers purchase wine and the reasons for selecting that environment; (2) consumers' perceptions of retail managers' expertise in the different types of retail environment; (3) retail managers' perceptions about their own level of expertise in the different types of retail environment; and (4) the evidence of the managers' expertise. The participants' knowledge and perceptions of wine competitions were used as a proxy for specialist wine knowledge.

3.5   Findings and discussion

3.5.1   Findings from the focus groups with wine consumers

3.5.1.1   The supermarket

This study determined the retail environment where consumers purchase wine and the reasons for selecting that particular environment. Participants across all focus groups purchased their wine in both the supermarket and wine specialty stores. However, participants predominantly purchased wine at the supermarket. The majority of participants indicated they preferred the convenience of being able to buy wine when doing their other grocery shopping. This is consistent with the finding of Hawkes (2008) that consumers liked the convenience of purchasing all groceries at one place. This was described as being "easy" (FG4). Many of the participants also mentioned that their preference toward wine shopping at the supermarket was due to the low prices and frequent sales available:

> *At the supermarket if the wine has come down on price I will certainly buy it . . . (FG1).*

> *It's not super specialist and there might be a lower price range (FG4).*

Participants also discussed the promotional activities used by supermarkets to encourage purchase behavior (Ritchie, Elliott, & Flynn, 2010). Such activities include cheap deals on international and national top wine brands for which the supermarkets intend to build and maintain customer loyalty. These sales were noticed by consumers when passing the wine aisle and encouraged purchase behavior. Participants responded to wine branding in conjunction with sales:

> *Sometimes when I'm walking through the supermarket and see one that I know that it is a quite good brand is on special, I would grab it (FG1).*

The majority of the participants noted that supermarkets were also smart in their promotional strategies (Bäckström & Johansson, 2006). In agreement with Ritchie et al. (2010), in-store tastings in the supermarket were often appreciated by consumers:

> *Supermarkets sometimes have someone with the tiny little plastic cups and you can taste it (FG2).*

> *...and quite often they will make tastings there as well, so people can always go and try it before you buy it, which is nice (FG4).*

> *If you are in the supermarket where there isn't somebody to recommend, they have the shelf talkers and various things (FG3).*

In addition, it appeared that the selection of national and international wine at the supermarket has now expanded to rival that of specialty wine stores, providing further incentives for respondents. For example, one of the participants said that she was buying wine at "[supermarket name] because of the broad selection" (FG4).

3.5.1.2   The wine specialty store

While wine was predominantly purchased in the supermarket, many respondents also talked about the reasons for occasionally going to a wine specialty store. Those who enjoyed drinking wine and who wanted to "buy a more expensive bottle" or "a new case [of wine]" (FG1) would go to a wine specialty store. Participants thought that the wine specialty store was the right place to get additional information about wine, particularly when considering purchasing quality or expensive wine, or a larger quantity of wine. The supermarket was not the right place to make bigger purchases as explained by one of the participants:

> *If I wanted to buy a new case [of wine] I would go and see a specialist, I wouldn't buy a good wine at the supermarket ... I am not going to go to the supermarket to stock up (FG1).*

Other participants emphasized how they felt that in the wine specialty store they could talk to a knowledgeable salesperson. This interaction was particularly important when participants felt they needed answers to wine-related questions regarding production, vintage, flavor, and pairing. Those who worked in the wine specialty store and who had many years of work experience were perceived as being more knowledgeable than those in the supermarket:

> *I quite enjoy wine so I generally go to a specialist shop because if you want to ask questions, people tend to have more specialist knowledge so they can tell you where the wine comes from, about the different flavors and things like that ... generally you get a good answer (FG4).*

> *I really enjoy shopping for wine in the wine shops and there is one really good one [in the city]. The staff there [are] really knowledgeable and I think that they have been doing their job for many years ... If I want to try something new then I would go and talk to [the sales person in the wine specialty store] about what I would want it for and who do I want it for ... I wouldn't buy an expensive and unknown wine from the supermarket (FG1).*

Other participants emphasized how they would go to a wine specialty store to buy wine for a special occasion (e.g., as a gift). Talking to a salesperson was particularly important when wanting to buy a particular type of wine. Participants also talked about how the salesperson in the wine specialty store was generally able to match the wine according to what they had described to them and how much they would trust them:

> *The two times I had to buy wine for a gift for someone who really liked red wine, I went to [the wine specialty store] and I said, "Listen, this is what I want to spend" ... and I really trusted them. We all trust the man at [the wine specialty store] (FG2).*

> *I just couldn't believe how well it was how much I wanted ... I keep going back to get that wine and to try some more (FG1).*

3.5.2 Findings from the in-depth interviews with wine retail managers

The findings of the second part of this study focused on retail managers' expertise in different wine retail environments. It was found that the expertise of wine retail managers in the wine specialty store is greater than that of those who worked at the supermarket or off-license stores. Out of the seven participants (P1–P7), four had more than 20 years of work experience in the wine retail industry (P4–P7). These experienced participants included all those at the specialty wine stores, and one of the managers of a chain off-license. The remaining three participants (from the supermarkets and the other chain off-license manager) had worked much less in the wine retail industry: one had worked for 2.5 years (P1), one for 7 years (P2), and the other one for 10 years (P3, who was the chain off-license manager). At least one of the specialty wine store managers had also undertaken formal wine training. In this section, empirical findings from retail managers' perspectives are presented.

3.5.2.1 Supermarket and chain off-license

Retail managers who worked in the supermarket said that they did not consider themselves to be experts. Nevertheless, one retail manager admitted that although his knowledge about wine was not extensive, he knew enough about "the differences between all the sort of variants of wine and that sort of thing" (P2). Another indicated that despite a lack of in-depth expertise, he was able to "match the consumer with the type of bottle of wine that will suit them" (P1). The other supermarket and off-license retailers matched this level of competence, knowing basic differences between the different wine varietals:

> *I know the difference in variance ... between white and red and Shiraz and Merlot and Merlot Cab ... I definitely wouldn't say I'm an expert. I know ... the procedure and how they're made, obviously with the skins and stuff, but I don't go round... tasting wines and know what's a good wine compared to a bad wine ... Probably just intermediate [knowledge] (P3).*

The level of expertise was reflected in the amount of product used by the retailers. For instance, although working in an off-license retail shop where customers are reliant on retail managers, one respondent admitted he did not have the knowledge to advise on the quality of the wine and drank little:

> *I've probably tasted a few of the wines in the store, but again, I probably wouldn't know what is good compared to another one ... I don't really drink wine that much (P3).*

All the retailers in the supermarket and off-license chain emphasized that it was their basic knowledge (often gained from reading the back of the wine bottles while stacking the shelves) that allow them to provide customers with general and limited advice on the wine characteristics and food pairings:

> *I point them in the right direction and tell them what sells and what is a good wine to go with ... what sort of dish ... a red wine with a beef ... or white wine with chicken... I just ... read the back and ... whatever it says on the bottle, I'll ... memorize through putting it on the shelf a thousand times (P2).*

Our results also suggest that this group of wine retailers thought that they were perceived by their customers as being knowledgeable about wines, despite their actual limited wine knowledge and brief wine retail experience:

> *I think they probably think we're experts. So we don't tell them we're not ... you've got to remember here people just really want bang for their buck ... you still get the people that are looking for really nice wines and things but we don't sell those. We're mainstream. It's 10 percent of our income so it's not something we put a great store on now (P4).*

To assess the retail managers' specialist knowledge, the topic of wine awards was used as a proxy for the participants' wine knowledge. Most supermarket and chain off-license managers displayed a lack of knowledge of wine awards. When asked to talk about their knowledge of wine awards, one of the participants said he didn't "really know a lot about them" (P2). Another said that he didn't have "any experience with them" (P3). In their personal wine consumption, awards played no role in their purchase decisions:

I like to try the wine and make my own decision (P4).

I don't look for awards. I just ... look for something that I've drunk and I've liked (P3).

Results also suggest that their interest in wine awards largely concerned consumer reaction to stickers:

Well people like bottles of wine with medals on them. The more medals, the more likely they will pick them up by themselves. I have my doubts sometimes on how true they are (P4).

No, don't really notice what awards. I mean you get New Zealand awards and then overseas awards. All the supplier does is really just stick a gold thing on it if they've won it (P3).

Other participants had similar opinions that the stickers were only important as a visual attraction.

I also have an interest in which wines are winning which awards. When a sales rep comes in ... and if they want to sell me a wine ... one of the selling points they'll use, "It's been awarded these awards", and one question I'll ask is, "When you deliver these wines, are they going to be stickered?" (P1).

3.5.2.2 Wine specialty store

In contrast to the supermarket and chain off-license managers, the wine specialty store managers regarded themselves as wine experts. One participant said:

I would think I'd be pretty good in that respect I would say ... We've plenty of experience over the years tasting wines so I don't know whether I'd be up to a top wine judge's level, but certainly I would say I was more than competent in that respect (P5).

The specialist managers thought consumers saw them as either expert or reasonably expert: "I think they would tend to regard what we have to offer and the product knowledge that we have as being at the expert level" (P6). Surprisingly, two out of three specialist managers were slightly modest about their perceived image. For example, one of the managers discussed the perceived ranking by consumers when coming to the wine specialty store by saying, "In this particular forum ... they need to see it as higher than theirs so ... 1−9, it would have to be 6 or 7" (P7).

In contrast to the supermarket and chain off-license managers, the wine specialty store managers were knowledgeable about their individual products and had personally tasted most of them.

> *We like to try and have the smaller, more boutique vineyards and ... we try probably about 80 percent of the wines that we stock before we stock them, and we try because we've got to hand sell them, we've got to know what the wines are like (P5).*

The consumers shopped at the wine specialty stores for a number of reasons including the perceived expertise of the wine retailer:

> *I'd like to think people come in here because they can get a bit of advice ... The wine industry ... we're awash with wine from, not just ... here but from all over the place, and I think to most people who aren't that savvy with wine, when they're confronted with rows and rows of wines and some of them have medals on them, some of them don't ... some of them have got rather fancy labels ... people are a little bit overawed by that and ... they will come in somewhere like here and say, "Look, I want to spend X amount, what have you got, what could you recommend?" ... People would come in here specifically ... to get something that maybe they would be a bit wary about spending NZ$40 in the supermarket and not really knowing what they're getting (P4).*

In contrast to the other retailers, the specialty store retail managers had knowledge about the wine awards process, and strong opinions on the awards procedures and standards. These opinions were generally strongly critical:

> *I think they're a complete waste of time. Because again, they've divorced themselves from the reality of drinking a wine ... a person tasting 1,000 wines in a wine tasting has no more ability at tasting the third or fourth label sensibly, simply because as a human being his taste receptors are completely shot after the third glass (P6).*

This detailed knowledge and strong personal convictions were coupled with a concern about the impact and number of awards, based on a strong involvement with the industry:

> *It's changed over the years. I think if you'd asked me maybe 10 years ago what the impact of wine awards was in this country, I would have said it was quite significant and I would say now it is quite different ... I'm certainly in the industry and I can't keep up with it all, so I don't know how the consumer is supposed to be able to keep with it all (P5).*

The specialist retailers believed that the wine awards were used in supermarkets where the staff had a lack of personal knowledge and expertise, and that their own expertise rendered wine awards redundant in specialty stores:

> *I think they perform a very important message or communication at a particular level of the market. That's a level where ... a wine stands alone on a shelf without any external advice or communication or recommendation around it (P7).*

> *It's probably more beneficial for a wine to have medals on it in a supermarket setting just because there isn't anybody there generally speaking to advise people (P5).*

The specialty store retailers' opinions that awards were more influential in supermarkets were consistent with the observations of the supermarket and general off-license retail managers that consumers were clearly influenced by awards in their stores. The specialty retailers, however, were influenced by some particular wine awards, which they felt retained credibility:

> *... less and less but still they must have some wine awards I still respect because the criteria is still integrity (P7).*

> *If we did have a wine that ... had a trophy of a big wine award and we thought it was particularly good, I might sort of mention that (P5).*

The wine specialty store managers only took the presence of particular awards as a weak indication to try the wine themselves before buying for stock. They then used their personal taste to corroborate or discard award information:

> *We don't believe the distributors and the winemakers when they tell us that, "This wine is fabulous, it's won gold medals, we should buy this and sell it for that." We have to take the wine, look at it, taste it and decide whether it is in actual fact worth the money that they are actually charging, and if it is not we reject it (P6).*

3.6 Conclusions

The goal of this chapter was to investigate the effect of retail concentration with respect to "experience goods." Based on a case study of supermarkets and wine retailers in New Zealand, this study has demonstrated that there is a lack of expertise in bulk retailers. The majority of the goods are sold in the bulk retailers with a small market for specialist stores with expert retail staff.

Overall, specialist retailers' expert knowledge affects the stores' inventory, as the retailers use their knowledge to purchase high-quality and specialty wine products. This knowledge also affects the ability of staff to communicate with consumers about the wine products, with specialist retailers offering detailed advice reflecting their high involvement with the industry. These retail managers could be classified as "expert-influencers." The staff of supermarkets and general off-licenses still thought they were perceived as experts, but might more accurately be classified as "pseudo-expert influencers," offering basic advice gleaned from the labeling on wine products.

Most importantly, results presented in this chapter indicate that the concentration of power may lead to a loss of expertise of retail staff with respect to the products they are selling, as large supermarket chains out-compete those specialty retailers that often hold specialist knowledge. However, although the one-stop-shopping concept is increasingly seen as being convenient, there is still currently enough demand from consumers for expert advice that a small number of specialized retailers still survive.

Although most theoretical attention has been devoted to price and variety as a function of retail power (Inderst & Shaffer, 2007), there has been a recent recognition of the need to include service in the theoretical models (Vargo & Lusch, 2004). Such service includes the on-going relationships between skilled and knowledgeable retail staff and consumers. This study's findings contribute to the theoretical framework of retail power and "experience goods" by highlighting need to include the role of expert knowledge in marketing and economic models.

In order to improve customer experience and to build long lasting relationships and customer loyalty, it is necessary that "pseudo-expert influencers" become "expert-influencers." Supermarkets can achieve this by providing product training and tastings on a regular basis that will increase product familiarity and sales. Then consumers will retain the benefit of expert advice despite retail power concentration.

References

ANZ Bank New Zealand Limited. (2016). *New Zealand wine industry: Full bodied growth.* Wellington: ANZ Bank New Zealand Limited.

Aqueveque, C. (2006). Extrinsic cues and perceived risk: The influence of consumption situation. *Journal of Consumer Marketing, 23,* 237−247.

Bäckström, K., & Johansson, U. (2006). Creating and consuming experiences in retail store environments: Comparing retailer and consumer perspectives. *Journal of Retailing and Consumer Services, 13,* 417−430.

Binkley, J. K., & Connor, J. M. (1998). Grocery market pricing and the new competitive environment. *Journal of Retailing, 74,* 273−294.

Braun, V., & Clarke, V. (2006). Using thematic analysis in psychology. *Qualitative Research in Psychology, 3,* 77−101.

Dixon, J., Hattersley, L., & Isaacs, B. (2014). Transgressing retail: supermarkets, liminoid power and the metabolic rift. In *Food Transgressions: Making Sense of Contemporary Food Politics,* (pp. 131−153). Routledge, Abingdon.

Dobson, P. W., Waterson, M., & Davies, S. W. (2003). The patterns and implications of increasing concentration in European food retailing. *Journal of Agricultural Economics, 54,* 111−125.

Foodstuffs New Zealand. (2017). *About foodstuffs* [Online]. Available: https://www.food-stuffs.co.nz/about-foodstuffs/ [Accessed 20 January 2017].

Foodstuffs North Island Ltd. (2017). *Concise annual report for foodstuffs North Island Limited. Year ending 2 April 2017.*

Gubrium, J. F., Holstein, J. A., Marvasti, A. B., & Mckinney, K. D. (2012). *The Sage handbook of interview research: The complexity of the craft.* Thousand Oaks, CA: Sage.

Gustafsson, K., Jönson, G., Smith, D., & Sparks, L. (2006). *Retailing logistics & fresh food packaging: Managing change in the supply chain.* London. Philadelphia, PA: Kogan Page Limited.

Hawkes, C. (2008). Dietary implications of supermarket development: A global perspective. *Development Policy Review, 26,* 657−692.

Hovhannisyan, V., & Bozic, M. (2016). The effects of retail concentration on retail dairy product prices in the United States. *Journal of Dairy Science, 99,* 4928−4938.

Inderst, R., & Shaffer, G. (2007). Retail mergers, buyer power and product variety. *The Economic Journal, 117,* 45−67.

Ivey, I. (2004). *Looking over the wall. NZ business.* Auckland: Adrenalin Publishing Ltd.

Johnson, J. M., & Rowlands, T. (2012). The interpersonal dynamics of in-depth interviewing. In J. F. Gubrium, J. A. Holstein, A. B. Marvasti, & K. D. Mckinney (Eds.), *The Sage handbook of interview research. The complexity of the craft* (2nd ed.). London: Sage.

Koch, S. L. (2013). *A theory of grocery shopping: Food, choice and conflict.* Bloomsbury, London.

Krueger, R., & Casey, M. A. (2015). *Focus groups: A practical guide for applied research.* New Delhi: Sage.

Neuninger, R. (2017). Hospitality marketing methodologies: Qualitative marketing methodologies. In D. Gursoy (Ed.), *Routledge handbook of hospitality marketing.* Abingdon: Routledge.

Neuninger, R., Mather, D., & Duncan, T. (2017). Consumer's skepticism of wine awards: A study of consumers' use of wine awards. *Journal of Retailing and Consumer Services, 35,* 98−105.

OECD. (2013). *Competition issues in the food chain industry.* Available: https://www.oecd.org/daf/competition/CompetitionIssuesintheFoodChainIndustry.pdf

Progressive Enterprises Limited. (2017). *Welcome* [Online]. Available: http://www.progressive.co.nz [Accessed February 10 2017].

Ritchie, C., Elliott, G., & Flynn, M. (2010). Buying wine on promotion is trading-up in UK supermarkets: A case study in Wales and Northern Ireland. *International Journal of Wine Business Research, 22,* 102−121.

Vargo, S. L., & Lusch, R. F. (2004). Evolving to a new dominant logic for marketing. *Journal of Marketing, 68,* 1−17.

TazeDirekt.com: Branding charm or operational basics?

4

Selcen Ozturkcan[1] and Deniz Tuncalp[2]
[1]Linnaeus University, Kalmar, Sweden, [2]Faculty of Management, Istanbul Technical University, Turkey

4.1 Introduction

In December 2015, Hasan Aslanoba, the founder of TazeDirekt, a loved online grocery e-tailer brand (Batra, Ahuvia, & Bagozzi, 2012), announced that plans for the year 2016 involved 700% growth in Turkey (Demirel, 2015). However, only a few months later, in February 2016, news broke of the sudden closure of the company (Kutsal, 2016), and subsequently of its acquisition by the Migros retail store chain in November 2017. This case study describes the rise and fall of TazeDirekt, analyzes the reasons for its failure, and explains how the scaling up of a venture is never enough. All start-ups, including e-tailers—and contrary to what many start-up entrepreneurs seem to believe—are dependent on favorable unit economics and require sound business practice for survival.

TazeDirekt had "cloned" aspects of US online retailer FreshDirect.com's operations to Turkey, bearing a very similar name and faced with growing e-commerce demand from the Turkish market. Like FreshDirect, TazeDirekt were e-tailing organic, genetic-modification-free food from small, local producers, like Makarna Lütfen, Gündoğdu Peynirleri, and Tire Süt Kooperatifi, providing them directly to a medium-to-high income consumer segment, consisting of health-conscious and gourmet individuals in three big Turkish cities (Istanbul, Ankara, Bursa). Besides sourcing from local producers, TazeDirekt was also selling eggs, chickens, and other meat products from Mr. Aslanoba's massive farm in Boğazköy, Bursa (Fig. 4.1). The ownership of the farm provided TazeDirekt with an excellent opportunity to better manage the quality of its meat products, as they were coming from the naturally grown and grass-fed animals of this farm. As e-commerce in Turkey grows, TazeDirekt's business value had the potential to achieve significant growth as an early mover.

4.2 Online retailing in Turkey

Regarding internet penetration, Turkey was lagging behind the United States, where the penetration was around 80% among households as early as 2001. There were

Case Studies in Food Retailing and Distribution. DOI: https://doi.org/10.1016/B978-0-08-102037-1.00004-9

Figure 4.1 Aslanoba Farm in Boğazköy, Bursa, Turkey.

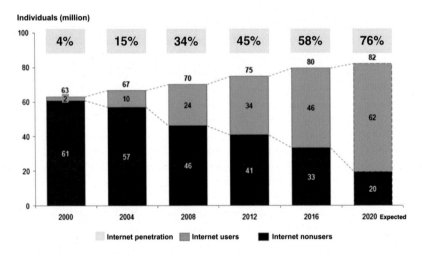

Figure 4.2 Penetration of the internet in Turkey.
Source: BKM (2017).

around 46 million internet users in Turkey in 2016, which amounted to only 58% penetration (Fig. 4.2), indicating considerable potential into the future. e-Commerce in Turkey reached USD4.96 billion [17.5 billion Turkish Lira (TL)] in 2016, growing 34% in the 3 years to this point (Fig. 4.3). However, the ratio of the online retailing remained only at 3.5% of the overall retail market, which fell behind the world average of 8.5%. By contrast, the 65% penetration of smartphones in Turkey exceeded the world average of 60%. Despite the high smartphone penetration, the share of mobile commerce lingered around 19%, while the world average was 44%. The major e-commerce verticals in the Turkish market included pure-play retailing,

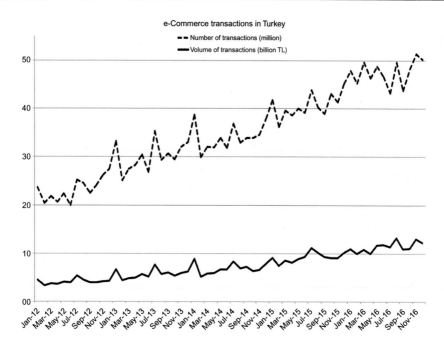

Figure 4.3 e-Commerce transactions in Turkey.
Source: BKM (2017).

travel and vacation, clicks-and-mortar retailing, and online gambling, in their respective order of size (Fig. 4.4). Among these four sectors, pure-play retailing captured the highest annual growth of 7% and 20% in the years 2015 and 2016, respectively.

Shoppers across the globe value online-shopping due to its convenience; however, most Turkish consumers value online-shopping primarily and predominantly for the availability of lower prices. Only one out of every three Turkish consumers chose to complete her/his shopping online, while 25% of those online shoppers often complain about after-sales service (Kantarcı, Özalp, Sezginsoy, Özaşkınlı, & Cavlak, 2017). The most common problems that shoppers encounter are prolonged delivery times and the delivery of damaged or wrong goods. When internet users are considered, again only one in every three internet users has shopped online in Turkey (TUIK, 2016). According to Interbank Card Center, 36% of online shoppers are between 25 and 34 years of age, which is followed by 32%, 24%, and 18% of online shoppers that are 18−24, 35−44, and 45−54 years of age, respectively Canko (2015). While lack of trust has been listed as a primary concern in choosing not to shop online (Canko (2015), there is significant potential for growth for e-tailers, both food and non-food, if they can solve these problems in the eyes of their customers.

Previous research indicated that the average food expense of consumers in Turkey was USD306 (TL555) per month in 2012. As the tendency to spend more

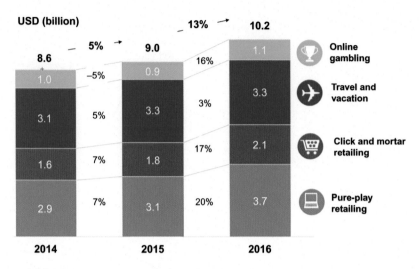

Figure 4.4 Major e-commerce verticals in Turkey.
Source: TÜBİSAD (2017).

money on food has increased, the inclination to perceive e-retailing as a reliable option has also improved (Sayılı & Büyükoğlu, 2012), giving a positive outlook for an online grocery offer.

4.3 Scale up to survive?

In 2014, using funds from the 2006 sale of the successful bottled water production brand that his family created in the 1960s, Mr. Aslanoba established a new brand named TazeDirekt for online food retailing (Fig. 4.5). He had initiated organic farming, an online logistics auction and optimization infrastructure, and livestock production in the years 2008, 2011, and 2013, respectively. These preparations had provided the necessary steps toward TazeDirekt's launch in 2014.

Mr. Aslanoba hoped that his new e-tailing business could revolutionize the food retail and distribution industry in Turkey, delivering a wide variety of food products directly to consumers (Fig. 4.6) and offering personalization through the intensive use of technology. Previous research indicated that online grocery shopping could involve several advantages, such as: convenience, access to the broader product range, and price-based shopping—despite the fear of receiving inferior quality delivered goods from online grocery providers and a reduced leisure dimension of shopping that might be inherent in bricks-and-mortar retail alternatives (Ramus & Nielsen, 2005). In line with these research findings, TazeDirekt offered a comfortable online shopping experience with the promise of next day delivery of a wide variety of perfectly conditioned food products.

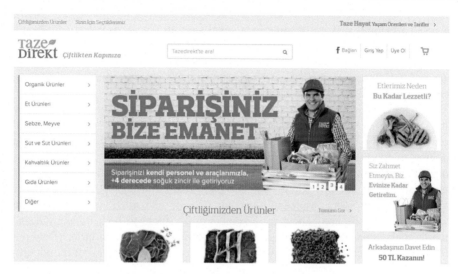

Figure 4.5 TazeDirekt.
Source: https://www.tazedirekt.com/.

Figure 4.6 A selection of TazeDirekt.com products.

Mr. Aslanoba began to undertake fully certified organic farming with plans to launch TazeDirekt in the next 3−4 years. Livestock production at the 15,000 m^2 Bursa facility and the gigantic warehouse and operations center in Mustafakemalpasa (Bursa) were all planned steps toward realizing TazeDirekt. Last, but not least, a fully equipped cold logistics chain was built for TazeDirekt. All in all, Mr. Aslanoba had undertaken an investment of USD35 million to provide most integrated elements of the value chain in-house, which included breeding farms as well as meat shredding and packaging facilities (Gökbayrak, 2016). Hence, TazeDirekt was not starting as a small start-up but as an investment-heavy venture that ensured high service levels and superior product quality from the start. Part of this effort went into preparing a fully automated and integrated order processing and fulfillment system based on SAP enterprise resource planning software, across various entities of its value chain.

TazeDirekt had aimed to remove all intermediaries and to become a vertically integrated online retailer, controlling the end-to-end value chain by bringing food products directly from its farms to its customers' kitchens (Descartes, 2014). In addition to food production and logistics, almost every function was handled in-house, including software design, search engine optimization, social media management, and call center operations. The company had more than 300 employees to handle deliveries every weekday. Around 40% of all orders were placed via TazeDirekt's mobile app. There were some 1,500 visitors to both its website and mobile application, and, on average, 5% of visitors had placed an order (Demirel, 2016). Most new customers came with a friend referral, making word-of-mouth crucial for TazeDirekt's customer base expansion. TazeDirekt had reached approximately 1,300 daily orders by around 1 year into its launch. It had reached monthly revenues of USD617,000 (TL1.8 million) by December 2015 and was growing 30% in revenue every month.

4.3.1 The magic or the curse

Although growth is considered the ultimate ideal for many start-ups, the fast-growing operations of TazeDirekt were indeed its curse. There was long-term high growth for TazeDirekt, which may have attracted an endless venture capital (VC) investment in another start-up ecosystem like San Francisco, where scaling seems to be the ultimate objective. However, in Turkey, there was no such VC money. Mr. Aslanoba was the top "angel" with the highest number of investments in the country. An endless stream of entrepreneurs was trying to reach him to get an opportunity to pitch. Likewise, he required more than a hopeful story to continue investing in his venture.

The market demand for online grocery was rising, and it had to be fulfilled either by TazeDirekt or by its many competitors, like tazemasa.com or memlekettengelsin.com. Hence, TazeDirekt could not risk ignoring the fast-growing demand in the market. However, fast-scaling also brought higher costs to the company. For sustainable growth, TazeDirekt had to provide organic food certifications and enhance its cost structure by improving its product variety, geographical expansion, logistics, and digital marketing.

4.3.2 Organic food certifications: costs and benefits

Most Turkish consumers assumed that buying food directly from a producer was the best way possible to ensure good quality, which often included significant assumptions about organic farming. Perception-wise, buying from remotely located producers, or if not from the weekly producer markets in the nearby neighborhoods, was considered the best alternative one could have for accessing organic and healthy food products traditionally. The tradition brought several challenges for organic producers including TazeDirekt. The certification process for organic produce was costly. Organic food certification had to be explained to consumers in detail, to address their misconceptions and misinterpretations of organic food. TazeDirekt, on the other hand, had a minimalist design for its private-label product packages which opted for brief product descriptions that limited any efforts in conveying information about the certifications obtained. Therefore, even though certification-related costs were already incurred, the associated benefits were mostly underutilized, particularly in justifying the marginally higher price tags. Small product labels that include little information could not include details of certifications. Labeling with certification information could have proved particularly important for encouraging hesitant first-time shoppers to purchase the product. On some online customer communities, some of the customers started questioning whether the products on TazeDirekt were organic, with valid certification, despite their high prices (SikayetVar, 2015b). The company was at first reluctant to respond to these issues, but in the end they provided detailed accounts of audit reports and organic certificates, particularly when the complaints grew and the company started to receive more extensive attention.

4.3.3 Product variety

TazeDirekt began its operation by offering fresh vegetables, fruits, and dairy and meat products. All these products were produced, sold, and delivered by TazeDirekt. To expand its product variety, TazeDirekt also began selling other products that were not necessarily organic. For example, they added local producers' selected products and vegetable dishes to the product range, which did not directly contradict the value proposition. However, to be able to offer full service, like competing supermarkets, they also added regular products that could have been found in any supermarket, potentially weakening their market position.

Increasing product variety with third-party products had also brought numerous managerial difficulties, further increasing the costs of quality control, service, and supply-chain management. It also put the company in a comparable position with supermarket chains like Macro Center (part of the Migros group), Gourmet Garage, Eataly, and also the medium-to-high segment Migros chain. Availability of select third-party products on TazeDirekt made price comparisons more accessible for consumers; and TazeDirekt were 20%−30% more expensive, due to the company's smaller size, higher cost structure, and affluent positioning.

4.3.4 Geographical expansion and logistics

Many grocery e-tail start-ups have been faced with "the last mile" problem, which refers to difficulties transporting goods to their ultimate destination. This has resulted in eventual failure for such companies. Laseter, Houston, Chung, Byrne, and Turner, (2000) identify four key challenges: "the limited online potential; the high cost of delivery; the selection-variety trade-offs; and existing entrenched competition." Ring and Tigert (2001), on the other hand, identified picking and delivery costs as the "killer costs" of online grocery. Also, existing consumer behavior regarding store shopping was robust, requiring costly promotions to attract them toward the online options.

FreshDirect.com avoided some of these problems, as its core market, New York City, suggested a sufficient level of sales density and consumers were open to flexible delivery windows (Laseter & Rabinovich, 2011, p. 103). Taking a direct-business model like Dell Inc., FreshDirect was able to remove intermediaries in the food value chain and served the customer's requests with custom-prepared food at lower prices (Laseter, Berg, & Turner, 2003). FreshDirect has not been offering precision in delivery timing and has made the most of its deliveries in the evenings and on weekends due to better traffic conditions and to drive down delivery costs in urban areas. In suburban areas, its refrigerated trailers come to the parking lots of train stations or office parks, where customers can collect their orders themselves. With a highly automated food processing facility just outside the New York City, the company has been able to decrease inventory costs without sacrificing speed. FreshDirect also implemented a prudent geographical expansion strategy, neighborhood-by-neighborhood, to keep avoiding the last mile problem as the company grows (Laseter & Rabinovich, 2011), and thereby optimizing its order densities and delivery policies (Boyer, Prud'homme, & Chung, 2009). Could TazeDirekt attain a similar level of effectiveness and efficiency while reaching its customers and expanding in Turkey?

TazeDirekt had first begun its market penetration in Bursa (denoted with a red star in Fig. 4.7), where its production and main farming facilities were also located.

Figure 4.7 TazeDirekt.com geographical expansion.
Source: Compiled by the authors.

As the fourth largest city in Turkey with 3 million people, Bursa seemed to provide an available market opportunity, where the proximity of delivery addresses to production sites enabled better logistics operations, especially for perishable products. However, nearby Istanbul also represented an outstanding opportunity with 15 million residents, hosting nearly 20% of the Turkish population. TazeDirekt's management decided to expand its operations not only to Istanbul, but also to Ankara, Izmir, and Balıkesir; cities with populations of 5.3 million, 4.2 million, and 1.2 million, respectively.

However, this geographic expansion involved numerous challenges for TazeDirekt. First of all, while these cities are big, their populations have been dispersed across a more extensive geographical area, unlike New York City. For example, Istanbul's urban area holds 15 million people in 1,539 km^2, whereas New York City has some 8.5 million people in just 789 km^2. Sometimes, finding an address becomes a daunting and time-consuming task (SikayetVar, 2016). Second, while Balıkesir is relatively close, Ankara and Istanbul are far away from the main operation center of TazeDirekt in Bursa. Therefore, to expand into these cities, the company had to incur the costs and other burdens of long-haul logistics. The distance between these cities not only increased operational costs and lowered delivery speed but also limited the shelf life of freshly prepared products (SikayetVar, 2015a). Thirdly, there is also intense competition for attracting organic food demand. In general, many consumers choose to connect directly with rural farmers from different parts of the country. Some farmers and producers were also running online websites that shipped fresh produce to consumers by post to big cities weekly. Especially in Ankara and Istanbul, gourmet supermarkets with organic food options and organic produce shops have been located in many neighborhoods. Most of these supermarket chains and produce shops also provide home delivery for their customers. Furthermore, farmers' markets and organic food markets were available in many different urban locations periodically. Therefore, attracting demand was not an easy task.

4.3.5 Digital marketing

As a pure-play online retailer, TazeDirekt heavily relied on digital marketing practices to reach its target customers. It was crucial to appear—if possible with a high ranking—on the listed results of the search engine inquiries involving specific keywords that were related to TazeDirekt's offers. Google Trend analysis of the keyword "TazeDirekt" suggests that steady growth was captured in internet searches from Turkey (Fig. 4.8). As of February 25, 2016, a few days after the shutdown of business, Tazedirekt.com ranked 87,209 globally and 2,945 nationally on Alexa.com, a popular website ranking service. In the last 3 months, it had 445,000 total visitors that had spent around 3.05 minutes on the site to view 3.9 pages, on average. Among these visitors, 98% of them were from Turkey. Most desktop visitors landed on TazeDirekt's homepage through a search engine (Fig. 4.9) (39%) or directly (36.6%). Traffic due to referrals and social media were 12.9% and 9.1%, respectively. Facebook

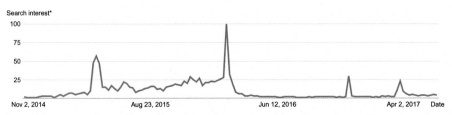

* Numbers represent search interest relative to the highest point on the chart for the given region and time. A value of 100 is the peak popularity for the term. A value of 50 means that the term is half as popular. Likewise a score of 0 means the term was less than 1% as popular as the peak.

Figure 4.8 Tazedirekt keyword search on Google Trends.
Source: Compiled by the authors via https://trends.google.com/.

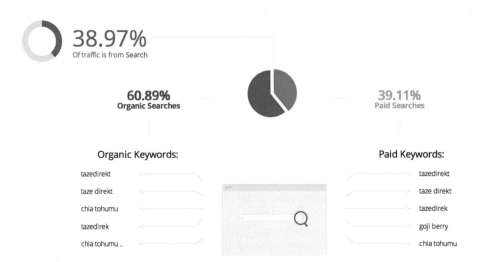

Figure 4.9 Tazedirekt internet search breakdown.
Source: Compiled by the authors via https://trends.google.com/.

(88.3%) drove the majority of TazeDirekt's social media traffic, while Twitter (9.9%), YouTube (1.1%), Google Plus (0.6%), and Pinterest (0.1%) only marginally contributed.

4.4   An abrupt closure

FreshDirect.com had been a successful business in its home market in the New York metropolitan area. Details of its business model had been widely discussed and shared within business circles (Bruno, 2010; Cruz, 2013; Forbes, 2006; Laseter et al., 2003). Previously, cloning and translating working business models from the United States to Turkey had proven to be successful for many others, such as

Trendyol.com's adaptation of private shopping site Gilt.com (BusinessInsider, 2011) since 2010, or Gittigidiyor.com's cloning of the peer-to-peer marketplace Ebay.com (Rao, 2011) since 2000. However, the economics were not working out for TazeDirekt. The company's prices were, of course, higher than the market average, as the products were of higher quality. However, the price markup was not enough to cover the additional cost of providing superior quality products and prime customer service. They tried to increase the prices of some of the items further to compensate for overheads and other indirect costs. However, this had limited success.

TazeDirekt had tried to scale up with aggressive marketing promotions for customer acquisition and retention. For example, the company was providing a USD20 (TL50) welcome promotion for all new registrants on the website and sending large rebate coupons with every order. It was also offering free shipping together with unusual gifts like small pots of strawberry plants or lemongrass for returning customers in order to increase satisfaction and "stickiness." The drop experienced in operational unit costs was not enough to satisfy Mr. Aslanoba, and there were no signs of near-term improvement in sight (Kara, 2016). The maintenance and opportunity cost of the vast Boğazköy farm of the founder, as well as the production and logistics costs of farming produce, were not decreasing significantly with the scale of operations. The costs of aggregating local producers, imposing quality controls, and delivering them to customers' doors in three major cities as early as the next day was not a simple business. The revenues were increasing approximately 30% monthly in the last 5 months of trading (Hürriyet, 2016), but costs were also increasing almost as fast, leading to an unsustainable financial bottom line. As Mr. Aslanoba stated:

> If I saw we could make our monthly USD2.84 million (TL10 million) revenue target for December 2016 without burning too much money, I would continue. However, there was a great difference between our budgeted and realized loss, and it is not getting any smaller. Especially in our order processing, especially for grocery preparation, we had high levels of grocery waste, eating up our margins. To achieve manageable unit costs for distribution, we needed to be 7–8 times larger scale (Kara, 2016).

Hence, TazeDirekt suddenly announced the closure of the business at 12:31 p.m. on February 18, 2016, declaring a significant failure of the business model. This was followed almost immediately by closure of the e-tailer website at 1:10 p.m., ceasing its operations overnight (Kutsal, 2016). The statement released to news agencies read:

> We launched TazeDirekt in December 2014 as we followed our dream of providing access to healthy food for everyone. Since that first day, we believed in what we have worked out. We became a large family with you and the TazeDirekt team. We held each other. We grew, learned as we grew, and shared as we learned.

Our belief and heartfelt love for TazeDirekt grew every day. We want to be sincere with you in sharing a recent decision. Today we are going to close TazeDirekt due to market conditions and operational balances. We are deeply saddened as we end pursuing our dream, and get to leave you. We thank you for sharing our dreams, holding our hands. We wish to meet somewhere again. Always stay fresh. Thank you so much for keeping us on this path! Hope to meet somewhere again. Always stay fresh. Sincerely, TazeDirekt Family (Akgün, 2016).

4.5   Reopening: a revival or a masquerade?

In November 2016, Migros, one of the largest multibrand bricks-and-mortar retail store chains in Turkey officially announced that they had purchased the TazeDirekt brand and its online assets. It was big news, as Migros was one of the most important brands in organized retailing in Turkey, with 1,527 "Migros" supermarket stores in 73 cities and 39 "Macro" gourmet supermarket stores in 5 cities, as of December 2016. It had been selected as a "lovemark" (Roberts, 2006) nine times in the supermarket category in 2016 by Ipsos and the Quality Association of Turkey, and was reported as having the highest customer loyalty in its category (Migros, 2016). The Migros ownership also meant strong financial backing, as it belonged to the Anadolu Group, a prominent Turkish industrial conglomerate with operations in 80 companies, across 19 countries, and with 61 production facilities and more than 50,000 employees (AnadoluGroup, 2017).

With this acquisition, Migros was able to complement its existing online retail presence on sanalmarket.com.tr (established in 1997) and online.macrocenter.com.tr (established in 2005) with TazeDirekt (Migros, 2016). Migros ownership could mean an essential logistics advantage to TazeDirekt. For example, the Migros online brand, sanalmarket.com, had been serving 24 cities integrated with 102 of the Migros stores (Migros, 2016). The same strategy could also apply to TazeDirekt for Macro and Migros stores. The relaunched TazeDirekt was expected to limit its product range to items sold in the Migros and Macro supermarkets, significantly decreasing the logistics variability and production costs that formerly pushed TazeDirekt out of business. Macro had been selling some organic and gourmet products as well, but it was not its sole offering.

The site was relaunched to customers on March 21, 2017. The renewed TazeDirekt would not be able to focus on organic food and superior meat products strictly. Hence it would necessarily lose its original value proposition. The Boğazköy farm's products would no longer be available on the site, but instead, the brands of global producers would be offered, as the product range needed to match the existing Macro offer. The distribution would be scaled down to cover only Istanbul, and shipping would no longer be free for orders below USD55 (TL200). Despite the limited relaunch of TazeDirekt, Macro already had excellent accessibility to the consumer grocery market, especially in wealthy neighborhoods with its 40 stores in the 5 most prominent cities of Turkey, together with its online e-tail operations. Since Macro Online would continue its existing operations, TazeDirekt's relaunch would primarily be an additional sales channel.

As the value chain of Macro would assume operational costs, TazeDirekt may become sustainable with favorable bottom line results this time around. However, some questions remain:

- What will customers make of the revised value proposition?
- Will the quality and the experience of Macro suffice to fulfill the previously pampered TazeDirekt customers?
- How many previous TazeDirekt customers will return to the relaunched site?
- To what extent will Macro Center Online and TazeDirekt cannibalize each other?
- When earlier TazeDirekt customers, who were fully aware of the relaunch story and the difference between the old site and the new site, are considered, will they order from the relaunched TazeDirekt instead of preferring the online or physical stores of Macro or Migros?

4.6 Conclusion

The causes, factors, and people to blame for such failures have long been subject to scrutiny (Pal, Medway, & Byrom, 2011; Shaver, 1985). When did TazeDirekt management know of the forthcoming failure? How could TazeDirekt interpret the weak financials and the high-cost figures during the scale-up execution? To what extent did TazeDirekt's location selection and expansion strategy lead to the failure? Could this failure have been identified earlier with specific online retail key performance indicators (KPIs) and have been avoided with different policies?

TazeDirekt's closure in 2016 has been a monumental failure for the food retailing and distribution industry and the start-up ecosystem of Turkey. It proved that food e-tailers are also prone to the basic rules of economics and sound business practice. Creating a "lovemark" of superior quality that has differentiated marketing and a "high-touch" consumer experience would not be enough for sustained performance, without creating a supportive operational infrastructure and a matching business model. Cloning a successful start-up would never be enough to ensure success. A business model and a retail marketing strategy that are successful in one specific context may not be readily transferable to other locations.

TazeDirekt served as a "poster child" for issues involved with the get-big-fast philosophy. This case has aimed to show how establishing sound business processes is critical for start-ups in the medium-to-long run. Effective and efficient logistics and other vital business processes are crucial in realizing an intended value proposition. For food e-tailing, logistics, quality management, the warehousing of perishables, and the analytics of product range, delivery routes and scheduling, are all critical for success. Optimization of these factors and their mutual fit to company policies during the execution phase is essential if problems like "the last mile" are to be avoided and the relevant costs of growth and customer service are to be managed.

Tanskanen, Yryola, and Holmstron (2002) argue that e-grocery should be a complementary channel for supermarkets, rather than a pure-play substitute. TazeDirekt's acquisition by Migros evolved parallel to this recommendation. Naturally, when two business models are combined, store-based order fulfillment becomes feasible, and the e-tailer can avoid investing in new logistical and operational facilities near customers, which can help retailers to utilize their existing resources better enabling them to expand quickly to larger geographies (Fernie, Fernie, & Kinnon, 2014). However, the long-term sustainability of this model is also questionable, as the priorities of the two channels and two consumer groups conflict, especially in order picking and fulfillment.

At the stage of final delivery to the home, significant challenges continue that require both pure and hybrid e-tailers to find a balance between convenience and cost. For example, Amazon plans to deliver purchases to consumers inside their homes even when they are away, whereas Walmart even plans to put groceries inside customers' fridges (Harris, 2017), thereby providing superior value and looking to overcome part of the last mile problem. Besides these home access alternatives, there are now widespread examples of delivery to reception boxes and collection points across the world (Fernie et al., 2014). As e-tailers are providing new technological innovations and optimized business processes, and whilst the next generation of internet-native consumers are entering the market, pure online e-grocery retailing may become widespread in different parts of the world.

References

Akgün, M. (2016). Tazedirekt bir yılda pes etti! *Radikal*, 19 February.

AnadoluGroup. (2017). *Genel Bakış*. [Online]. Available: http://www.anadolugrubu.com.tr/genelbakis/1/9/biz [Accessed 12 June 2017].

Batra, R., Ahuvia, A., & Bagozzi, R. P. (2012). Brand love. *Journal of Marketing*, 76, 1–16.

BKM. (2017). *e-Commerce transactions*. The Interbank Card Center, Istanbul.

Boyer, K. K., Prud'homme, A. M., & Chung, W. (2009). The last mile challenge: Evaluating the effects of customer density and delivery window patterns. *Journal of Business Logistics*, 30, 185–201.

Bruno, K. (2010). *Inside FreshDirect's expansion* [Online]. *Forbes*. Available: https://www.forbes.com/2010/06/29/freshdirect-online-grocery-braddock-cmo-network-grocery.html - 1f3c17505d3d [Accessed 29 June].

BusinessInsider. (2011). *Trendyol is Turkey's gilt and is actually more popular in one important area* [Online]. Available: http://www.businessinsider.com/trendyol-is-turkeys-gilt-and-is-actually-more-popular-in-one-important-area-2011-8 [Accessed 12 December 2017].

Canko, S. (2015). *2015 Webrazzi e-Commerce conference, Istanbul*. Available: https://www.slideshare.net/webrazzi/demelerde-dijital-dnm-webrazzi-eticaret.

Cruz, J. (2013). *FreshDirect vs. peapod: New York's online food fight* [Online]. *Business Week*. Available: https://www.bloomberg.com/news/articles/2013-03-14/freshdirect-vs-dot-peapod-new-yorks-online-food-fight [Accessed 12 December 2017].

Demirel, F. (2015). *Aslanoba: Aylık yüzde 30 büyüyen Tazedirekt, 2016'da 7 kat büyümeyi hedefliyor* [Online]. *Webrazzi*. Available: https://webrazzi.com/2015/12/17/aslanoba-aylik-yuzde-30-buyuyen-tazedirekt-2016da-7-kat-buyumeyi-hedefliyor/ [Accessed].

Demirel, F. (2016). *Tazedirekt'in Aralık 2015 cirosu 1.8 milyon TL, Aralık 2016 hedefi 10 milyon TL* [Online]. *Webrazzi*. Available: http://webrazzi.com/2016/01/09/tazedirektin-aralik-2015-cirosu-1-8-milyon-tl-aralik-2016-hedefi-10-milyon-tl/ [Accessed 25 May 2017].

Descartes. (2014). *Tazedirekt.com selects Descartes' home delivery solution for next-day online grocery operations* [Online]. Available: https://www.descartes.com/news-events/general-news/tazedirektcom-selects-descartes-home-delivery-solution-next-day-online [Accessed 25 May 2017].

Fernie, J., Fernie, S., & Kinnon, A. (2014). The development of e-tail logistics. In J. Fernie, & L. Sparks (Eds.), *Logistics and retail management: Emerging issues and new challenges in the retail supply chain* (pp. 205−235). London: Kogan Page.

Forbes. (2006). *Will work with food* [Online]. Available: https://www.forbes.com/global/2006/0918/041.html - dd8ee6a1c717 [Accessed 10 May 2017].

Gökbayrak, U. (2016). *Tazedirekt'e ne oldu?* [Online]. *Medium*. Available: https://medium.com/@umutgokbayrak/tazedirekt-e-ne-oldu-41365e8a140 [Accessed 10 May 2017].

Harris, S. (2017). *Amazon wants to leave packages in your home, Walmart will put groceries in your fridge* [Online]. *CBC News*. Available: http://www.cbc.ca/news/business/amazon-key-walmart-in-home-delivery-security-1.4383049 [Accessed 13 December 2017].

Hürriyet. (2016). Tazdirekt'i Migros Aldı. *Hürriyet*, 22 November 2016.

Kantarcı, O., Özalp, M., Sezginsoy, C., Özaşkınlı, O., & Cavlak, C. (2017). *Dijitalleşen Dünyada Ekonominin İtici Gücü: E-Ticaret*. Istanbul: TUSIAD.

Kara, M. (2016). *Hasan Aslanoba Tazedirekt'i neden kapattığını anlattı* [Online]. *Webrazzi*. Available: https://webrazzi.com/2016/03/02/hasan-aslanoba-tazedirekti-neden-kapattigini-anlatti/ [Accessed 25 May 2017].

Kutsal, A. (2016). *Tazedirekt.com kapanıyor!* [Online]. *Webrazzi*. Available: http://webrazzi.com/2016/02/18/tazedirekt-com-kapaniyor/ [Accessed 10 May 2017].

Laseter, T., Berg, B., & Turner, M. (2003). What FreshDirect learned from Dell. *Strategy and Business*, *30*, 20−25.

Laseter, T., Houston, P., Chung, A., Byrne, S., & Turner, M. D. A. (2000). The last mile to nowhere: Flaws & fallacies in internet home-delivery schemes. *Strategy and Business*, *20*, 40−49.

Laseter, T., & Rabinovich, E. (2011). *Internet retail operations: Integrating theory and practice for managers*. London: CRC Press.

Migros (2016). *Migros 2016 faaliyet raporu [Migros 2016 annual report]*. Available at https://www.migroskurumsal.com/userfiles/file/faaliyet_raporu/Migros_Faaliyet_Raporu_2016.pdf.

Pal, J., Medway, D., & Byrom, J. (2011). Deconstructing the notion of blame in corporate failure. *Journal of Business Research*, *64*, 1043−1051.

Ramus, K., & Nielsen, N. A. (2005). Online grocery retailing: What do consumers think? *Internet Research*, *15*, 335−352.

Rao, L. (2011). *eBay acquires Turkish marketplace GittiGidiyor* [Online]. *TechCrunch*. Available: https://techcrunch.com/2011/04/12/ebay-acquires-turkish-marketplace-gittigidiyor/ [Accessed 12 December 2017].

Ring, L. J., & Tigert, D. J. (2001). Viewpoint: The decline and fall of Internet grocery retailers. *International Journal of Retail & Distribution Management*, *29*, 264−271.

Roberts, K. (2006). *The lovemarks effect: Winning in the consumer revolution*. Seattle: Mountaineers Books.

Sayılı, M., & Büyükoğlu, A. M. (2012). Analysis of factors affecting the attitudes of consumers on food purchasing through E-commerce. *Journal of Agricultural Sciences, 18*, 246–255.

Shaver, K. (1985). *The attribution of blame: Causality, responsibility and blameworthiness.* New York, NY: Springer.

SikayetVar. (2015a). *Tazedirekt.com Sitenin Adı Taze Direkt Ama Taze Değil* [Online]. Available: https://www.sikayetvar.com/tazedirektcom-sitenin-adi-taze-direkt-ama-taze-degil [Accessed 25 May 2017].

SikayetVar. (2015b). *Tazedirekt.com Ürünlerin Hiçbiri Organik Değil!* [Online]. Available: https://www.sikayetvar.com/tazedirektcom-urunlerin-hicbiri-organik-degil [Accessed 25 May 2017].

SikayetVar. (2016). *Tazedirekt.com Siparişim Bana Ulaşmadı* [Online]. Available: https://www.sikayetvar.com/tazedirektcom-siparisim-bana-ulasmadi [Accessed 25 May 2017].

Tanskanen, K., Yrjölä, H, & Holmström, J. (2002). The way to profitable Internet grocery retailing—six lessons learned. *International Journal of Retail & Distribution Management, 30*(4), 169–178.

TÜBİSAD. (2017). *Türkiye'de E-Ticaret 2016 Pazar Büyüklüğü.* İstanbul: TÜBİSAD Bilişim Sanayicileri Derneği.

TUIK. (2016). *Hanehalki Bilişim Teknolojileri Kullanım Araştırması.* Ankara: Turkiye Istatistik Kurumu.

Factors influencing consumers' supermarket visitation in developing economies: The case of Ghana

5

Schmidt H. Dadzie[1] and Felix A. Nandonde[2]
[1]Niels Brock Copenhagen Business College, Denmark, [2]Department of Business Management, Sokoine University of Agriculture, Morogoro, Tanzania

5.1 Introduction

The emergence of modern retail stores in Africa with self-service consumer goods has increasingly attracted the attention of scholars and researchers. Studies which have been conducted in the continent thus far have focused on the linkages of local food suppliers with supermarkets (e.g., Louw, Vermeulen, Kirsten, & Madevu, 2007; Nandonde, 2016; Nandonde & Kuada, 2016), supermarkets and gender (Sehib, Jackson, & Gorton, 2013), and consumers' behavior in supermarkets (Neven, Reardon, Chege, & Wang, 2006). These studies, however, examined supermarkets in Kenya and South Africa, since the sector in these countries is much more advanced. Furthermore, these studies paid little attention to consumers' motives for visiting modern stores. Africa has witnessed the emergence of supermarkets, but little is known of the factors influencing consumers' visits to these retail outlets (Meng, Florkowski, Sarpong, Chinnan, & Resurreccion, 2014).

Previous studies on Ghanaian supermarkets focused on consumers' demography and purchase behavior (Meng et al., 2014), the food retail environment, and consumer access to various food retail formats (Oltmans, 2013). Other works looked at consumer motivations for visiting malls (Hinson, Anning-Dorson, & Kastner, 2012). Most studies on supermarket visitation usually focus on analyzing consumer choices between traditional food retailers and modern food retailers but do not pay specific attention to supermarkets as a different format of grocery shopping (Oghojafor, Ladipo, & Nwagwu, 2012; Aryeetey, Oltmans, & Owusu, 2016). According to Ahmed, Ghingold, and Dahari (2007) and Nandonde and Kuada (2014), there is a need for future research to include more demographic variables in understanding motivational factors that influence consumers into buying food items from supermarkets. The current study, therefore, takes this specific focus.

Case Studies in Food Retailing and Distribution. DOI: https://doi.org/10.1016/B978-0-08-102037-1.00005-0

5.2 Supermarkets in Ghana

Retail business is increasing in Ghana and is forecast to rise from US$8 billion in 2015 to US$11 bilion in 2019 (USDA, 2017). Furthermore, Ghanaians have witnessed the emergence of local and international supermarkets, including South African supermarket chains, such as Shoprite (see Table 5.1), which are located in the malls in the capital, Accra. However, there are also local retailers that have, over the years, established themselves as "mega markets" throughout the country. A good example is Melcom Group, which has over 30 shops across the country. There are also a growing number of local supermarkets that serve specific regions in the country. However, supermarkets are still estimated to account for only 4% of the country's annual retail sales (USDA, 2017). This calls for a need to understand what motivates buyers into purchasing goods in the supermarkets of developing economies, and in Ghana specifically, with a view to understanding the potential for growth within the sector.

At the other end of the food retail spectrum are huge numbers of small-scale food retailers which are located in convenient places and offer food products to customers at varied prices and quality. These businesses, which are typically located in open-air markets in various towns and cities in Ghana, are normally referred to as *traditional* food retailers. Most of the retail needs of Ghanaians are served by the informal sector, usually by vendors using small tabletop shops and street stalls, all of which account for approximately 90% of the country's retail activity (AtKearney, 2014). Although supermarkets are not largely entrenched in Ghana, they give a degree of choice to consumers. They provide convenient parking spaces, a clean environment, sales promotions, return policies, and air-conditioned shopping areas for customers. This is a direct contrast to what traditional shops can provide; which is, better bargaining and fresh foods. As the supermarket retail format begins to take hold and evolve in Ghana, consumers' motives for buying food are likely to change. It is therefore vital to understand the factors that influence consumers' choice of these new retail formats.

Table 5.1 Some of the supermarkets that operate in Ghana

Name of the company	Number of stores	Country of origin
Melcom Group	35	Ghana
Shoprite	5	South Africa
Maxmart	5	Ghana
Marina supermarket	1	Ghana
Eakaza Limited	5	Ghana
Lara mart	5	Ghana
Shop-n-Save	5	Ghana
Game	1	South Africa
A life supermarket	3	Ghana
All needs	7	Ghana

Source: USDA (2017).

5.3 Literature review

Supermarkets need to evaluate consumers' behavior based on their visitations because this is a proxy for store patronage (Hutcheson & Moutinho, 1998). The current study draws from the theory of planned behavior (TPB) in understanding consumers' store visitation motives, as proposed by Ajzen (1985, 1991). The theory holds that consumers' behavioral intention is influenced by external factors, which are attitudes, norms, and perceived behavior control. The theory holds further that the effects of these factors may vary across different consumers' behaviors and situations.

According to Azjen (1991), motivation is a central factor of TPB. Thus, it can be argued that there must be a motivation for a buyer to visit supermarkets. This motivation could either be economic or social gains. As Fishbein and Azjen (1975) argue, a person's attitude towards an object is measured by his or her beliefs of salient attributes that the object has, and his or her evaluation of each attribute. Therefore, for a person to engage in a certain behavior, he or she must believe that there is a benefit from his/her engagement in that behavior (Lam & Hsu, 2004).

In general, previous studies show that food has a relatively higher price in modern supermarket stores than is the case in traditional stores (Minten & Reardon, 2007; Neven et al., 2006). Indeed, in developing economies, including Ghana, supermarkets are perceived to be shops for high-status end consumers. Thus, shopping in supermarkets in Ghana can be considered as a marker of social class and is perceived as a luxury for most consumers. That means if a customer visits a supermarket in Ghana, then he/she is likely to have a reason that motivates this visit beyond straightforward economics.

Tauber (1972) asserts that people's shopping motives go beyond satisfaction and he classifies these into personal and social motives. Personal motives include the need for role-playing, diversion, self-gratification, learning about new trends, physical activity, and sensory stimulation. Social motives, on the other hand, relate to satisfaction from shopping that is socially created. It includes satisfaction from communicating with other shoppers and the social experience of shopping, along with the pleasure of bargaining. A person will shop when their needs for these motives are strong and can be satisfied through shopping activity.

Recent studies in developing economies have shown that there are a number of factors that motivate consumers into visiting supermarkets. Such factors include an increase in income (Reardon & Hopkins, 2006), ownership of the means of transport such as cars, and ownership of storage facilities such as refrigerators (Gorton, Sauer, & Supatpongkul, 2011). However, recent developments in African shopping trends have shown that even at the level of the village, supermarkets are also emerging and that consumers from low-income brackets will visit (Nandonde, 2016). On the other hand, a number of supermarkets are not performing well in the continent. For example, the failure of Shoprite in Uganda and Tanzania was associated with the failure of the retailers to understand consumers' needs.

As The Economist (2015) argues, one of the challenges for those firms that invest in Africa is a lack of understanding of consumer behavior there.

5.4 Methodology

5.4.1 Research design

A cross-sectional research design was carried out in two Ghanaian cities, Accra and Kumasi, in order to explore the factors that influence consumers into shopping in modern food retail supermarket stores. Accra is the capital city of Ghana and, with a population of 4 million, accounts for 16.3% of the country's total population. Kumasi also has a largely urbanized population of about 1,730,249 according to the 2010 population and housing census (Ghana Statistical Service, 2012). Modern food retail establishments such as supermarkets are springing up in these urban areas and are increasingly serving as an alternative to more traditional food retail outlets. However, the spread of these supermarkets has at best been uneven. Supermarkets that are located in the city centers of Accra and Kumasi are mostly in shopping malls, with some of them also opening chains of stores in high-income suburbs. Accordingly, the selected supermarkets for this study also followed this geographical pattern.

5.4.2 Sample and sampling procedure

The respondents for this study were selected from a convenience sample of 388 individuals. A questionnaire was administered at the shopping sites for shoppers to fill out after they had completed their shopping. Five supermarket chains were selected from across 13 supermarket locations for the study. Efforts were made to undertake the research at supermarkets that sell or deal with similar products and have similar physical attributes and similar levels of convenience for shoppers. For instance, all the selected supermarkets in the study locations had designated parking areas, large shopping areas, and all had self-service. The malls ranged in size from 6,230 to 27,500 square meters (USDA, 2017). Furthermore, the supermarkets have rented spaces, each of which is above 100 square meters (USDA, 2017). These features distinguished the supermarkets from other food retail outlets such as open-air markets.

5.4.3 Data collection

Following the agreement with supermarket managers, the data collection process began. The data were collected from May to June 2015 using a questionnaire that had previously been pilot tested on 30 respondents and revised accordingly. Questionnaires were administered by university students, who were supervised by teaching and research assistants with experience in data collection. Since the questionnaire was developed in English, issues such as translation of terms into local languages were considered carefully by the researchers to avoid misinterpretation. The questionnaire was administered on different days during the study period, and at various times of the day, to avoid bias. The target sample size of viable returns was set at 300. This was deemed to be an adequate number of respondents for multivariate analysis techniques such as factor analysis (Hair, Black, Babin, & Anderson, 2009). During data collection, some 388 questionnaires had to be administered to reach this target, with 88 of the 388 returns discounted as they were either incomplete or had errors.

The questionnaire was developed with inspiration from prior studies on consumer shopping typologies (Buttle, 1992; Jamal, Davies, Chudry, & Al-Marri, 2006; Kenhove & De Wulf, 2000). The questionnaire comprised 38 questions. Twenty-six questions measured shopping influencers on a seven point Likert scale, ranging from 1 ("strongly agree") to 7 ("strongly disagree"). The remaining questions collected information on demography and the food shopping behavior of respondents. Data analysis involved an initial description of demographic factors relating to shoppers. A factor analysis was then carried out using data from the 26 Likert questions, with the aim of ascertaining those factors that influence consumers to shop for food in supermarkets. The identified components were then combined using summated scales. Finally, the relationship between demographic variables and the factors influencing consumers to shop in supermarkets was analyzed.

5.5 Results

5.5.1 Respondents' profile

A descriptive analysis of the respondents ($n = 300$) is presented in Table 5.2. The findings show that females comprised 54% and males 46% of the total sample. With regards to age distribution, the findings show that respondents below 35 years accounted for 78% of the shoppers. Out of this, 43.7% were under the age of 25 years, and 34.3% fell between the ages of 26 and 35 years. This trend broadly reflects Ghana's national demographic distribution in age, whereby younger age groups dominate (Ghana Statistical Service, 2012).

Approximately 3% of the respondents indicated they did not have had any formal education, with around 23% and 46% having secondary education or a diploma/bachelor degree, respectively. Furthermore, 19% indicated they had completed postgraduate education. This shows that the respondents were generally highly educated and confirms previous research (see e.g., Neven et al., 2006) indicating that customers who visit modern retail formats in emerging markets tend to be well educated. Income levels were normally distributed with approximately 11% of the respondents having a monthly income of up to 100 Ghanaian cedis, and 42% of respondents' incomes falling between 101−500 and 501−1000 Ghanaian cedis a

Table 5.2 **Respondent demographics**

	n	*%*
Gender		
Female	162	54
Male	138	46
Total	**300**	**100**
Age		
Under 25	131	43.7
25−35	103	34.3
36−45	32	10.7
46−55	32	10.7
Above 56	2	0.7
Total	**300**	**100**
Monthly income in Ghanaian cedis[a]		
0−100	32	10.7
101−500	85	28.3
501−1000	41	13.7
1001−1500	22	7.3
1501−2000	10	3.3
Above 2000	5	1.7
Total	**195**	**65**[b]
Education		
No formal education	10	3.3
Basic education	26	8.7
Secondary education	69	23.1
Diploma/bachelor	137	45.8
Postgraduate	57	19.0
Total	**299**	**99.7**[b]
Employment		
Student	114	38.0
Employed	143	47.7
Unemployed	20	6.7
Housewife	20	6.7
Total	**298**	**99.3**[b]

[a]US$1 = 4.39 Ghanaian cedis, as at 27 October 2017.
[b]Missing values resulted in less than 100% response for variables.

month. The annual average wage of 2,623 Ghanaian cedis (Ghana Statistical Service, 2016) implies that supermarkets in the two studied Ghanaian cities are accessed not only by middle to high income customers, but also by other customers at lower income levels.

5.5.2 Exploratory factor analysis

Exploratory factor analysis was used as a data reduction technique that compresses data by looking for groups that have very strong intercorrelations within a set of large variables. The test was run several times to give more efficient components. Items/statement reduction was carried out and thereby factors with low factor loadings and communalities were deleted in order to improve the analysis (Costello & Osborne, 2005; Dhurup, 2008). Out of the original 26 questions relating to factors that might influence consumers into shopping in supermarkets, 7 items were deleted, leaving 19 items for the final analysis. The final analysis gave a four component solution and ignored all factor loading of below 0.40. The decision was made using the comparison of the eigenvalues.

The final run factors were tested to ascertain intercorrelation or the suitability of factorability with Bartlett's Test of Sphericity and the Kaiser-Meyer-Olkin test. The approximated chi square value of the Bartlett's Test of Sphericity was 2212.309 (df = 171) and significance was at 0.000. The Kaiser−Meyer measure of sampling adequacy was 0.729, which is considered as "middling" (see Hair et al., 2009). The four component results are presented in Table 5.3.

The 19 variables in the four factors were subjected to an internal consistency reliability analysis with the computation of the coefficient of alpha (Cronbach α). The Cronbach α provides a measure of the internal consistency of test scales. Internal consistency implies that all the test scales measure the same concept or construct (Tavakol & Dennick, 2011). Cronbach measures whether or not the various variables in the four factors measure similar motives (respectively). The alpha values must lie between $+0$ and $+1$ with alpha values ranging from 0.70 to 0.95 as acceptable. All the four factors recorded adequate levels of alpha with a total of $\alpha = 0.791$; number of items 19 with 300 sample size. The results are reported in Table 5.3.

Factor one was named "curious economic shopper" and comprised five variables including "hunt for a bargain," "I came to supermarket to chat with other shoppers," "I came to the supermarkets to compare prices with other options", "supermarkets have higher quality goods", and "I lose sense of time when I am in the supermarkets". Factor two was named "quality and safety": this comprised variables such as

Table 5.3 Internal reliability analysis: Cronbach α coefficients

Factors	Cronbach α
Curious economic shopper	0.781
Quality and safety	0.780
Aesthetic motives	0.750
Social motives	0.718
Total α	0.791

"it is always very easy to find what I want", "foods which are bought here are of higher quality than those from other locations", "I feel very secure in this supermarket", "the supermarket serves as a one-stop shopping place for me", and "I visit this supermarket for its complimentary services, better management, and promotion". Factor three was named "Aesthetic motives" and included the following: "this supermarket is beautifully designed to attract people like me", "the interior design of the supermarket usually attracts my attention", "I feel excited whenever I visited the supermarket", and "the environment inside the supermarkets attracts me". The fourth factor was named "social motives" and included variables such as "shopping would provide me with social experiences outside home", "when am in the supermarket I feel like I am in another world", "I visit supermarkets to meet new friends", and "I enjoy talking to other customers and sales people".

5.5.3   Demographic factors and shopping influencers

A one-way analysis of variance (ANOVA) and t-test were used to access the relationship between shopping influences and demographic factors. These statistical tools have been widely used to study shoppers' motives and demographic factors and have revealed varied evidence of relationships (Jin & Kim, 2003). The choice of which type of analysis to use depends on the kinds of data that makes up the demographic factors. A one-way ANOVA was used for interval or ratio variables (i.e., age group, income levels, educational levels), while the t-test was used to access the relationship for categorical demographic variables (i.e., gender, employment status) (Bryman & Bell, 2011). In all instances, Levene's test was used to access the level of significant relationship. A P-value of less than 0.05 is considered as a significant relationship. Significant relationships are further explored using a post hoc test to discover at which levels the differences occur. The results are presented in Tables 5.4 and 5.5.

The one-way ANOVA tests between three demographic variables showed different results from those factors which emerged from the factor analysis. The study shows that age and income are the factors that classify curious economic shoppers. Perhaps the study findings suggest that a good number of consumers who are visiting supermarkets are looking for products which are on promotion or on sale. No statistically significant differences were noted in income levels, quality and safety, social and Aesthetic motives. A significant difference was, however, observed between income levels and curious economic shoppers at the <0.05 significance level. Post hoc analysis was carried out to find at what level of income the differences occurred. The results show that the differences in the means between various levels of income are not significant. The study findings showed further that Ghanaian consumers visit supermarkets due to social motives. This includes a chance to meet friends. Similar findings are reported by Ahmed, Ghingold, and Dahari (2007) who showed that consumers in Malaysia visit supermarkets to socialize with friends.

Table 5.5 shows that employment has a significant influence on Aesthetic motives. This implies that respondents who are unemployed may have different

Table 5.4 ANOVA results (between groups); Levene's test and decision

Independent variable	Dependent variable	Sum of squares	Mean square	F-Ratio	Sig.	Levene's test	Decision
Income	Curious economic shopper	10.693	1690.466	1084.708	0.000	0.004	Accept
	Quality and safety	6.905	1.381	2.033	0.076	0.009	Reject
	Aesthetic motives	24.490	4.898	6.319	0.000	0.161	Reject
	Social motives	51.414	10.283	6.035	0.000	0.397	Reject
Age	Curious economic shopper	35.044	8.761	5.924	0.000	0.000	Accept
	Quality and safety	7.362	1.840	2.078	0.084	0.537	Reject
	Aesthetic motives	8.158	2.040	3.031	0.018	0.780	Reject
	Social motives	15.654	3.914	2.684	0.032	0.141	Reject
Educational level	Curious economic shopper	10.693	2.139	1.372	0.237	0.000	Reject
	Quality and safety	6.905	1.381	2.033	0.076	0.000	Reject
	Aesthetic motives	24.490	4.898	6.319	0.000	0.522	Reject
	Social motives	51.414	10.283	3.035	0.000	0.004	Accept

Table 5.5 *t*-Test with Levene's test independent variable

Independent variable	Dependent variable	Levene's test	Sig.	Decision
Employment	Curious economic shopper	0.000	0.001	Accept
	Quality and security shopper	0.097	0.463	Reject
	Aesthetic motive	0.027	0.000	Accept
	Social motive	0.936	0.020	Reject
Gender	Curious economic shopper	0.937	0.000	Reject
	Quality and security shopper	0.098	0.129	Reject
	Aesthetic motive	0.439	0.000	Reject
	Social motive	0.049	0.000	Accept
Vehicle ownership	Curious economic shopper	0.002	0.864	Reject
	Quality and security shopper	0.001	0.004	Accept
	Aesthetic motive	0.939	0.006	Reject
	Social motive	0.076	0.000	Reject

sensory satisfaction from respondents who are employed. Employed respondents may also show different economic motives for buying food from supermarkets. As Hinson et al. (2012) argue, hectic working conditions are likely to influence employees into shopping at places with a good environment, and this encourages them to migrate from traditional markets to supermarkets.

Being a male or a female has no significant effect on respondents' supermarket visitation motives. Our findings are in contrast with those of Kotze, North, Stols, and Venter (2012), whose study in South Africa found that gender has a significant impact on mall visitation.

Car ownership is not a significant factor in terms of being a curious economic shopper or Aesthetic motives. Neither does it have any significant influence on social motives. That means car ownership is not a factor that motivates consumers in Ghana to visit supermarkets. Our finding is in contrast with the findings in a study by Gorton et al. (2011) who showed that supermarkets in developing economies emerged due to car ownership. This finding correlates with the emergence of curious economic shoppers who are influenced by promotion and discounts which are offered by supermarkets.

The post hoc test was run for educational level and social motives. The results show that respondents with basic level education differ significantly from the respondents with a diploma or bachelor degree in terms of social motives. Secondary school certificate holders also differ from diploma or bachelor degree holders, as shown in Table 5.6. This means that respondents with low education have social motives for shopping at supermarkets which are different from those individuals with higher education.

Table 5.6 Post hoc multiple comparisons: educational levels and social motives

Educational level	Educational level	Mean difference	Std. error	Sig.	95% Confidence interval	
					Lower bound	Upper bound
No formal education	Basic education	1.0974	0.44637	0.114	−0.1280	2.3229
	Secondary education	0.9295	0.40590	0.119	−0.1148	1.9738
	Diploma/bachelor level	0.4165	0.39294	0.917	−0.5442	1.3773
	Postgraduate	0.5450	0.41127	0.807	−0.5281	1.6182
Basic education	No formal education	−1.0974	0.44637	0.114	−2.3229	0.1280
	Secondary education	−0.1680	0.27605	0.999	−0.9250	0.5891
	Diploma/bachelor level	−0.6809[a]	0.25661	0.046	−1.3543	−0.0075
	Postgraduate	−0.5524	0.28389	0.389	−1.3383	0.2335
Secondary education	No formal education	−0.9295	0.40590	0.119	−1.9738	0.1148
	Basic education	0.1680	0.27605	0.999	−0.5891	0.9250
	Diploma/bachelor level	−0.5129[a]	0.17708	0.035	−1.0052	−0.0206
	Postgraduate	−0.3844	0.21471	0.533	−0.9892	0.2203
Diploma/bachelor level	No formal education	−0.4165	0.39294	0.917	−1.3773	0.5442
	Basic education	0.6809[a]	0.25661	0.046	0.0075	1.3543
	Secondary education	0.5129[a]	0.17708	0.035	0.0206	1.0052
	Postgraduate	0.1285	0.18907	0.999	−0.3927	0.6496
Postgraduate	No formal education	−0.5450	0.41127	0.807	−1.6182	0.5281
	Basic education	0.5524	0.28389	0.389	−0.2335	1.3383
	Secondary education	0.3844	0.21471	0.533	−0.2203	0.9892
	Diploma/bachelor level	−0.1285	0.18907	0.999	−0.6496	0.3927

Based on observed means. The error term is Mean Square (Error) = 1.439.
[a]The mean difference is significant at the 0.05 level.

5.6   Discussion and conclusions

This study aimed to understand the factors that influence Ghanaian consumers into buying goods in supermarkets. The findings indicate that demographic factors influence consumers' decisions in purchasing commodities from supermarkets. In general, the study shows that consumers who visit supermarkets are mainly young and this correlated with the findings of Gorton, Sauer, and Supatpongkul (2009). The exploratory findings suggest that educated people dominate supermarket visitation in developing economies. Contrary to previous findings by Traill (2006), which shows that consumers who visit supermarkets in developing economies do so by using their own vehicles, in general, the study shows that most consumers who visit supermarkets are those who are going to experience a new mode of distributing consumer goods in Ghana.

The findings of this study can be used by supermarket management to comprehend the reality that consumers who visit supermarkets come from different demographic segments. For example, the study shows that visitors are mostly shoppers who are looking for sales promotions. This suggests that in order to attract all types of buyers in developing economies, retailers should provide low-cost products. In general, previous studies show that customers in developing countries are price-conscious (Mai & Zhao, 2004).

The current study shows that TPB theory can be used to understand behaviors that influence consumers in Ghana. Specifically, the study contributes to the TPB; in that what motivates consumers is the outcome of their decisions. Consumers who are visiting supermarkets are not those from the "have" class alone, but also come from the "have not" class. In general those from the "have not" class are visiting supermarkets having decided not to go to traditional retail stores due to the availability of sales promotions at supermarkets. The study shows that despite modern retail stores being perceived as areas for high-end consumers in Africa, recently low-income consumers are observed to have been migrating to the stores from traditional markets. This indicates that the prospects of having more consumers purchasing commodities from supermarkets is promising.

Supermarkets are emerging in Africa and this indicates there will be more competition amongst retailers who operate in the continent. Our study has two specific implications for supermarket management: First, esthetic factors influence African consumers to shop in supermarkets as opposed to shopping in the traditional market. This implies that in order to compete, retailers should design a more attractive environment for consumers, including car parking and merchandise display, and should invest more in these areas. Second, the study shows that Ghanaian consumers also visit supermarkets so that they can socialize. This implies that supermarket managers should create more areas within their store premises for recreation purposes, to attract more customers and be competitive in Ghana.

In spite of the contribution of the study, there are some limitations. The study used convenience sampling due to budget and time constraints. It is recommended that future research should use random sampling to understand how consumers perceive different facilities provided by retailers. Furthermore, Africa has witnessed urbanization of its cities in different countries, but little is known on how consumers in different cities behave according to such geographical differences. In general, studies from China show that consumers' behaviors are different across cities (Wong & Yu, 2003). In this regard, more studies are needed to raise our understanding of what motivational factors seem important to consumers in different cities in developing economies.

References

Ahmed, Z. U., Ghingold, M., & Dahari, Z. (2007). Malaysian shopping mall behavior: An exploratory study. *Asia and Pacific Journal of Marketing and Logistics*, *19*(4), 331–348.

AtKearney, (2014). *The 2014 African retail development index: Seizing Africa's retail opportunities.* Available at https://www.atkearney.com/documents/10192/4371960/ Seizing + Africas + Retail + Opportunities.pdf/730ba912-da69-4e09-9b5d-69b063a3f139.

Aryeetey, R., Oltmans, S., & Owusu, F. (2016). Food retail assessment and family food purchase behavior in Shongman estates, Ghana. *African Journal of Food, Agriculture, Nutrition and Development*, *16*(4), 11386–11403.

Azjen, I. (1991). The theory of planned behaviour. *Organization Behavior and Human Decision Processes*, *50*, 179–211.

Azjen, I. (1985). From intentions to actions: A theory of planned behaviour. In: Kuhl J, & Beckmann J (Eds.), Action control (pp. 11–39). Berlin: Springer.

Bryman, A., & Bell, E. (2011). *Business research methods* (3rd ed.). Oxford: Oxford University Press.

Buttle, F. (1992). Shopping motives constructionist perspective. *The Service Industries Journal*, *12*(3), 349–367.

Costello, A. B., & Osborne, J. W. (2005). Best practices in exploratory factor analysis: Four recommendations for getting the most from your analysis. *Practical Assessment, Research and Evaluation*, *10*(7), 1–9.

Dhurup, M. (2008). A generic taxonomy of shopping motives among hypermarkets (hyper-stores) customers and the relationship with demographic variables. *ActaCommercii*, *8*(1), 64–79.

Fishbein, M., & Azjen, I. (1975). *Belief, attitude, intention, and behavior: An introduction to theory and research.* Addison-Wesley, Reading, MA.

Ghana Statistical Service (2012). *2010 Population and Housing Census.* Ghana Statistical Service, Accra.

Ghana Statistical Service (2016). *Ghana living standards survey round 6.* Ghana Statistical Service. ⟨http://www.statsghana.gov.gh/docfiles/glss6/GLSS6_Main%20Report.pdf⟩ (accessed 20 October, 2017).

Gorton, M., Sauer, J., & Supatpongkul, P. (2009). Investigating Thai shopping behaviour: Wet-markets, supermarkets and food quality. In: *83rd Annual Conference of the Agricultural Economics Society, Dublin* (Vol. 30).

Gorton, M., Sauer, J., & Supatpongkul, P. (2011). Wet markets, supermarkets and the "big middle" for food retailing in developing countries: Evidence from Thailand. *World Development*, *39*(9), 1624–1637.

Hair, J. F. J., Black, W. C., Babin, B. J., & Anderson, R. E. (2009). *Multivariate data analysis* (7th ed.). Upper Saddle River, NJ: Prentice Hall.

Hinson, R. E., Anning-Dorson, T., & Kastner, A. N. A. (2012). Consumer attitude towards shopping malls in sub-Saharan Africa: An exploration of the new retail format in Ghana. *African Journal of Business and Economic Research*, 7(2−3), 97−134.

Hutcheson, G. D., & Moutinho, L. (1998). Measuring preferred store satisfaction using consumer choice criteria as a mediating factor. *Journal of Marketing Management*, 14(7), 705−720.

Jamal, A., Davies, F., Chudry, F., & Al-Marri, M. (2006). Profiling consumers: A study of Qatari consumers' shopping motivations. *Journal of Retailing and Consumer Services*, 13(1), 67−80.

Jin, B., & Kim, J. O. (2003). A typology of Korean discount shoppers: Shopping motives, stores attributes, and outcomes. *International Journal of Services Industry Management*, 14(4), 396−419.

Kenhove, P. V., & De Wulf, K. (2000). Income and time pressure: A person-situation grocery retail typology. *The International Review of Retail, Distribution and Consumers Research*, 10, 149−166.

Kotze, T., North, E., Stols, M., & Venter, L. (2012). Gender differences in sources of shopping enjoyment. *International Journal of Consumers Studies*, 36, 416−424.

Lam, T., & Hsu, C. H. C. (2004). Theory of planned behavior: Potential travelers from China. *Journal of Hospitality and Tourism*, 28(4), 463−482.

Louw, A., Vermeulen, H., Kirsten, J., & Madevu, H. (2007). Securing small scale farmers participation in supermarket supply chains in South Africa. *Development Southern African Journal*, 24(4), 539−551.

Mai, L., & Zhao, H. (2004). The characteristics of supermarket shoppers in Beijing. *International Journal of Retail and Distribution Management*, 32(1), 56−62.

Meng, T., Florkowski, W. J., Sarpong, D. B., Chinnan, M. S., & Resurreccion, A. V. A. (2014). Consumer's food shopping choice in Ghana: Supermarket or traditional outlets? *International Food and Agribusiness Management Review*, 17, A, 107−130.

Minten, B., & Reardon, T. (2007). Food prices, quality and quality's pricing in supermarkets versus traditional markets in developing countries. *Applied Economic Perspectives and Policy*, 30(1), 480−490.

Nandonde, F.A. (2016). *Linkages of local food suppliers in modern food retail in Africa: The case of Tanzania*. Unpublished PhD thesis, Aalborg University, Denmark.

Nandonde, F.A. and Kuada, J. (2014). Empirical studies of food retailing in developing economies. In: International Food Marketing Research Symposium, Aarhus Business School, Denmark, 19−20 June.

Nandonde, F. A., & Kuada, J. (2016). Modern retail buying behavior in Africa: The case of Tanzania. *British Food Journal*, 118(5), 1163−1178.

Neven, D., Reardon, T., Chege, J., & Wang, H. (2006). Supermarkets and consumers in Africa: The case of Nairobi. *International Food and Agribusiness Marketing*, 18(1−2), 103−123.

Oghojafor, B. E. A., Ladipo, P. K. A., & Nwagwu, K. O. (2012). Outlet attributes as determinants of preference of women between supermarkets and traditional open market. *American Journal of Business Management*, 1(4), 230−240.

Oltmans, S.J. (2013). *A case study on the food retail environment of Accra, Ghana*. Unpublished Masters dissertation, Iowa State University, Ames, Iowa.

Reardon, T., & Hopkins, R. (2006). The supermarket revolution in developing countries: Policies to address emerging tensions among supermarkets, suppliers, and traditional retailers. *The European Journal of Development Research, 18*(4), 522–545.

Sehib, K., Jackson, E., & Gorton, M. (2013). Gender, social acceptability and the adoption of supermarket: evidence from Libya. *International Journal of Consumer Studies, 37,* 379–386.

Tauber, M. (1972). Marketing notes and communication: Why do people shop? *Journal of Marketing, 36*(4), 46–49.

Tavakol, M., & Dennick, R. (2011). Making sense of Cronbach's alpha. *International Journal of Medical Education, 2,* 53–55.

The Economist (2015). *Africa's middle class: Few and far between,* 22nd October. ⟨https://www.economist.com/news/middle-east-and-africa/21676774-africans-are-mainly-rich-or-poor-not-middle-class-should-worry⟩ (accessed on 20 July 2016).

Traill, W. B. (2006). The rapid rise of supermarkets? *Development Policy Review, 24*(2), 163–174.

USDA (2017). *Ghana: Retail food report, 2017.* ⟨https://gain.fas.usda.gov/Recent%20GAIN%20Publications/Retail%20Foods%20Report%20_Accra_Ghana_5-22-2017.pdf⟩ (accessed on 20 May 2017).

Wong, G. K. M., & Yu, L. (2003). Consumers' perception of store image of joint venture shopping centres: First tier versus second-tier cities in China. *Journal of Retailing and Consumers Services, 10,* 61–70.

The home as a consumption space: Promoting social eating*

Donatella Privitera[1] and Rebecca Abushena[2]
[1]Department of Educational Sciences, University of Catania, Italy, [2]Department of Marketing, Retail and Tourism, Manchester Metropolitan University Business School, United Kingdom

6.1 Introduction

This chapter discusses the food economy and the way it falls into a collaborative consumption model based on sharing, assisted by the internet. Starting with a review of the sharing economy, the chapter evaluates food-sharing platforms that are taking the social dimensions of dining to new levels.

We use the term "food sharing," as it is possible to sell meals that are home cooked to someone in your city via internet-facilitated platforms. It is claimed that this has positive environmental and social effects and the act of sharing could bring people together. Food sharing comprises diverse concepts, indicating a new phenomenon that people in urban spaces, particularly in cities, share land to produce food (e.g., community gardens). Such practices blur the boundaries between work and home. Given the novelty of this phenomenon, the case study of food sharing platform BonAppetour.com is presented as an exploratory approach to this relatively new and unique topic.

The number of people involved in the so-called sharing economy is growing rapidly as alternatives to more traditional approaches to consumption emerge. The focus on consumption is no longer simply on ownership of a commodity, but increasingly involves a consideration of *the means* of acquisition of commodities, along with the experience of acquiring them. There is an emphasis on the sharing or pooling of resources, where products and services are redefined via the use of technology and the peer communities who use it.

The sharing economy has become a term used to define configurations of economic activity and collaborative consumption based on sharing. A solid definition of the sharing economy that reflects common usage is almost impossible; however, it is said to be "economic activity that is peer-to-peer, or person-to-person, facilitated by digital platforms" (Schor, 2015, p. 13). One of the mechanisms which

---

*Part of this study was presented at the 2016 International Conference "Economic Science for Rural Development," held in Jelgava, Latvia, from 21 to 22 April. This chapter is part of a wider research project on the sharing economy carried out at the University of Catania, Italy.

Case Studies in Food Retailing and Distribution. DOI: https://doi.org/10.1016/B978-0-08-102037-1.00006-2

enhances the sharing economy is that of social networks, where consumers actively participate in online communities to share information, products, knowledge, and suggestions about a new initiative and/or brand. In reality, there is significant debate concerning conceptions of sharing as part of economic or social practices (Eckhardt & Bardhi, 2015). Examples of sharing ventures that fit this definition are "Airbnb" for apartment sharing, "Enjoy" for car sharing, and "BlaBlaCar" for ride sharing. Sharing has become a global phenomenon, due to the expansion of platforms to other countries, as well as the diffusion of information and communication technologies (ICT), and because the idea of sharing has caught on around the world, following and aided by the economic crisis.

The internet is intrinsically linked to the rise of the sharing economy, providing the means by which sharing can and often does occur. The internet and new technologies have become an integral part of our lives making it easier to communicate, research information, and purchase any kind of product and service. Online transactions are becoming simpler and faster and have definitely made people's lives easier, as with just one click consumers are able to acquire whatever they want in a matter of days. In 2017, there were 3.7 billion internet users in the world with significant penetration ratios in Asia (50.1%), as well as European countries like the United Kingdom (62%), Germany (72.2%), and France (56.3%) (Internetworldstats, 2017). The increase in usage can be explained by the proliferation of content accessible through the internet, mobile apps, and platforms, which has grown exponentially. Beyond new technologies, consumers using digital platforms tend to be motivated by economic, environmental, and social factors, along with a mixture of internal motivations, all of which cause them to use these technologies. Consumers are also increasingly looking for ways to earn or save money, which is why they are currently more receptive to peer-to-peer business models centered on consumer needs, as both potential suppliers and buyers.

The rise of the sharing economy via the medium of the internet has had far-reaching impacts, including in the field of food. Within this domain, there are crucial challenges and opportunities. Food has been identified as a key area for consideration in the challenge of sustainable consumption, due to increasing evidence of the impact of the prevailing food system on the environment, local communities, and social justice. Social eating is seen as a positive thing in order to enhance sustainability and understanding of community action in the development of new food systems. Social eating can be reconsidered and seen as a form of grassroots innovation in itself (Seyfang & Smith, 2007).

Sharing models now exist at all stages of production and distribution for food: from the land to the plate, and from restaurant to home. A good example is food that falls into the collaborative consumption model that is based on sharing, with facilitation by the internet. Whilst cooking can be a burden for some people, for many, cooking and eating is a relaxing leisure pursuit. The kitchen is a place where work and consumption are inseparable. It can involve the development and utilization of websites, social media and apps in order to share skills, spaces or "stuff" (e.g., food itself, meals, devices, tools, etc.); related to growing, preparing, or eating food that dominates consumption in terms of food sharing. These mechanisms offer the possibility of sharing food with wider communities. Essentially, ICT is stretching the spaces over which food sharing can occur.

"Private cuisine" in the home can be viewed as a place where meals are served, just like in restaurants. Cooking is an activity that so often appears to be outside the relations of paid work and consumption (Cox, 2013); instead, individuals cooking in their own kitchen entails a new sharing activity. Home restaurants have become a competitive area of consumption in the hospitality sector. Although dining out means searching for experiences outside the home, this can be misleading. The home can be an important part of dining-out experiences; and the co-production and co-consumption of food services with customers in the physical setting of the home is a key aspect of home restaurant services (Honggang & Qunchan, 2015). There are many influences on what makes a home restaurant successful; ambiance for example, is different in every home. Other things that may vary are word-of-mouth customer-to-customer marketing; customer service beyond simple product advice; community embeddedness; and informal but meaningful interpersonal relations between owners and customers of the home restaurant. These are some of the key pillars of the strategic marketing approach pursued by the "new home restaurant" concept on the web. This could indicate a counterbalance to globalization because the approach promotes home-cooked meals with a localized feel. The role of the home is fundamental to the success of such businesses and to promoting a localization agenda.

Food tourism has also increasingly attracted the attention of practitioners, consumers and researchers over the past decade (Robinson & Getz, 2014). The search for culinary authenticity, in a cultural context, is an important topic related to motivation and helps broadly explain food tourism experiences (Beer, 2008; Sims, 2009), particularly in event settings (Robinson & Clifford, 2012). Food is an inherent part of culture and national identity and as such, it has a connecting power, bringing people, communities and businesses together. The attitude of restaurants toward using and championing local food is crucial in promoting local food production and bringing authenticity to the experience that tourism destinations deliver to the market (Presenza & Del Chiappa, 2013). Restaurants can play a strategic role in transmitting new information to diners about food and possibly influencing their tastes and preferences. The home restaurant links the hospitality of home with appreciation of local food and traditional cuisine.

BonAppetour.com is a social dining platform that allows you to organize meals and gastronomic events at home as a social dining marketplace. It connects travelers with local hosts for home-dining experiences, including dinner parties, cooking classes, etc. Potential diners can see the menu in advance, and read details of the venue and the host. The following case study clarifies the interrelationships between the social contexts in which people learn food practices. It points out that different social spheres may sometimes apply contradictory influences and that food learning involves emotional and social experiences, with the aim of helping people create healthy, economically vibrant neighborhoods through the development of local food systems.

Currently, research on sharing business models is still in its infancy but, without doubt, certain sectors, e.g., mobility and tourism, have already adopted such models. However, there are no rigorous methodologies for success at the moment and further studies are needed. This case study is a first attempt to explore the field and to provide a starting point for future studies. The case study emerges as an interesting example and offers some insights into the role of food in connecting people who want to make dining a time for sharing, conviviality, and discovery. The overall aim of this chapter is to link theory on the sharing economy and food sharing, through an exploration of the case study as an example of collaborative consumption in practice using new technologies.

6.2   Forms of the sharing economy: a literature review

The sharing economy (or collaborative economy or collaborative consumption) was first defined by Benkler (2004). It is a type of economic activity, as connoted by the term "sharing economies," which can involve production or consumption. Social sharing and exchange is becoming a common modality in meeting valuable needs at the very core of the most advanced economies in the information, culture, education, computation, and communications sectors (Benkler, 2004). As Belk (2014) has argued, sharing implies interdependence, an obligation of care, and responsible use by those involved; it creates a social bond and gives rise to an (implicit) debt.

It is a collaborative economy which has accessibility at its core, where a continuation of the ongoing economic crisis means consumers are looking to supplement their income with new revenue opportunities to save or purchase at low cost and for short periods (Belk, 2014). The sharing economy may help consumers achieve this aim because they actually just want cheaper services and less hassle. The sharing economy also involves online platforms that help people share access to assets, resources, time and skills (Wosskow, 2014). The European Parliament (2015) defines it as "a new socioeconomic model that has taken off thanks to the technological revolution, with the internet connecting people through online platforms on which transactions involving goods and services can be conducted securely and transparently."

With sharing models, the goal is not always to maximize profits. This calls into question the current capitalist model and replaces it with a cooperative model based on shared objectives that will allow redistribution of wealth. The central point appears to be that emphasis on the *possession* of property or goods appears to be shifting to the *use* of them and the quality of services themselves (Botsman & Rogers, 2010). Goods and services are offered to the consumer, but it is necessary to distinguish how they access them. In many cases, the use of shared goods is financially compensated, thus resembling a simple market transaction, where the only difference is the means through which the exchange is intermediated (but in other cases, financial compensation is substituted by a nonmonetary exchange).

A clear example is that of music streaming (access) and music downloaded (possession). Whilst there have been numerous literature reviews discussing the negative impact of file sharing on music sales (Connolly & Krueger, 2006; Liebowitz, 2006; 2016), this phenomenon has eliminated many of the intermediary organizations usually involved in the consumption process, such as commercial structures, financial and traditional institutions. It has thus setup new patterns of consumption without ownership. This trend toward *access* rather than *possession* of goods and services can be seen as an extension of a rental or leasing model, where ownership is not the focus of consumption activity (Goudin, 2016).

Energy, transport, communications, and tourism seem, at present, to be the sectors most affected by the sharing economy. Airbnb is one of the most cited examples in this regard, but the spectrum of conveyed goods or services via the web is not limited to tourism (Guttentag, 2013) though it is based on the idea of relational goods (i.e., goods which are enjoyed through the establishment of interpersonal relationships) (Forno & Garibaldi, 2015). More rarely discussed is the way such services replace less formally regulated activities, along with ideas about the legal obstacles or barriers that prevent the sharing economy from reaching its full development potential (EPRS, 2016). Besides selling and renting, other forms of sharing include lending, donating, and bartering, where non-anonymous agents share something more than just goods, namely, solidarity and a sense of belonging.

Sharing may be one possible alternative market structure that can be adopted by anticonsumption proponents (Ozanne & Ballantine, 2010). Sharing is obviously not an extreme form of active market rebellion (Dobscha, 1998), but possibly more similar to minimization behaviors such as the downsizing of consumption practices and needs (Fournier, 1998). Sharing could be a new services model; a platform that changes the way companies manage their value chain and address changing customer needs and wants. As sharing is a communal act, it may connect us to others and create feelings of solidarity and bonding (Belk, 2007, 2010). Four forms seem to emerge, but they are fast changing (see Fig. 6.1).

In summary, the main issues or dimensions that the literature associates with the sharing economy are surrounding collaboration and access. Celata, Yungmee Hendrickson, and Sanna (2017) tell of three crucial dimensions: connectivity, reciprocity and trust. This implies intimacy between people and means social trust, i.e. between consumers and workers as they need to trust both the platform they are using and the people they are connecting with (Heinrichs, 2013). The primary and most formal way to increase trust is via insurance and/or refund systems, which have been implemented by almost all platforms. Many online social networking sites have also implemented reputation systems which are becoming an increasingly important component of online communities. They are based on members' digital footprints and feedback received, which encourages good behavior, collaboration and new mechanisms for trust between individuals anywhere in the world. Tripadvisor is a good example of a platform regulated by such an implicit reputation system.

The most diffused mechanisms for increasing trust are digital reputation systems or feedback, which give participants time and space to think and learn from each

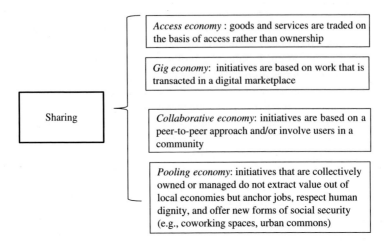

Sharing

Access economy : goods and services are traded on the basis of access rather than ownership

Gig economy: initiatives are based on work that is transacted in a digital marketplace

Collaborative economy: initiatives are based on a peer-to-peer approach and/or involve users in a community

Pooling economy: initiatives that are collectively owned or managed do not extract value out of local economies but anchor jobs, respect human dignity, and offer new forms of social security (e.g., coworking spaces, urban commons)

Figure 6.1 Forms of sharing initiatives.
EU Commission for Economic Policy (2015).

other, and to take into account reputational scoring and rankings (Wosskow, 2014). Sharing platforms emphasize the idea of being a "community marketplace," and even part of some sort of social movement based on an ethos of sharing, solidarity and alterity with respect to traditional, impersonal, and standardized markets (Celata et al., 2017). Indeed, community spirit is also of pivotal importance to sharing platforms (Albinsson & Perera, 2012), as are economic and environmental benefits, emphasized by the fact that local jobs can be created by the localization agenda of sharing (Hamari & Ukkonen, 2013); cost saving and convenience; enjoyment motives (Bardhi & Eckhardt, 2012); application of high technology; and the need for a certain level of digital access. In terms of the latter, it is argued that low levels of digital access could be a significant obstacle to the geography of the sharing economy (Goudin, 2016).

It is important to specify that there are an increasing number of different actors involved in this model of sharing. Actors act simultaneously as consumers and as operators of services. They create new firms where possible, finding large entrepreneurs offering web platforms with a high number of contacts. Given its rapid development, this kind of economy and social norms are yet to be fully established or adapted to the changing reality. It appears that sharing platforms will not be a temporary phenomenon. As Martin (2016) affirms, the sharing economy can be viewed as a niche of sociodigital experiments, with the paradoxical potential to promote more sustainable consumption and production practices, and to reinforce the current unsustainable economic paradigm. The sharing economy has therefore started to transform many aspects of our current social economic system by allowing individuals, communities, organizations and policy makers to rethink the way we live, grow, connect, and sustain.

Recent debates about the future of the sharing economy also address the development of "platform cooperativism" (Scholz, 2016). Sharing platforms are turned into cooperative structures where the former platform "users" also become owners and decision-makers and gain parts of their livelihoods from these platforms. This is a reaction against the global, contemporary, for-profit sharing economy platforms, which build on peer-to-peer production and sharing of resources (Bradley & Pargman, 2017).

6.3 Understanding the context of food sharing

There is a real interest in the culture of food on the part of both consumers and the media, with celebrity chefs, for example, seen as "heroes of the kitchen." Food is the subject of television programs, movies, social networking, sharing on social networks and other imagery. Many actors play different roles to create or change new models of consumption, new lifestyles, or different food networks. In the emerging systems of food, the reincorporation of production processes and local consumption is spreading quickly, taking on different forms such as the so-called Alternative Food Networks (AFNs) or Food Community Networks (FCNs) (Feenstra, 2002; Goodman, 2004). Forms of AFNs are, for example, farmers' markets and box schemes. This form of participation practiced by consumers and producers could contribute to confirmation of a new paradigm of development based on agricultural, eco-compatible, multifunctional models; and on sustainable consumption. FCNs originate in a specific social and cultural background, aimed at increasing social and democratic equity among all members of the community (Feenstra, 2002). FCNs are a particular type of self-organized collective action whose goal is to find a cooperative form for sustainability based on the active participation of actors involved in agriculture and food production and consumption. These experiences represent a possible solution where consumers and producers actively build an economic model founded on human relations (Migliore, Forno, Dara Guccione, & Schifani, 2014). Thus, consumption of food can be tied to the community of the web, where the preference is to favor different solutions and create new entertainment options online. In this domain, there is complete freedom regarding access to multiple foods and nutritional patterns and styles. Online consumption communities are therefore "affiliative groups whose online interactions are based upon shared enthusiasm for, and knowledge of, a specific consumption activity or related group of activities" (Kozinets, 1999, p. 254).

As for the literature in general, a scarcity of studies on the food-sharing marketplace is apparent. The development of case studies that illustrate different degrees of organizational, technological and social change for more sustainable eating, including community eating containing more social and communal elements, is limited. Social innovations (e.g., slow-food events and online food-distribution communities) are organized and initiated through bottom-up, citizen-led approaches (Davies, 2013). The impact of web platforms is shaped by both their market orientation (for-profit vs nonprofit) and market structure (peer-to-peer vs business-to-peer). These dimensions shape the platforms' business models, logics of exchange, and potential for competing with conventional businesses. While all sharing economy platforms effectively create "collaborative markets" by facilitating exchanges, the imperative for a platform to generate a profit influences how sharing takes place and how much revenue devolves to management and owners.

Sharing platforms, particularly nonprofit ones that provide a public benefit, can also function as "public goods" (Schor, 2014). Food sharing can take the form of selling as well as donating and bartering initiatives (Falcone and Imbert, 2017). Food sharing is not a discrete empirical object; it is something that is emerging through a combination of practice and performance (Davies et al., 2017). Food sharing initiatives have also been rising in most developed societies through a variety of forms such as web food networks, underground restaurants, public refrigerators, or simply private initiatives within specific households consisting of unrelated people like students (e.g., Kera & Sulaiman, 2014). Consumers choosing sharing economy initiatives are mostly driven by economic rather than environmental reasons. However, food sharing practices can be viewed within the classifications of "alternative food initiatives" demonstrating practical and varied connections to food. One dimension established by food sharing is that of "facilitating access" to experiential learning, which it is argued can have more enduring or transformational impacts on participants (Sharp, Wardlow, & Lewis, 2015). The principal forms of food sharing are summarized in Fig. 6.2.

The most diffused models of on-line food sharing are donation-based, especially within the distribution marketplace where consumers swap food. In some cases, this empowers communities to transform food waste, surplus and loss into new value and resource efficient (Davies & Doyle, 2015) platforms that bring together people who have a passion for home-grown food. In this sense, food sharing practices can help with the creation of sustainable food (Heinrichs, 2013); encouraging healthy, sustainable food choices, and offering direct environmental benefits such as

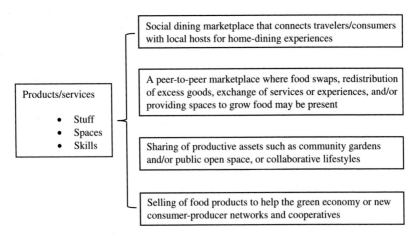

Figure 6.2 Forms of food sharing initiatives.
Adapted from Davies et al. (2017).

"servicizing": "a business model that holds the potential to support a shift toward more sustainable production and consumption by selling to the consumer the product's function, rather than the material product itself" (Plepys, Heiskanen, & Mont, 2015, p. 117).

Sharing models now exist at all stages of production and distribution for food, from the land to the plate. Examples include:

- Pop-up restaurant platforms such as Grub Club, which connect food lovers and creative gourmet chefs in temporary home restaurants.
- Supper clubs and meal sharing platforms such as Casserole Club to help tackle the growing social problems of loneliness and malnutrition among older people, whilst at the same time helping connect people with their neighbors.
- I Food Share, a web platform allowing users, retailers or manufacturers to offer food surplus for free.
- VizEat, which is a social dining platform that allows individuals to organize meals and gastronomic events at home. Potential diners can see the menu in advance and read details of the venue and the host.

While some new food sharing platforms represent substitutes for existing food distribution and retailing practices, others are likely to cover sectors, geographical settings and consumer groups where the reach of food sharing is limited. Indeed, in many cases the goal is to build something completely new and separate from existing food systems (DiVito Wilson, 2013). The aim is also to help people create workable, healthy, economically vibrant neighborhoods through the development of local food systems.

6.4 Methodological approach

BonAppetour is a real-world example that provides additional insights into social eating. It is a platform for connecting people who want to make the meal a moment for sharing, conviviality and discovery. It aims to teach how to prepare favored dishes, and open up new food experiences. The BonAppetour case was chosen in order to have sufficient robustness to capture the distinct characteristics of the observational units of analysis (i.e., the different food sharing platforms, and the drivers and activities of related service organizations), and because the aim of the study was to understand collaboration and sharing practices in a real-life context. Data were collected through observations and interviews with members of an online sharing community that involved social eating between guests, which was useful for gaining a better understanding of the community's physical, social, cultural and economic dimensions. A survey of guest groups was carried out in Italy during early 2017. A pilot questionnaire was created and conducted with 10 online respondents. The recruitment of the sample was realized by word-of-mouth. Data were collected via 33 completed questionnaires. Various aspects of respondents' behavior were explored, especially the emotions arising from the experience of social eating and their leisure preferences. Another section of the questionnaire examined the type of social eating experiences respondents carried out and meals selected. At the same time, the study also involved secondary data taken from the BonAppetour website. This represents a "real-world" case study that provides additional insights into social dining but also into food tours, cooking classes, tastings, etc.

Lastly, documentary sources such as newspapers, books and online sources have been considered as additional sources of information. Given the novelty of the phenomenon under investigation, and the inductive nature of the research questions, we employed a qualitative, case-based approach. At the same time, to improve the understanding of food sharing, the study integrated community development theory (Christenson & Robinson, 1989) to establish a more holistic, conceptual framework to support research and advocacy efforts. Community development refers to a group of people in a community reaching a decision to initiate a social action process to change their economic, social, cultural and environmental situation. Sharing enacted through a community is therefore contingent upon participation and the ability of individuals to contribute to something that will not just benefit themselves. Development is a process that increases choices and means new options, diversification, thinking about issues differently, and anticipating change (Christenson & Robinson, 1989).

6.5 BonAppetour: A food community marketplace

BonAppetour is a platform for connecting people who want to make the meal a moment of sharing, conviviality and discovery. It connects hosts and guests from all over the world in order for them to experience new flavors and meet new people, whether they are visiting a foreign country and looking to escape the tourist traps or are locals looking to have a new and unique experience.

BonAppetour is a community marketplace that connects travelers with local home chefs for a unique home-dining experience, anywhere around the world. In fact, the mission is to make home dining an integral ingredient of every remarkable travel experience. In reality, social eating is the desire to be together around the table; a transversal concept moving between notions of public and private, and professional and amateur (Fig. 6.3). Especially in travel and tourism, local food consumption behavior is considered as a social experience to learn about other cultures: "from a local food paella making workshop on a terrace in Spain, to an exquisite Italian feast with a "nonna" in Rome and a traditional tea ceremony experience in Tokyo. You can choose from a range of unique dining experiences hosted by a carefully selected host community" (BonAppetour, 2017).

Founded in 2014, with an initial focus on Italy, BonAppetour has expanded to include dining experiences across Africa, Asia, North America, Oceania, and South America. It is a start-up in the booming sharing economy with an international team. The platform is available in four languages and reaches 30 different cities worldwide. An innovative company, it employs the latest and most up-to-date technology in food systems. Accelerating its growth, the two parties to a food

Figure 6.3 The mission of BonAppetour.
www.bonappetour.com.

experience exchange—hosts and guests—must register on the website. Hosts list online their available menus and detailed profiles, where they introduce themselves, their cooking styles, their preferred places to travel and things to do, while simultaneously making money to fund their passion. Guests can then browse and book meals. The platform allows access to meals in every price range. BonAppetour takes a 15% guest service fee every time the platform invitation is booked and receives remuneration in the form of a commission in exchange for its contact service.

As with the majority of sharing economy platforms, BonAppetour does not own any of the "spaces" in which it operates; rather the organization acts as a facilitator, matching hosts with guests. Thus, the website serves as a platform for listings and the exchange of information. It has been established to create an alternative option to the traditional food service provided by restaurants.

Testimonials from the BonAppetour website still make compromises between the hedonistic values of food and the location of the home restaurant and friendliness of the host: "The atmosphere was relaxed, authentic and cheerful, the owner was very hospitable and welcoming. The clean and familiar atmosphere makes you feel at home. Excellent food at zero kilometer. Plentiful dishes. Personally I feel I recommend a visit" (BonAppetour, 2017).

As a safety measure, BonAppetour performs background checks on all users and offers a private messaging system for both parties to learn more about each other before agreeing to a transaction. The platform has a number of safety-related advantages built in, including:

- No handling of cash, with payment transferred using PayPal. Payment for the experience is transferred to the host 24 hours after the dining experience. The money is held in order to protect both hosts and guests in cases of any unforeseen circumstances. Transactions are secure and data are protected.
- Profiles of hosts are attached to a reputational mechanism, emphasizing the importance of examining clients' trust-building processes, as well as host attributes and strategies in

facilitating the delivery of products to alleviate any "trust" tension. Trust is clearly important therefore, and is built into the technology on the BonAppetour platform. In the vocabulary of the sharing economy, the function of this platform is that a host's profile acts like a curriculum vitae for other users, and guests in particular, to verify it as credible. The profile is also a medium to market oneself, to elaborate about hobbies, interests, as well as languages spoken. A further dimension is added through the accumulation of profile reviews that are visible to other guests and hosts. The platform still operates as a form of technological assurance as the intermediary of the financial transaction.

- Guests and hosts both verify their identity by verifying their official user account and confirming personal details.
- Hosts also have the option of requiring a fixed amount payable before the BonAppetour invitation to cover costs.
- Hosts are covered by up to €100,000 in damages for every meal through the Host Guarantee, for instance when guests damage a host's property.
- A message system occurs between host and guest before an experience. BonAppetour is a social platform so the relationship starts by guests introducing themselves, explaining why they choose a particular event, and identifying who is coming with them. Guests can ask at which time they should come, but also about any special requests, e.g., vegetarian options.

BonAppetour is a place for travelers and foodies who want to find new ways to explore countries and for hosts willing to promote their food culture to make money at the same time. Key elements of BonAppetour's business model are the emphasis on building communities and encouraging social interaction. Platforms with new cuisine advice comprise the most diverse concepts by highlighting various conceptualizations of collaborative consumption, means of use, accessibility, ownership and internet facilitation. This is achieved by introducing and recommending local food to tourists or residents. This business model must endeavor to build a community that desires these types of food experiences. In general, local food is a destination attraction that motivates tourists to visit because it reflects local culture and provides a connection with the place (Cavicchi, Ciampi, & Stancova, 2016; Cohen & Avieli, 2004). For example, a key message on the platform from a host in Florence (Italy) is as follows: "What about enjoying good Tuscan food with a real wonderful view of Florence?" (BonAppetour, 2017). Guests find it important, not only to obtain information from the experience of food, but also to give something back, in the form of reporting their progress and sharing recipes. For instance, a guest reveals: "During the last week, I have managed to cook such delicious food that I have almost burst into tears" (female respondent A). The host in this instance offers tourists and travelers traditional foods, using local products which are strictly seasonal and cooked traditionally; with the aim of showing guests local Italian hospitality and authentic cuisine — all home-cooked of course.

Another option is to join in a food tour that brings the food to you — minus the queues. BonAppetour offers "The Great Singapore Food Tour" which promises the following:

> plenty of tasty Michelin-starred dishes, a professional tour guide who shares the stories behind the dishes and Singapore culture, meals enjoyed in the company of fellow foodies and hosts who stand in line on your behalf — so you get the food without having to endure those agonisingly long queues (BonAppetour, 2017).

At the same time, the presence of the community aspect also means that residents and tourists can coexist and come into contact with each other, thereby sharing services and entertainment and leisure locations. A guest interviewed on the BonAppetour website says:

> Had my first experience with BonAppetour in March 2015 in Florence. The food was delicious! I am pretty open-minded to food and just asked to try any traditional Tuscan dish... The hosts were good about explaining the history of each dish with an Italian food rookie like myself (BonAppetour, 2017).

What is important is building a relationship with the community and that users are engaged in sharing their experiences with others. One interviewee, an adult male, revealed:

> Currently I'm looking for eating with friends to address the issue of social isolation; and I'm looking for my own style and the foods that suit me... so becoming a continuing, living process (male respondent B).

6.6 Discussion and conclusions

Our results show that food sharing dynamics can be effectively studied by analyzing communities' shared food practices. Sharing of this type has the potential to compete with many traditional food distribution and retailing business models, like the traditional restaurant for example. In fact, there are opportunities to expand the principles of sharing to the provision of physical, social and recreational infrastructure. Colocation reduces the need for infrastructure and enables more to be achieved with less resource use and at lower cost. For example, land not being used by one government agency may be transferred and used more effectively by another. It is possible to connect with customers anywhere in the world to cut transaction costs and shrink the advantages of economies of scale that larger chains hold. To ignore the sharing phenomenon could indicate skepticism about the autonomy and social aspects of virtual communities (Labrecque, Esche, Mathwick, Novak, & Hofacker, 2013) as a tool for social development that can bring surprising results. However, the sharing economy is best understood as a series of performances rather than a coherent set of economic practices (Richardson, 2015). Equally, the sharing economy presents opportunities to promote innovation and entrepreneurship, which might help create jobs, strengthen community resilience and drive economic growth. Old businesses, firms, and occupations disappear and new ones emerge, enabled by new technologies. A benefit is that users can earn and work more flexibly and operators can complete existing services with sharing models, learn new skills, or support ethical causes.

Our results provide insights into the dynamics of food sharing that revolve around complex and nuanced lifestyles. Food needs to be not only good to eat but also have social potential. The advent of social communities through food sharing is mobilized as a framework to facilitate novel forms of participation in the economy. This case study aims to outline innovative solutions in an area that is evolving rapidly in terms of customer behavior in the food industry. But barriers to such development include trust: compared to impersonal market exchanges, sharing implies a number of problems with regard to this, which are directly related to its social depth. As Belk (2007, 2010) has argued, sharing implies interdependence, an obligation of care and responsible use from those involved; it creates a social/communal bond and gives rise to an (implicit and social) debt. The risk of opportunistic, incorrect, mistrustful behaviors is high, especially in the case of nonreciprocation. Indeed, it is becoming clear that the platforms themselves exist in a regulatory grey area where they cannot be held liable as employers or asset holders, nor do they function like many of their traditional competitors.

At the same time, sharing as an economic phenomenon is technology dependent. But technology, whilst being an enabler, can also be an obstacle, as not everyone is familiar or has access to it. This issue could be addressed not only by communication, but also by experts who could focus on designing discrimination-free platform solutions which establish trust (Plewnia & Guenther, 2017). Currently, research on sharing business models is still in its infancy, but without doubt certain sectors (e.g., transportation, tourism) have already seen changes. The trend is toward development of new products and processes located on expert platforms, with work organized on a project-by-project basis and carried out by transitory teams. It is also important to recognize the existence of different economic practices in the marketplace, as DiVito Wilson (2013) affirms, it is a step to rejecting the hegemony of capitalism and a way to rethink and recreate new economic and social realities as the strengthening of social capital within a community. However, it is unclear for the time being how a sharing economy redefines the roles of tourists and locals compared to the conventional market economy (Cheng, 2016).

The sharing economy has clearly not reached its full potential in food marketplaces but continues to grow in scope. The entrance of sharing platforms to the market has led to the introduction of quality products and services at a much lower price. Peer-to-peer sharing can also prioritize utilization of, and accessibility to, products and services over ownership. Online sharing platforms are used to access a global pool of workers on a just-in-time basis. They allow people to eat well at the price of a mid-range restaurant. Furthermore, Web 2.0 and social media have the lowered transaction costs for connecting providers and buyers to almost zero (Allen & Berg, 2014).

Overall, our results imply that food sharing communities can reveal information about consumer preferences regarding ingredients, food consumption habits, and what is considered fake or inauthentic. This suggests that online communities should be considered major partners in developing and marketing new products. The BonAppetour experience represents an exchange of potential tangible and intangible effects, leading to consumer opportunities for a better travel experience. The service is for those people who want to connect, to open their homes, whilst acting as both host and guest at the same time. On the tourism side, BonAppetour works mainly on the basis of cultural exchange: helping visitors to access a wide range of products and services at a more affordable price; facilitating authentic encounters between tourists and locals; and contributing to the employment and income of locals. It also provides good memories of a vacation for the tourist, with the knowledge of having tasted the traditional dishes of a country. Bon Appetour can also be an important tool to make new friends among people with common interests, and/or where there is a sense of community and belonging, even though this may be mainly virtual.

Despite all these benefits, food sharing has arguments against it: home restaurants increase the casualization of labor in tourism, for example, they can often avoid government regulations designed to protect both consumers and employees. In terms of hygiene and risk of infection, Gullstrand Edbring, Lehner, and Mont (2016), have investigated the attitudes of young consumers to different sharing economy consumption models, affirming this to be an obstacle in all consumption domains because of the great uncertainty there is about what is safe or good to eat. Consumers may prioritize values differently in this regard; for instance, some think that improvising and inventing dishes is better than the safety of their ingredients.

In conclusion, there are no rigorous methodologies for success in online food sharing platforms at present. This case study is a first step in exploring the field. It has reviewed relevant literature on the topic, and has provided a starting point for further studies in the area. Future research could extend our insights with primary quantitative data, comparing profit vs nonprofit online food sharing initiatives, and identifying those factors that impact on their success or failure.

References

Albinsson, P. A., & Perera, Y. B. (2012). Alternative marketplaces in the 21st century: Building community through sharing events. *Journal of Consumer Behaviour, 11*(4), 303–315.

Allen, D., & Berg, C. (2014). *The sharing economy. How over-regulation could destroy an economic Revolution*. Melbourne: Institute of Public Affairs. Retrieved <www.ipa.org.au> Accessed 12.12.2015.

Bardhi, F., & Eckhardt, G. (2012). Access based consumption: The case of car sharing. *Journal of Consumer Research, 39*, 881–898.

Beer, S. (2008). Authenticity and food experience—Commercial and academic perspectives. *Journal of Foodservice, 19*(3), 153–163.

Belk, R. (2007). Why not share rather than own? *Annals of the American Academy of Political and Social Science, 611*(1), 126–140.

Belk, R. (2010). Sharing. *Journal of Consumer Research, 36*(5), 715–734.

Belk, R. (2014). You are what you can access: Sharing and collaborative consumption online. *Journal of Business Research, 67*, 1595–1600.

Benkler, Y. (2004). Sharing nicely: On shareable goods and the emergence of sharing as a modality of economic production. *The Yale Law Journal, 114*, 273–358.

BonAppetour, 2017. Retrieved <https://www.bonappetour.com> Accessed 06.07.2017.

Botsman, R., & Rogers, R. (2010). *What's mine is yours. The rise of collaborative consumption.* New York: Harper Business.

Bradley, K., & Pargman, D. (2017). The sharing economy as the commons of the 21st century. *Cambridge Journal Regions, Economy and Society, 10*(2), 231–247.

Cavicchi, A., Ciampi., & Stancova, K. (2016). *Food and gastronomy as elements of regional innovation strategies.* Spain: European Commission, Joint Research Centre, Institute for Prospective Technological Studies. EUR 27757. Retrieved <https://ec.europa.eu/jrc>.

Celata, F., Yungmee Hendrickson, C., & Sanna, V. S. (2017). The sharing economy as community marketplace? Trust, reciprocity and belonging in peer-to-peer accommodation platforms. *Cambridge Journal of Regions, Economy and Society, 10*(2), 349–363.

Cheng, M. (2016). Current sharing economy media discourse in tourism. *Annals of Tourism Research, 60,* 111–114.

Christenson, J. A., & Robinson, J. W. (1989). *Community Development in Perspective.* Ames, IA: Iowa State University Press.

Cohen, E., & Avieli, N. (2004). Food in tourism attraction and impediment. *Annals of Tourism Research, 31*(4), 755–778.

Connolly, M., & Krueger, A. B. (2006). Rockonomics: The economics of popular music. In: V. A. Ginsburg, & D. Throsby (Eds.), *Handbook of the Economics of Art and Culture* (pp. 667–719). Elsevier.

Cox, R. (2013). House/Work: Home as a space of work and consumption. *Geography Compass, 7*(12), 821–831.

Davies, A. R. (2013). Food futures: Co-designing sustainable eating practices for 2050. *EuroChoices, 12*(2), 4–11.

Davies, A. R., & Doyle, R. (2015). Transforming household consumption: From backcasting to HomeLabs experiments. *Annals of the Association of American Geographers, 105*(2), 425–436.

Davies, A. R., Edwards, F., Marovelli, B., Morrow, O., Rut, M., & Weymes, M. (2017). Creative construction: Crafting, negotiating and performing urban food sharing landscapes. *Area, 49*(4), 510–518.

DiVito Wilson, A. (2013). Beyond alternative: Exploring the potential for autonomous food spaces. *Antipode, 45*(3), 719–737.

Dobscha, S. (1998). The lived experience of consumer rebellion against marketing. *Advances in Consumer Research, 25,* 91–97.

Eckhardt, G. M., & Bardhi, F. (2015). The sharing economy isn't about sharing at all. *Harvard Business Review,* 1–28. Retrieved <http://bit.ly/1EWNdmb> Accessed 10.1.2016.

EPRS, (European Parliamentary Research Services). *Cost of non-Europe in the sharing economy: Legal aspects.* (2016). Retrieved <http://www.europarl.europa.eu/thinktank> Accessed 25.01.16.

EU Commission for Economic Policy. *The local and regional dimension of the sharing economy. ECON-VI/005, 2015.* (2015). Retrieved <https://webapi.cor.europa.eu/.../COR-2015-02698-00>.

European Parliament. *The sharing economy and tourism, tourist accommodation, briefing, September.* (2015). Retrieved <www.europarl.europa.eu/RegData/etudes/BRIE/2015/568345/EPRS_BRI(2015)56%208345_EN.pdf> Accessed 01.11.2015.

Falcone, P. M., & Imbert, E. (2017). Bringing a sharing economy approach into the food sector: The potential of food sharing for reducing food waste. In: P. Morone, F. Papendiek, & V. E. Tartiu (Eds.), *Food Waste Reduction and Valorisation: Sustainability Assessment and Policy Analysis* (pp. 182–198). Cham: Springer International.

Feenstra, G. (2002). Creating space for sustainable food systems: Lessons from the field. *Agriculture and Human Values, 19*, 99−106.

Forno, F., & Garibaldi, R. (2015). Sharing in travel and tourism: The case of home-swapping in Italy. *Journal of Quality Assurance in Hospitality & Tourism, 16*(2), 202−220.

Fournier, S. (1998). Consumer resistance: Societal motivations, consumer manifestations, and implications in the marketing domain. *Advances in Consumer Research, 25*, 88−90.

Goodman, D. (2004). Rural Europe redux? Reflection on alternative agro-food and paradigm change. *Sociologia Ruralis, 44*(1), 3−16.

Goudin P. *The cost of non-Europe in the sharing economy.* (2016), Retrieved <www.europarl.europa.eu/thinktank> Accessed 22.1.2016.

Gullstrand Edbring, E., Lehner, M., & Mont, O. (2016). Exploring consumer attitudes to alternative models of consumption: Motivations and barriers. *Journal of Cleaner Production, 123*, 5−15.

Guttentag, D. (2013). Airbnb: Disruptive innovation and the rise of an informal tourism accommodation sector. *Current Issues in Tourism, 18*(12), 1192−1217.

Hamari, J., & Ukkonen, A. (2013). The sharing economy: Why people participate in collaborative consumption. *SSRN Electronic Journal.* Retrieved <http://papers.ssrn.com/sol3/Papers.cfm?abstract_id = 2271971>.

Heinrichs, H. (2013). Sharing economy: A potential new pathway to sustainability. *Gaia, 22* (4), 228−231.

Honggang, X., & Qunchan, F. (2015). The production and consumption of home in the home restaurants in Guangzhou. *Journal of China Tourism Research, 11*(1), 105−120.

Internetworldstats. (2017) *Internet usage and world population statistics.* Retrieved <http://www.internetworldstats.com>.

Kera, D., & Sulaiman, N. (2014). FridgeMatch: Design probe into the future of urban food commensality. *Futures, 62*(B), 194−201.

Kozinets, R. V. (1999). E-tribalized marketing? The strategic implications of virtual communities of consumption. *European Management Journal, 17*(3), 252−265.

Labrecque, L. I., Esche, J., Mathwick, C., Novak, T., & Hofacker, C. F. (2013). Consumer power. Evolution in the digital age. *Journal of Interactive Marketing, 27*, 257−269.

Liebowitz, S. J. (2006). File-sharing: Creative destruction or just plain destruction? *Journal of Law and Economics, 49*(1), 1−28.

Liebowitz, S. J. (2016). How much of the decline in sound recording sales is due to file-sharing? *Journal of Cultural Economic, 40*, 13−28.

Martin, C. J. (2016). The sharing economy: A pathway to sustainability or a nightmarish form of neoliberal capitalism? *Ecological Economics, 121*, 149−159.

Migliore, G., Forno, F., Dara Guccione, G., & Schifani, G. (2014). Food community networks as sustainable self-organized collective action: A case study of a solidarity purchasing group. *New Medit, 4*, 54−62.

Ozanne, L. K., & Ballantine, P. W. (2010). Sharing as a form of anti-consumption? An examination of toy library users. *Journal of Consumer Behaviour, 9*, 485−498.

Plepys, A., Heiskanen, E., & Mont, O. (2015). European policy approaches to promote servicizing. *Journal of Cleaner Production, 97*, 117−123.

Plewnia, F., & Guenther, E. (2017). Advancing a sustainable sharing economy with interdisciplinary research. *Umwelt Wirtschafts Forum, 25*(1−2), 117−124.

Presenza, A., & Del Chiappa, G. (2013). Entrepreneurial strategies in leveraging food as a tourist resource: A cross-regional analysis in Italy. *Journal of Heritage Tourism, 8* (2−3), 182−192.

Richardson, L. (2015). Performing the sharing economy. *Geoforum, 67*, 121−129.

Robinson R.N.S., & Clifford C. (2012). Authenticity and festival foodservice experiences, *Annals of Tourism Research 39*(2), 571–600.

Robinson, R. N. S., & Getz, D. (2014). Profiling potential food tourists: An Australian study. *British Food Journal, 16*(4), 690–706.

Schor J.B. (2014) *Debating the sharing economy.* Retrieved <http://www.greattransition.org/publication/debating-the-sharing-economy> Accessed 17.1.2016.

Schor, J. B. (2015). Getting sharing right. *Contexts, 14*(1), 12–19.

Scholz, T. (2016). *Platform Cooperativism. Challenging the Corporate Sharing Economy.* New York, NY: Rosa Luxemburg Stiftung.

Seyfang, G., & Smith, A. (2007). Grassroots innovations for sustainable development: Towards a new research and policy agenda. *Environmental Politics, 16*(4), 584–603.

Sharp, E., Wardlow, F., & Lewis, N. (2015). Alternative framings of alternative food: A typology of practice. *New Zealand Geographer, 71*, 6–17.

Sims, R. (2009). Food, place and authenticity: Local food and the sustainable tourism experience. *Journal of Sustainable Tourism, 17*(3), 321–336.

Wosskow D. *Unlocking the sharing economy. An independent review.* (2014). Retrieved <www.gov.uk/bis> Accessed 13.12.2015.

Supply chain analysis of farm-to-restaurant sales: A comparative study in Vancouver and Christchurch

7

Hiran Roy[1], C. Michael Hall[2] and Paul W. Ballantine[2]
[1]Hospitality Management, Acsenda School of Management, Vancouver, BC, Canada,
[2]Department of Management, Marketing and Entrepreneurship, University of Canterbury, Christchurch, New Zealand

7.1 Introduction

The demand for local food is growing and creating new economic opportunities for farmers and restaurants who participate in local food systems (Hall & Gössling, 2016a). However, the growing interest in local foods has been explored largely from the perspective of consumers (Martinez et al., 2010) and the role of farmers' markets (Hall, 2013), rather than for other elements in the culinary system and supply chain such as restaurants and food wholesalers (Gössling & Hall, 2013; Hall & Gössling, 2016b). Previous studies have indicated that the direct marketing of farm products to restaurants would increase farm sales and broaden consumers' exposure to local farming operations (Thilmany, 2004). Other ancillary benefits (e.g., changing consumer demand for local food products, and enhancing brand development and product differentiation for producers) have also been identified in selling to restaurants (Curtis & Cowee, 2009; Gössling & Hall, 2016). However, very little empirical research has been conducted on the benefits and barriers for farmer/producer direct-marketing efforts and supply relationships to local restaurants (Nilsson, 2016).

This chapter presents the results of a comparative study of local farmer/producer and restaurant relations in Vancouver (Canada) and Christchurch (New Zealand) in the context of the promotion of local foods. The study identifies strategies for successful local food selling by farmers/producers and provides a more comprehensive perspective than earlier studies in the field, noting especially the roles of network relationships and intermediaries, whose significance in the culinary system has often been ignored.

Case Studies in Food Retailing and Distribution. DOI: https://doi.org/10.1016/B978-0-08-102037-1.00007-4

7.2   Background

There is no consensus on defining "local" and what constitutes a local food system (Pearson et al., 2011). Research indicates that the definition of "local food" is complex, as are its implications for small-scale producers (Trivette, 2015), and is "based on a general idea of where local food is coming from" (Dunne, Chambers, Giombolini, & Schlegel, 2011, p. 50). Nevertheless, although there is no consistent definition of what constitutes "local food," it remains an important component of food promotion and purchase (Hall, 2013).

7.2.1   Local food movements

A number of socioeconomic and environmental movements have converged around the idea of a local food system. A local food system refers to deliberately formed food systems that are characterized by "a close producer-consumer relationship within a designated place or local area" (Hall & Gössling, 2016b, p. 10). The concept of a local food system symbolizes a paradigm shift from the globalized and industrialized food system toward local or relocalized food systems (Allen, FitzSimmons, Goodman, & Warner, 2003; Hinrichs, 2003; McMichael, 2009; Wilhelmina, Joost, George, & Guido, 2010), which are often regarded as an alternative to conventional food production (e.g., Feagan, 2007; Higgins, Dibden, & Cocklin, 2008). Indeed, many localized communities have initiated alternative food and agricultural systems (Feenstra, 2002). As Anderson and Cook (2000, p. 237) note:

> The major advantage of localizing food systems, underlying all other advantages, is that this process reworks power and knowledge relationships in food supply systems that have become distorted by increasing distance (physical, social, and metaphorical) between producers and consumers ... [and] gives priority to local and environmental integrity before corporate profit-making.

Local food movements have therefore been regarded as offering new economic benefits for small and medium-sized farms, reductions in the environmental footprint of food, and closer relations between consumers and producers while providing good nutrition to consumers (Hall & Gössling, 2016b; Kloppenburg, Lezberg, De Master, Stevenson, & Hendrickson, 2000). As a result, since the mid-1990s, consumer interest in using and purchasing local foods has increased substantially, a trend also supported by the growth of the slow food movement and a food media focus on local food (Hall, 2012; 2013). This interest is evidenced in a variety of "alternative" forms of food retail and distribution, including farm-to-school programs, farmer and producer direct marketing, farmers' markets, and community supported agriculture (Allen et al., 2003; Izumi, Wright, & Hamm, 2010; Hall, 2013; Tregear, 2011). However, restaurants are also a very important part of the culinary system, especially given the extent to which people "eat out." Therefore, the way in which restaurants support local foods is a significant factor influencing the food supply chain in such systems.

7.2.2 Benefits and obstacles perceived by restaurants and chefs

Restaurant decisions to purchase local food are based on a number of factors, such as product taste; perceived higher quality, freshness, and safety; access to unique or specialty products; satisfaction of consumer requests; increase in bottom-line profits of the establishment; public relations; supporting the local economy and communities; possibility of purchasing smaller quantities; lower transportation costs; competitive pricing; and the dependability of farmer suppliers (Inwood, Sharp, Moore, & Stinner, 2009; Kang & Rajagopal, 2014; Murphy & Smith, 2009; Reynolds-Allie & Fields, 2012; Roy, Hall, & Ballantine, 2016). Significantly, Sharma, Strohbehn, Radhakrishna, and Ortiz (2012) reported that restaurant customers are willing to pay extra for menu items sourced from local farmers.

A number of barriers to restaurants purchasing local food have also been identified, particularly relating to cost factors; payment procedure conflicts; product availability; dealing with multiple suppliers; complicated ordering processes; packaging and handling; inadequate distribution systems and service; ineffective communication; and higher product costs (DeBlieck, Strohbehn, Clapp, & Levandowski, 2010; Inwood et al., 2009; Nilsson, 2016; Peterson, Selfa, & Janke, 2010; Pillay & Rogerson, 2013; Reynolds-Allie & Fields, 2012; Roy et al., 2016; Sharma, Gregoire, & Strohbehn, 2009). Lack of knowledge with respect to local food sources can clearly be a significant issue in the food chain, while local food purchasing patterns can vary significantly by the type of restaurant (Curtis & Cowee, 2009).

7.2.3 Benefits and obstacles perceived by farmers/producers

Although the promotion of local food on restaurant menus appears to be increasingly popular, little systematic empirical research has been conducted on the food producer's perspectives of the benefits and barriers to marketing their products directly to local restaurants. Where this has occurred, the reported perceived benefits of direct marketing and selling included supporting local farmers and the local economy; providing fresher, flavorful, and higher quality products for customers; convenience; personal commitment to environment and food safety; reduction in food miles; price premiums; product knowledge; and personal relationships with restaurants (Curtis, Cowee, Havercamp, Morris, & Gatzke, 2008; Green & Dougherty, 2008; Gregoire, Arendt, & Strohbehn, 2005; Nilsson, 2016; Sharma et al., 2012). Nevertheless, a number of obstacles to selling local foods to restaurants have also been identified: failure to match supply and demand; limited product ranges; restaurant problems in dealing with multiple sellers; price; seasonality issues and lack of year round production; high transportation and delivery costs; higher costs of production; delivery issues; and unexpected changes in buyer demands (Dougherty, Brown, & Green, 2013; Green & Dougherty, 2008; Sharma et al., 2012).

The extent of knowledge transfer between restaurants and suppliers appears essential for understanding each other's needs and challenges (Self et al., 2016). For example, research suggests that many producers appear unsure how to enter the market; there is often a lack of knowledge by both producers and buyers about regulations surrounding food distribution, a lack of knowledge on the part of producers in developing relationships with restaurants, and a lack of time to invest in building these relationships (Curtis et al., 2008; Gregoire et al., 2005; Self et al., 2016). Schmit, Lucke, and Hadcock (2010) also found that the fact that restaurants were already dealing with multiple sellers created a significant obstacle for farmers to directly market to restaurants. The issue in many such cases is not that a restaurant is not interested in using local foods, but more so that they already have existing suppliers in place and they do not have sufficient time to modify their supply chains by finding new suppliers (Peterson et al., 2010; Roy et al., 2016).

7.3 Method

This chapter is based on individual semi-structured interviews conducted from September to November 2014 in Vancouver, and February to April 2015 in Christchurch, with farmers and/or farmers' market vendors. A convenience sampling approach was used in order to complete a large number of interviews as quickly and as cost-effectively as possible (Neuman & Robson, 2009). The interviews were also used to develop a survey of restaurants with respect to local food purchase (see Roy et al., 2016 for a discussion of the Vancouver results). A total of 12 farmers and/or farmers' market vendors from Vancouver and 8 farmers and/or farmers' market vendors from Christchurch that currently sold local products to local foodservice establishments were identified and recruited for the interview sessions.

A letter of purpose presenting the objectives of the study and a consent form was sent to all the farmers and/or farmers' market vendors. The interview date and venue was then arranged. The average length of the interview discussions with farmers and/or farmers' market vendors was 60 minutes and they were conducted at the farm and at farmers' markets. All interviews were conducted face-to-face in English on a one-on-one basis and audio recorded with the permission of the interviewees for later transcription. Transcripts were compiled verbatim as soon as possible after each interview by the researchers. The responses were then analyzed and coded; for example, Respondent 1 was labeled F1, Respondent 2 as F2, and so on. Content analysis was undertaken based on the textual data derived from the transcripts of semi-structured interviews to gain knowledge, new insights, and understanding of phenomena through valid inferences from text data to the context of the study (Krippendorff, 1980). The data was then extracted manually under the thematic headings (Braun & Clarke, 2006).

7.4   Results

7.4.1   Interview respondents' profile

The Vancouver sample consisted of 50% (six) male and 50% (six) female participants, while all the interviewees (eight) from Christchurch were male. The majority
of the respondents from both samples were experienced farmers (average 17.9 years
in Vancouver and 15.9 years in Christchurch). About 42% of the farms used some
form of sustainable farming technique (e.g., integrated pest management, organic)
in Vancouver and about 37% in Christchurch. In both samples, restaurants were the
most popular choice of outlets for the producers' products followed by farmers'
markets, while farms also utilized wholesale distributors to market their products.

7.4.2   Respondent definition of local food

Respondents were asked what local food meant to them. The definitions offered by
the majority of respondents were based on geographical or political boundary lines
(province or region), such as products "grown" within the region or within a political boundary, rather than by a distance measure. In contrast, very few (one respondent in Christchurch and two respondents in Vancouver) defined their "local food"
in terms of the mileage or distance they would travel to sell. The actual number of
miles they would travel varied considerably, ranging from 90 in Christchurch to a
100−210 mile radius from where they lived in Vancouver. Six respondents in
Vancouver and seven respondents in Christchurch defined "local food" according
to political boundaries rather than by a distance measure from restaurants. One
respondent in Vancouver went even further to explain that he would prefer food to
be grown closer, rather than simply within the geographical or political boundary
lines or distance measures, and wanted it to be from as close as possible:

> For me, local means 'just up the road'. When I was in California, I bought
> strawberries from a surplus stand outside an enormous farm that supplies berries
> all over North America. I considered the roadside ones local (Vancouver, F1).

These perspectives among the respondents in defining and describing local food
reflects a wide variety of definitions found in the academic literature (Peterson
et al., 2010; Sharma et al., 2012; Sims, 2010; Vecchio, 2010).

7.4.3   Benefits perceived by farmers/producers

Farmers were asked about perceived benefits and/or motivations for selling to restaurants and chefs. Farmers in both samples noted personal satisfaction, product
appreciation, higher prices, and personal relationships as the main reasons for this.

7.4.3.1   Personal satisfaction

Respondents in both samples indicated that they derive personal satisfaction from selling their products to restaurants and chefs. They see chefs creating beautiful dishes with their products, thus they feel they are valued for their products and their hard work is appreciated:

> Well we like the idea of our food being served to people who are our customers
> and now we really want the restaurant to have local produce. So, that's a big
> motivation and to see it is going to a chef who is creating a beautiful food with it.
> It is really satisfying. It's just like, you know, "Oh my God!" That means it paid off
> my hard work with my produce (Vancouver, F10).

Similarly, other respondents talked about wanting to sell to restaurants and chefs to keep their products local, with one stating:

> I like to see them going by using local products, and most of the restaurants and chefs
> we use are selling local products. So, our local products will be going there locally. It
> is just a good outlet and chefs can easily do the product justice sort of things and they
> do a pretty good job with that. So, it shows in a good light … (Christchurch, F19).

7.4.3.2   Product appreciation

For some respondents, their product's appreciation seemed to be a major motive to sell to the restaurants and chefs:

> I want to sell my products [to those] who entertain me and appreciate my products,
> and part of it is in terms of developing the brand. I want my products to be reached
> to the bigger audience through the chefs and restaurants and I think people are
> aware of it. It helps to bring people's awareness of my products and you know
> when it featured on the menu (Vancouver, F2).

Another respondent stated that he receives product appreciation, along with good prices, and that motivated him to sell the products to restaurants and chefs:

> The appreciation, getting good price, and the volume I am selling to them is good
> for me rather than selling to my retail consumers (Christchurch, F14).

However, the same respondent clearly mentioned that he would make more money by selling his products at the market, but due to the appreciation of his products by chefs, he pursued this marketing avenue for his business:

> …if I sell to the restaurants it is going to be larger amount of the products and
> selling to the consumers at the market will be smaller amount of the products. But
> the thing is that market makes more money for me, but for the restaurants I just
> price the stuff, pack it, and then send to the restaurants. Moreover, they appreciate
> my products too (Christchurch, F14).

7.4.3.3 Higher prices

Direct sales to restaurants can increase farmers' profits (Bendfeldt, Walker, Bunn, Martin, & Barrow, 2011; Nilsson, 2016). Several farmers in this study described the benefits of selling to restaurants and chefs in order to receive price premiums:

> They are totally supportive and they don't try [to say] it's low value and I like that. They are ready to pay above the market price. That's [what] I really like about restaurants. And you know, it is booming the restaurant thing, and it is growing and growing (Vancouver, F4).

> I often found that even dealing direct—in that charging higher price when you have to deal direct—and we are still cheaper than wholesalers (Christchurch, F20).

The result is consistent with previous studies (Green & Dougherty, 2008; Sharma et al. 2012). For example, Sharma et al.'s (2012) study found that most growers identified the benefits of selling to local restaurants; especially those that were able to pay a price premium for their products.

7.4.3.4 Personal relationships

Respondents described feelings of enjoyment and appreciation after building personal relationships with restaurants and chefs in both samples. In this study, three respondents (two from Vancouver and one from Christchurch) wanted to sell directly to restaurants and chefs due to personal relationships. Many farmers also engaged restaurants and chefs as individuals and see restaurants and chefs as a very valuable source of marketing guidance:

> The other thing we like to have the relationship with the chefs that we learn from them what people are cooking? What's out there? How can you do certain foods? These are the things we don't know out there. If we learn somebody is using the particular food in some other way, well then we know for the next year we can grow some of those items. Like chef [chef name] came here and says that you could sell those products, which I did not know ... We know lot of our stuff but we do not know the user end (Vancouver, F10).

In some cases, farmers see restaurants and chefs as their best source of information on prospective products. As one respondent from Vancouver remarked:

> We went to the restaurant at Richmond and [I] introduce myself and said we are starting out a new farm. One day he called to us right back at the beginning of this year and then he came to our farm. He is one of the few chefs ... because he is so close to our farm and he came to our farm an hour before the meal started and demanded by saying I need fennel, I need these mini tomatoes, and I need these and so on. So, we stopped what we were doing and did harvest the required products for him. It is all about relationship (Vancouver, F10).

However, support for local farmers was not necessarily shared by all restaurants and chefs:

Yeah, it is good to see ... some of the restaurant[s] using our products. Lot[s] of them talk about it but they do not actually do that (Christchurch, F15).

Two respondents from Christchurch were open to considering marketing their products to restaurants and chefs, as opposed to having the products sold to wholesalers directly, as it allows them to remain in full control of their products until the final point-of-sale and enables them to maximize value-added potential:

I have got no control with the wholesalers. If I sell it today, I do not know that reaches to kitchen and how they are handling my products. You know there is no key to love my products and products could be left in a box or ruining in the box or that could be two or three days old when it gets to the chefs and then it is going to have the bad reputation for my products. And you know, I do not feel comfortable with that (Christchurch, F18).

A different participant from Christchurch complained that dealing with markets or wholesalers was too unpredictable to market his products and he asserted that he would rather sell to restaurants and chefs:

...if we did not have lot of our direct supply customers then we wouldn't exist. Doing what we do even in a lot of horticulture I don't think you can survive just dealing with the market or the wholesalers, they are too unpredictable. The entire thing is that you can turn off one day with all of your stuff and then have no orders. And we claim, we turn up and you got all of your stuff then you need double the numbers because that is predictable. They just chase the price all the time and it is very hard to operate like that. I could sit down now just about for the restaurants and I could be 80% accurate with what ... all of my chefs will order tonight. For the wholesalers, some of them could do that, but most of them are not reliable to do that (Christchurch, F20).

Similarly, a respondent from Vancouver shared his personal values related to farming, and marketing seemed to be a major motive for selling to restaurants and chefs. The value he shared was the pride he has taken in growing the products. He offered this explanation:

Pride for who we are and our family's name and it is tied to us being a family business and at the end of the day it is our name on everything. And we are personally accountable to the people whom we deal with. So, we take pride in seeing our business name on chef's menus. We take pride in the chefs knowing us personally and knowing our families, and knowing details about us. And vice versa for us knowing about them and being comfortable with walking into a place, seeing a smiling face, and having a beer with those people that we are selling to ... because we mutually like and respect each other. So yeah, I mean my family and

... my family has ... we are very proud people and we have very strong egos and very strong personalities and that's what keeps us doing what we do as because we can sell cheap vegetables and crap but we will not do it. So, these are forces (Vancouver, F11).

7.4.4 Barriers perceived by farmers/producers

In the interviews with farmers, several barriers emerged that they perceived as hindering them in selling directly to restaurants and chefs. The barriers were very diverse in both samples. Several respondents from Vancouver reported that they were not able to supply a required quantity or volume of products that restaurants and chefs needed to purchase (F2, F7). A different respondent who sells his products at the farmers' market stated that because he had a limited volume it was difficult for him to supply the quantity that was necessary to meet the demands of a restaurant (F6). For the same respondent, the commitment to serve farmers' market consumers also created a barrier for him in providing the required volume of products to restaurants and chefs:

The farmers' market customers are big portions of our buyers, so yeah, it is hard for me to find the balance sometimes. I do not want to disappoint the chefs and restaurants and at the same time with my farmers' market customers. As you know, both segments are valuable to me (Vancouver, F6).

In another case, a respondent acknowledged that the uncertainty of weather conditions was the most challenging aspect of their farm work and that was one of the barriers for them to supply quality products to restaurants (see also Self et al., 2016):

To maintain the quality because my farm is far away and the weather conditions. For example, weather is not always same and it kind of affects the quality of my produce. If it rains then it destroys lot of the produce. If the temperature changes, or if it is too hot or cold then it is difficult for me to maintain the quality of the produce. So, restaurants have to consider that but they do understand that part. I just have to make them a phone call and say I don't have it and get it from somebody else. For sure, I cannot handle the Mother Nature (Vancouver, F8).

A respondent from Christchurch stated that the cost of production also presented difficulties to sell to restaurants and chefs:

Sometimes you sell the products and make huge amount of money and sometimes you do not make money out of that but still you have to do it to keep your customer happy. The products might take longer to produce, depending on the season, depending on the weather, and depending on the ground conditions you have got ... you know. The products could cost twice as much in spring time to produce than in summer time (Christchurch, F14).

This comment reflects Self et al.'s (2016) observations that small farmers in the local food system have smaller profit margins that could be affected by the higher cost of producing foods through sustainable growing practices. In some cases, several respondents from both samples complained that restaurants and chefs do not place orders on time and they recognized that there was a lot more work involved in this regard. As one respondent commented:

> I'd say 'timing'. Timing can be a significant challenge. You [the interviewer] have worked in restaurants for a long time and you know, sometimes chefs are pain in the butt to get hold of ... and [it's] hard to train them to get the orders out on time. So, that's definitely a challenge and basically setting boundaries around for when it's acceptable to place your order and if you're going to consistently place your orders ... (Vancouver, F5).

Another three respondents claimed that delivery costs were higher for selling directly to restaurants and chefs if they do not order enough volume/quantities of the products from them, particularly if they were located away from the city. This would imply that farmers would have to make frequent deliveries to the restaurants and chefs. This could help to create relationships but this cost was unavoidable for the farmers:

> The main problem is getting enough volume of order that makes [it] worthwhile for me. It is not worth for me to drive the products to the town and deliver the products which cost only $100. It costs me $100 for fuel to drive the vehicle to the town. By the time I use my labor and fuel cost then there is no worth for me to supply the products to them. So, it's mainly volume that is not enough for me (Christchurch, F15).

Barriers related to delivery costs are identified in previous studies (Dougherty et al., 2013; Schmit & Hadcock, 2012; Self et al., 2016; Sharma et al., 2012). Additionally, in line with previous research (Sharma et al., 2012), a lack of planning and foresight by restaurants can further fuel the uncertainty of product demand for farmers, as one respondent from Vancouver complained:

> Production time, because they are plants and they take a while to grow. Chefs will plan their menus without consulting us and expect products to be available on the drop of a hat or they will have special events booked for a long time in advance but they won't order and they won't give us enough lead time to ensure that there's enough products for them. So, as a grower and producer that is probably the biggest challenge for me (Vancouver, F11).

7.4.5 Other specific barriers (food safety and licensing concerns)

Many foodservice establishments require farmers to comply with regulatory requirements (licensing, certification, food safety protocols, and liability insurance) to protect against economic loss from food-borne illnesses attributed to the farmers' products. A number of respondents (seven from Vancouver and three from Christchurch) indicated that they had no concerns at all with licensing and regulations. This result is somewhat surprising, since food safety and liability concerns have often been noted as a challenge to small farmers (Gregoire et al., 2005; Peterson et al., 2010). Similarly, the majority of studies from foodservice perspectives cite food safety and/or liability insurance as a concern (Gregoire & Strohbehn, 2002; Gregoire et al., 2005; Pillay & Rogerson, 2013). However, several respondents did indicate their dissatisfaction with these licensing and regulation policies, with one remarking:

> *Compliance with food safety is often a nightmare. Not because of the compliance but obviously you do not make money out of poisoning your customers. But the amounts of paper work are enormous. You know, I mean the paper work and then I mean just developing the relationships and it's very easy for businesses that we rely on each other. And so that relationship is really important and you are not actually in an isolated business. You are in an actual business that relies on ... other relationships and we are just trying to maintain the relationships. For example, it is easy for me to cut a piece of meat but everything else is involved so that you know ... you are not doing [the] wrong thing (Christchurch, F13).*

Nevertheless, many respondents from both samples stated that having such certifications has been a positive impact for them in selling their products to restaurants and chefs:

> *I would feel that there is a positive impact as it does keep us to be honest that you have some regulations. It is also a source of confidence for the customers and for you too and that also keeps the grower honest. If anything goes wrong you can always check back, you know (Vancouver, F9).*

Only two respondents (one from each sample) mentioned that restaurants and chefs were not interested in these certifications.

7.4.6 Farmers/producers and fair prices

When farmers were asked whether they were paid a fair price for their products by restaurants, 11 out of 12 respondents in Vancouver and 7 out of 8 respondents in Christchurch stated that they were. The reasons for being paid fairly ranged from being able to set their own prices for products, higher quality products, and their efforts to grow the products:

> *I think the restaurants that we deal with, we are really comfortable with the price point. It is always challenging ... working with the restaurants because of their low price point and they do not want you to get [a] good price. And that's why only some restaurants does local food because they don't believe the price point and [that] they can still make money. So, the restaurant we have chosen ... I feel like our prices are very fair (Vancouver, F10).*

On the other hand, a respondent from Vancouver claimed that he was not paid a fair price because he believed people do not want to pay if he puts the real price on his products (F12). Another respondent from Vancouver discussed having problems with other farmers at the farmers' market who sold similar products for a much higher price and described his dissatisfaction with these farmers (F7).

In further discussion, respondents were asked what criteria they use to determine their products' prices. For both groups of respondents, the cost of production and a fair return on their work (i.e., wages, labor) plus a desired profit margin and matching other farmers' prices were stated (see also Schmit et al., 2010). As one respondent noted:

> I guess kind of mostly talking with the other farmers … It would be ridiculous to see the product price is high or low. I mean we kind of go sometimes like, "That is way too much" or, "That is too cheap". So, basically we go with the comparison of other farmers or vendors and market values sort of things, you know (Vancouver, F4).

While two respondents from Christchurch stated that the market sets the price and anything that can move the products is a fair price for their produce (F17, F20), another respondent from Vancouver had a strategy of setting prices for products that included an assessment of conventional and wholesale certified organic prices (F10).

7.4.7   Future prospects for selling local food products

Respondents were asked about their future plans for selling their food products locally to restaurants and chefs. All individuals interviewed wanted to continue this practice. Interestingly, some respondents want to increase the volume of food they supply to restaurants and chefs, while others want to stay about the same. Comparing the Vancouver and Christchurch samples, both were seen as being more interested in increasing (eight and five respondents, respectively) the amount of food they sell to restaurants and chefs than staying about the same. Respondents that expressed an interest in increasing the proportion of food they were selling to restaurants and chefs mentioned several reasons: an increase in production capacity; a decrease in sales through the wholesale distribution channel; an increase in specialty products in order to remain competitive in the market and an accompanying decrease in standardized products; and an increase in arable land. However, several respondents from Christchurch noted that lack of time and staff, and a limited product range, prevented them from being able to increase the number of restaurants and chefs they supply to. In addition, delivery and logistics was recognized by one respondent as a significant barrier to working with restaurants and chefs:

> I mean I get restaurants quite often asking us can you deliver the products, then I say, "Yes I do". And then you do not hear anything more from them. It would be easier for us if they collect the products from [a] farmers' market, but they do not want to do that. So, that's one of the reason[s] I do need to supply directly from my farm to their restaurants. (Christchurch, F15).

7.5   Discussion

7.5.1   Definition of "local food"

This study highlights that the term "local food" is a relatively fluid and dynamic concept (Duram & Cawley, 2012; Hall, 2013; Peterson et al., 2010; Sims, 2010; Trivette, 2015; Vecchio, 2010). There was no consensus on the definition of "local food" among the respondents from both samples. As Allen and Hinrichs (2007) noted, this reflects the extensive debate about the meaning of the term "local food," with farmers and/or farmers' market vendors adapting a range of definitions in accordance with their own interests and perceptions. In both Vancouver and Christchurch, farmers and/or farmers' market vendors primarily defined "local food" in terms of geographical or political boundaries than by a distance measure, even if, as in New Zealand, farmers' markets themselves use a distance measure (Hall, 2013). Nevertheless, these variations can lead to uncertainty surrounding the sourcing of local foods and challenges in labeling or branding products as "local" (Feagan, 2007).

7.5.2   Benefits of local food as perceived by farmers/producers

Personal satisfaction, product appreciation, and higher prices for products were the major perceived benefits and/or motivations reported by farmers for selling to restaurants and chefs in both samples. Vancouver and Christchurch respondents aim to maximize their share of the food dollar through marketing to foodservice businesses. As one farmer stated, "They are ready to pay above the market price. That's what I really like about restaurants, and you know it's booming the restaurant thing, and it is growing and growing" (Vancouver, F4). The current findings therefore reinforce research that suggests that farmers preferred to sell to restaurants to receive a price premium for their products and improve their cash flow (Sharma et al., 2012). Similarly, in a Swedish study, Nilsson (2016) also reported that farmers who sell direct to restaurants in the same region receive higher prices than through conventional sales channels.

7.5.3   Barriers to sale of local food as perceived by farmers/ producers

With regards to barriers to greater sales of local foods, the findings revealed several diverse marketing barriers that were reported by farmers in both Vancouver and Christchurch. Barriers reported by farmers included lack of quantity or volume of the products that restaurants and chefs needed to purchase; the uncertainty of weather conditions; placing orders on time; delivery costs; and cost of production. These barriers are echoed throughout the literature (Gregoire & Strohbehn, 2002; Gregoire et al., 2005; Nilsson, 2016; Sharma et al., 2012). However, among respondents, two financial barriers were consistently cited: cost of production and delivery costs. These findings also indicate that there were higher costs associated to farmers for the delivery of products to restaurants than when selling directly to wholesalers or at farmers' markets to consumers. In contrast, farmers' market sales require less transportation and are confined to one location, thus reducing delivery costs.

While several marketing issues were noted by farmers, the findings from both samples revealed that food safety and liability were not major concerns and all farmers felt that they generally received a fair price for their products. Farmers believed it was in their ability to set their own price for higher quality products.

7.5.4   Future prospects of selling local food products

Most of the interviewed farmers are interested in selling more of their products and want to decrease the wholesale distribution channel in order to maximize their revenue from direct selling to restaurants and chefs. However, for them, more consistent demand of local products and the need to reach a bigger and more stable market were the key factors that need to be met in order to expand their involvement in the local food system. This finding is similar to that of Gregoire et al. (2005), although as noted above, the issues of delivery and logistics are a significant barrier to working with restaurants and chefs.

7.6   Conclusions

The results of this chapter indicate that there is a positive attitude among farmers and/or farmers' market vendors toward increasing sales of their products to restaurants in both of the study sites (see also Pillay & Rogerson, 2013; Schmit et al., 2010; Sharma et al., 2012). Even though there is no clear definition of what local food means (Conner, Montri, Montri, & Hamm, 2009; Hall, 2013; Pearson et al., 2011; Trivette, 2015), there clearly remains a demand for it from restaurants, chefs, and consumers (Hall & Gössling, 2016a).

This study indicated that social networks are extremely important for local food systems and this has clear implications for farmers who wish to increase sales to local restaurants and chefs. Restaurants and chefs like the personal connections that can be developed with farmers through farmers' markets, direct sales with farmers, recommendations from fellow operations, and events (Roy et al.2016). The study therefore suggests that farmers need to communicate directly with restaurants and chefs to improve information flow. However, due to the fragmented nature of the local food value chain, many small-scale farmers may face obstacles to do so. Tactics such as workshop mingles, farm and restaurant tours, and locally sourced food events can be useful mechanisms to bring producers and restaurants together (Brain, Curtis, & Hall, 2015).

The findings also indicate that for farmers, the most important factor in explaining selling intentions to restaurants and chefs is the farmer's personal satisfaction, the chef's product appreciation, and the price premiums available for products. This is not consistent with the more general results of previous work (e.g., Dougherty et al., 2013; Kang & Rajagopal, 2014; Lillywhite & Simonsen, 2014;

O'Donovan, Quinlan, & Barry, 2012; Schmit & Hadcock, 2012). Nevertheless, the results indicate that economic considerations, specifically fair prices for farmers, were clearly central in both samples to the development of selling to restaurants and chefs. The study also highlighted that maintenance of personal relationships is a necessary step in creating a successful business with restaurants and chefs, yet there has been little empirical research of this in the literature (Duram & Cawley, 2012; Sharma, Moon, & Strohbehn, 2014).

In spite of the benefits, the findings indicate that several constraints may limit the expansion of this growing market channel for farmers and/or farmers' market vendors. Perhaps most critically this includes a better understanding of the way in which existing personal relationships between restaurants and chefs and their existing wholesalers and farm suppliers acts as a disincentive to the creation of new ones, unless a farmer is already a member of such a social network in some way. However, many of these barriers could be managed by better communication and supply channels between different actors in the food systems (e.g., cooperative marketing strategies) as well as endeavoring to create new relationships and networks (e.g., mingles, tours, and invitations). This study therefore highlights that the creation of food supply chain relations with respect to local food is grounded as much in the development of social networks as it is in economic relations, and that social capital can be both an enabler and a constraint in creating new distribution and sales opportunities.

References

Allen, P., & Hinrichs, C. C. (2007). Buying into buy local: Engagements of United States local food initiatives. In D. Maye, L. Holloway, & M. Kneafsey (Eds.), *Constructing alternative food geographies: Representation and practice.* (pp. 255–272). Oxford: Elsevier.

Allen, P., FitzSimmons, M., Goodman, M., & Warner, K. (2003). Shifting plates in the agrifood landscape: The tectonics of alternative agrifood initiatives in California. *Journal of Rural Studies, 19*(1), 61–75.

Anderson, M. D., & Cook, J. T. (2000). Does food security require local food systems? In J. M. Harris (Ed.), *Rethinking sustainability: Power, knowledge and institutions* (pp. 228–248). Ann Arbor: University of Michigan Press.

Bendfeldt, E. S., Walker, M., Bunn, T., Martin, L., & Barrow, M., (2011). *A community-based food system: Building health, wealth, connection, and capacity as the foundation of our economic future.* Virginia Cooperative Extension [online]. Available from: ⟨https://pubs.ext.vt.edu/3306/3306-9029/3306-9029-PDF.pdf⟩ (accessed 26 September 2016).

Brain, R., Curtis, K., & Hall, K. (2015). Utah Farm-Chef-Fork: Building sustainable local food connections. *Journal of Food Distribution Research, 46*(1), 1–10.

Braun, V., & Clarke, V. (2006). Using thematic analysis in psychology. *Qualitative Research in Psychology, 3*(2), 77–101.

Conner, D. S., Montri, A. D., Montri, D. N., & Hamm, M. W. (2009). Consumer demand for local produce at extended season farmers' markets: Guiding farmer marketing strategies. *Renewable Agriculture and Food Systems, 24*(4), 251–259.

Curtis, K. R., & Cowee, M. W. (2009). Direct marketing local food to chefs: Chef preferences and perceived obstacles. *Journal of Food Distribution Research, 40*(2),
26–36.

Curtis, K. R., Cowee, M. W., Havercamp, M., Morris, R., & Gatzke, H. (2008). Marketing
local foods to gourmet restaurants: A multi-method assessment. *Journal of Extension,
46*(6), 16–24.

DeBlieck, S., Strohbehn, C. H., Clapp, T. L., & Levandowski, N. (2010). Building food service staff familiarity with local food. *Journal of Hunger & Environmental Nutrition, 5*
(2), 191–201.

Dougherty, M. L., Brown, L. E., & Green, G. P. (2013). The social architecture of local food
tourism: Challenges and opportunities for community economic development. *Journal
of Rural Social Sciences, 28*(2), 1–27.

Dunne, J. B., Chambers, K. J., Giombolini, K. J., & Schlegel, S. A. (2011). What does 'local'
mean in the grocery store? Multiplicity in food retailers' perspectives on sourcing and
marketing local foods. *Renewable Agriculture and Food Systems, 26*(01), 46–59.

Duram, L., & Cawley, M. (2012). Irish chefs and restaurants in the geography of "local"
food value chains. *The Open Geography Journal, 5*, 16–25.

Feagan, R. (2007). The place of food: Mapping out the local in local food systems. *Progress
in Human Geography, 31*(1), 23–42.

Feenstra, G. (2002). Creating space for sustainable food systems: Lessons from the field.
Agriculture and Human Values, 19(2), 99–106.

Gössling, S., & Hall, C. M. (2013). Sustainable culinary systems: An introduction. In C. M.
Hall, & S. Gössling (Eds.), *Sustainable culinary systems: Local foods, innovation, and
tourism & hospitality.* (pp. 3–44). Abingdon: Routledge.

Gössling, S., & Hall, C. M. (2016). Developing regional food systems: A case study of restaurant–customer relationships in Sweden. In C. M. Hall, & S. Gössling (Eds.), *Food
tourism and regional development: Networks, products and trajectories* (pp. 76–89).
Abingdon: Routledge.

Green, G. P., & Dougherty, M. L. (2008). Localizing linkages for food and tourism: Culinary
tourism as a community development strategy. *Community Development, 39*(3),
148–158.

Gregoire, M. B., & Strohbehn, C. (2002). Benefits and obstacles to purchasing food from
local growers and producers. *Journal of Child Nutrition & Management, 26*(1), 1–7.

Gregoire, M. B., Arendt, S. W., & Strohbehn, C. (2005). Iowa producers' perceived benefits
and obstacles in marketing to local restaurants and institutional foodservice operations.
Journal of Extension, 43(1), 1–10.

Hall, C. M. (2012). The contradictions and paradoxes of slow food: Environmental change,
sustainability and the conservation of taste. In S. Fullagar, K. Markwell, & E. Wilson
(Eds.), *Slow tourism: Experiences and mobilities* (pp. 53–68). Bristol: Channel View.

Hall, C. M. (2013). The local in farmers' markets in New Zealand. In C. M. Hall, & S.
Gössling (Eds.), *Sustainable culinary systems: Local foods, innovation, tourism and hospitality.* (pp. 99–121). Abingdon: Routledge.

Hall, C. M., & Gössling, S. (Eds.), (2016a). *Food tourism and regional development:
Networks, products and trajectories.* Abingdon: Routledge.

Hall, C. M., & Gössling, S. (2016b). From food tourism and regional development to food,
tourism and regional development: Themes and issues in contemporary foodscapes.
In C. M. Hall, & S. Gössling (Eds.), *Food tourism and regional development: Networks,
products and trajectories* (pp. 3–57). Abingdon: Routledge.

Higgins, V., Dibden, J., & Cocklin, C. (2008). Building alternative agri-food networks: Certification, embeddedness and agri-environmental governance. *Journal of Rural Studies*, *24*(1), 15−27.

Hinrichs, C. C. (2003). The practice and politics of food system localization. *Journal of Rural Studies*, *19*(1), 33−45.

Inwood, S. M., Sharp, J. S., Moore, R. H., & Stinner, D. H. (2009). Restaurants, chefs and local foods: Insights drawn from application of a diffusion of innovation framework. *Agriculture and Human Values*, *26*(3), 177−191.

Izumi, B. T., Wright, D. W., & Hamm, M. W. (2010). Farm to school programs: Exploring the role of regionally-based food distributors in alternative agrifood networks. *Agriculture and Human Values*, *27*(3), 335−350.

Kang, S., & Rajagopal, L. (2014). Perceptions of benefits and challenges of purchasing local foods among hotel industry decision makers. *Journal of Foodservice Business Research*, *17*(4), 301−322.

Kloppenburg, J., Jr, Lezberg, S., De Master, K., Stevenson, G., & Hendrickson, J. (2000). Tasting food, tasting sustainability: Defining the attributes of an alternative food system with competent, ordinary people. *Human Organization*, *59*(2), 177−186.

Krippendorff, K. (1980). *Content analysis: An introduction to its methodology*. Beverly Hills: Sage.

Lillywhite, J. M., & Simonsen, J. E. (2014). Consumer preferences for locally produced food ingredient sourcing in restaurants. *Journal of Food Products Marketing*, *20*(3), 308−324.

Martinez, S., Hand, M. S., Da Pra, M., Pollack, S., Ralston, K., Smith, T., et al. (2010). Local food systems: Concepts, impacts, and issues. *Economic Research Report No. 97*. Washington, D.C: U.S. Department of Agriculture, Economic Research Service. [online]. Available from: ⟨http://www.ers.usda.gov/media/122868/err97_1_.pdf⟩ (accessed 30 September 2016).

McMichael, P. (2009). A food regime analysis of the 'world food crisis'. *Agriculture and Human Values*, *26*(4), 281−295.

Murphy, J., & Smith, S. (2009). Chefs and suppliers: An exploratory look at supply chain issues in an upscale restaurant alliance. *International Journal of Hospitality Management*, *28*(2), 212−220.

Neuman, W. L., & Robson, K. (2009). *Basics of social research: Qualitative and quantitative approaches*. Toronto: Pearson.

Nilsson, J.-H. (2016). Value creation in sustainable food networks: The role of tourism. In C. M. Hall, & S. Gössling (Eds.), *Food tourism and regional development: Networks, products and trajectories* (pp. 61−75). Abingdon: Routledge.

O'Donovan, I., Quinlan, T., & Barry, T. (2012). From farm to fork: Direct supply chain relationships in the hospitality industry in the south east of Ireland. *British Food Journal*, *114*(4), 500−515.

Pearson, D., Henryks, J., Trott, A., Jones, P., Parker, G., Dumaresq, D., & Dyball, R. (2011). Local food: Understanding consumer motivations in innovative retail formats. *British Food Journal*, *113*(7), 886−899.

Peterson, H. H., Selfa, T., & Janke, R. (2010). Barriers and opportunities for sustainable food systems in north eastern Kansas. *Sustainability*, *2*(1), 232−251.

Pillay, M., & Rogerson, C. M. (2013). Agriculture-tourism linkages and pro-poor impacts: The accommodation sector of urban coastal KwaZulu-Natal, South Africa. *Applied Geography*, *36*, 49−58.

Reynolds-Allie, K., & Fields, D. (2012). A comparative analysis of Alabama restaurants: Local vs non-local food purchase. *Journal of Food Distribution Research*, *43*(1), 65–74.

Roy, H., Hall, C. M., & Ballantine, P. (2016). Barriers and constraints in the use of local foods in the hospitality sector. In C. M. Hall, & S. Gössling (Eds.), *Food Tourism and Regional Development: Networks, products and trajectories* (pp. 255–272). Abingdon: Routledge.

Schmit, T. M., & Hadcock, S. E. (2012). Assessing barriers to expansion of farm-to-chef sales: A case study from upstate New York. *Journal of Food Research*, *1*(1), 117–125.

Schmit, T. M., Lucke, A., & Hadcock, S. E. (2010). *The effectiveness of farm-to-chef marketing of local foods: An empirical assessment from Columbia County, NY. (EB 2010-03)*. Department of Applied Economics and Management College of Agriculture and Life Sciences Cornell University. [online]. Available from: ⟨http://publications.dyson.cornell.edu/outreach/extensionpdf/2010/Cornell_AEM_eb1003. pdf⟩ (accessed 26 September 2016).

Self, J. L., Handforth, B., Hartman, J., McAuliffe, C., Noznesky, E., Schwei, R. J., … Girard, A. W. (2016). Community-engaged learning in food systems and public health. *Journal of Agriculture, Food Systems, and Community Development*, *3*(1), 113–127.

Sharma, A., Gregoire, M. B., & Strohbehn, C. (2009). Assessing costs of using local foods in independent restaurants. *Journal of Foodservice Business Research*, *12*(1), 55–71.

Sharma, A., Moon, J., & Strohbehn, C. (2014). Restaurant's decision to purchase local foods: Influence of value chain activities. *International Journal of Hospitality Management*, *39*, 130–143.

Sharma, A., Strohbehn, C., Radhakrishna, R. B., & Ortiz, A. (2012). Economic viability of selling locally grown produce to local restaurants. *Journal of Agriculture, Food Systems, and Community Development*, *3*(1), 181–198.

Sims, R. (2010). Putting place on the menu: The negotiation of locality in UK food tourism, from production to consumption. *Journal of Rural Studies*, *26*(2), 105–115.

Thilmany, D. D. (2004). Colorado crop to cuisine. *Applied Economic Perspectives and Policy*, *26*(3), 404–416.

Tregear, A. (2011). Progressing knowledge in alternative and local food networks: Critical reflections and a research agenda. *Journal of Rural Studies*, *27*(4), 419–430.

Trivette, S. A. (2015). How local is local? Determining the boundaries of local food in practice. *Agriculture and Human Values*, *32*(3), 475–490.

Vecchio, R. (2010). Local food at Italian farmers' markets: Three case studies. *International Journal of Sociology of Agriculture and Food*, *17*(2), 122–139.

Wilhelmina, Q., Joost, J., George, E., & Guido, R. (2010). Globalization vs. localization: Global food challenges and local solutions. *International Journal of Consumer Studies*, *34*(3), 357–366.

The new institutional economics (NIE) approach to geographical indication (GI) supply chains: A case study from Turkey

8

Pelin Bicen[1] and Alan J. Malter[2]
[1]Marketing, Sawyer Business School, Suffolk University, Boston, MA, United States,
[2]Marketing, Department of Managerial Studies, College of Business Administration,
University of Illinois at Chicago, Chicago, IL, United States

8.1   Introduction

Agrifood systems have been experiencing major changes worldwide, generating tensions and conflicts that are reflected in international trade negotiations (Goff, 2005; Menard & Valceschini, 2005). Some of the most significant indicators of these changes are the fading out of state interventions that played a key role in protecting and preserving agricultural interests, the increasing pressure of global competition in agricultural trade, the emergence and dominance of highly concentrated retail chain stores that impose quality standards and costs to farmers, technological disruptions related to the development and distribution of genetically modified organisms (GMOs), and rising concerns over food quality and safety (Bramley, Bienabe, & Kirsten, 2009; Menard & Valceschini, 2005).

One essential strategic response to these changes in agrifood systems is the institutionalization of food quality and safety (Allaire, 2004). Asymmetric information between consumers and producers, which leads to market distortions through adverse selection and moral hazard, combined with an increased perception of uncertainty, leads to growing demand for food products with higher quality and safer characteristics (Akerlof, 1970). Quality standards communicate information about the attributes of a product. These attributes can pertain to the product itself (e.g., the color of a cocoa bean, the taste of chocolate) or to production and process methods, which may include environmental and socioeconomic conditions (e.g., fair trade, labor rights, organic), safety (e.g., pesticide use), and authenticity of origin (e.g., geographic appellation). Changing features of agrifood systems, consumption in industrialized economies, as well as social and environmental concerns have led consumers and economic actors in supply chain networks to seek more control,

Case Studies in Food Retailing and Distribution. DOI: https://doi.org/10.1016/B978-0-08-102037-1.00008-6

not only over products but also over production processes as well (Menard & Valceschini, 2005; Reardon, Codron, Busch, Bingen, & Harris, 2001). Therefore, quality control and management issues, traceability, quality certifications, norms, and codes that intend to certify the origin of components and ingredients are now among the most essential parts of discussions in signaling food quality and guaranteeing safety strategies (Ponte & Gibbon, 2005).

In this book chapter, we see quality management as a question of cooperation among the economic actors of agrifood supply chain networks, by each having control over certain aspects of information on the product, as well as production and process methods. Then, solving quality management problems relates to the coordination and governance issues among the economic actors in the value chain. How to define quality and establish procedures to measure quality should be coordinated, organized, and implemented by the actors of the value chain; and effective governance is needed to streamline this process. Depending on the transactional attributes of the exchange process to economize transaction costs, governance modes can be tighter (closer to vertical integration) or looser (closer to market coordination) or somewhere in between, e.g., hybrid governance (North, 1990; Williamson, 1985).

We discuss issues related to the concepts of food quality and safety, their management, and the most efficient ways of achieving them by focusing specifically on institutional arrangements that facilitate developing quality conventions and safety standards. We discuss institutions that govern, can govern, and should govern agricultural and marketing activities of supply chain networks to ensure food quality. Drawing on the new institutional economics (NIE) literature, we elaborate on how institutional arrangements (e.g., mode of governance) are related to the institutionalization of food quality and safety. Our discussion will focus on the distinctive quality convention, geographical indications (GIs).

Briefly, GIs are intellectual property (IP) rights for agrifood products that highlight the unique tie between the quality of the GI product and the territory where it is produced. This tie encompasses both physical (i.e., soil, climate, local variety, and breed) and human-related factors (i.e., local know-how, specific skills, historical traces) (Belletti, Marescotti, & Touzard, 2015). GI supply chain systems tend to be collective in nature and quality standards result from the efforts of many individuals over the course of years; thereby, IP rights belong to a representative organization of the GI supply chain network rather than individuals (Arfini, Mancini, & Donati, 2012; Thevenod-Mottet & Marie-Vivien, 2011).

This chapter is structured as follows. First, we provide a brief review of GI systems and explore their hybrid and collective nature. We then discuss the NIE explanation of the institutional arrangements of GI supply chain systems. Then, we illustrate how the hybrid and collective nature of GI governance mode is formed using a notable GI case study from Turkey, Gemlik table olives, with a focus on practical implications of GIs in the Turkish domestic market including farmers and consumers. Finally, we draw implications for the literature on institutional arrangements and discuss implications for further research on agrifood supply chain networks.

8.2   Geographical indication (GI) systems

GIs are a type of IP defined and protected by the World Trade Organization Agreement on Trade Related Aspects of Intellectual Property Rights (TRIPS) (Josling, 2006). Article 22.1 of the TRIPS agreement defines GIs as a good originating in a territory or regional sub-area within that territory, where a given quality, reputation, or other characteristics of the goods are essentially attributable to its specific geographical origin, and the production and/or processing and/or preparation of which take place in the defined geographical area. Like trademarks and brands, GIs serve an identification function and are registered and recognized by a government authority. Unlike ordinary trademarks and brands, which distinguish the goods of one enterprise from those of another, GIs necessarily identify the location from where the good originates (Josling, 2006) and can apply to all participating producers within that area. GI protection can be granted for the name of a region (e.g., Eastern Crete Olive Oil, Darjeeling Tea), a specific place (e.g., Roquefort), or, in exceptional cases, an entire country (e.g., Café de Colombia, Greek feta cheese) used to describe a food product. Feta cheese is a special example of a location-related food name that is not strictly speaking a GI since there is, in fact, no geographic place called feta. However, under EU law, feta cheese is considered a traditional nongeographical name worthy of protection, similar to a GI (O'Connor, 2004).

In practice, there are two GI categories: protected GI (PGI) and protected designation of origin (PDO). PDO is the term used to describe foodstuffs which are completely produced, processed, and prepared within the designated geographical area using recognized local know-how and skills. In contrast, for PGI designation, the geographical link must occur in at least one of the stages of production, processing, or preparation. PGI imposes explicit sanitary constraints that contribute to the product differentiation in terms of quality, whereas PDO is concerned about product quality, mainly through characteristics related to the origin of the product and to specific processes in its production (Menard & Valceschini, 2005). PDO status requires a variety of factors to be met: the region (e.g., climate, soil, breed) must contribute to the quality of the product and proof of this link must be demonstrated as part of the GI registration process; the region should be delimited and specific ingredients required in the production must be available only in that specific region; and production should be according to a special manufacturing technique and must be based on the knowledge of local producers built up over generations. A PDO is granted only to groups; individuals cannot apply for PDO status. Therefore, to obtain a PDO status, producers of a good must form a representative association to manage the application and operate the eventual quality and branding scheme.

Goff (2005) discusses GIs as contemporary examples of economic nationalism (or regionalism), where policies seek to preserve and promote a set of shared meanings, cultural values, and social practices held dear by a significant portion of a national citizenry. Relatedly, she argues that the real winners from the GI system and regulations are the producers of the GI products and the consumers who assume that foodstuff names are indicators of "quality, origin, and authenticity" (p. 200).

By setting quality standards and enforcing their implementation, GIs promote rural livelihoods based on local resources and socioeconomic development, localize economic control by adding value to local production, and generate greater economic returns (Cacic, Tratnik, Kljusuric, Cacic, & Kovacevic, 2011; Folkeson, 2005; Jena & Grote, 2010). By providing a strong rural development tool, GI systems could constitute a strong rationale, especially for developing countries, to embrace and support origin-labeled products within a specific region (e.g., Bramley et al., 2009).

As a member of the developing countries list (e.g., IMF, 2015), Turkey joined the Paris convention, which was one of the first IP treaties, in 1925, and trademark law came into force in 1965. Turkish enactment of legal protection of GIs and implementation of regulations pertaining to the protection of GIs came into force in 1995 through Decree Law No. 555. The Turkish Patent Institute (TPI) is one of four public institutions in Turkey involved in certification of food quality and safety and acts as the certification and auditing body for trademarks and GIs (WIPO, 2014). As a candidate country for EU membership, Turkish food safety and environmental legislation is increasingly oriented toward EU standards (Koc, Asci, Alpas, Giray, & Gay, 2011).

The Turkish government has paid special attention to its GI system in recent years. GIs have been used as a strategic tool for rural development and competition (GTHB, 2013). Competitive pressure from globalized food production and retail companies prompted trade organizations, cooperatives, and producers of regional products to protect their brand identity and competitive advantage through GI registrations, which gives regional producers one product attribute that cannot be imitated by rivals outside the region. The increasing global race in agrifood industries resulted in increased numbers of GI registrations. There were only 25 Turkish GI applications in 1996, and by 2015, 180 products had been approved by the TPI and registered with a domestic Turkish GI certificate. Of these 180 products, 123 were agricultural and food products, including fruits, processed food, bakery items, oils, olives, and cheese. As of 2015, 196 applications for agricultural and processed food products were on the waiting list to obtain Turkish GI protection or designation of origin. Currently, only five Turkish GI products are in the EU's DOOR (Database of Origin and Registration) system [the DOOR database includes product names for foodstuffs registered with the EU as PDO, PGI, or TSG (Traditional Specialty Guaranteed), as well as names for which registration has been requested], four of which (Turkish apricots, figs, pastrami, and pepperoni) are still in the application process. Only one Turkish food product, Antep baklava, received PGI certification from the EU in 2013. Lack of control, leadership, and collective effort among farmers and organizations is cited among the top reasons for the current lack of EU-certified Turkish GI products (Dokuzlu, 2016; Gurkan, 2015).

8.3   The NIE approach to governance of GI supply chain networks

NIE is an approach that studies institutions and how they interact with modes of governance (Menard & Shirley, 2005). Institutions include both formal and informal rules and agreements with the purpose of reducing environmental uncertainty and unpredictability. Modes of governance include various forms of institutional arrangements, that is, the set of rules, laws, policies, customs, and norms that economic actors develop to facilitate transactions—i.e., contractual agreements that use formal structures and third-party enforcement to establish processes for joint actions and relational-based agreements that use social mechanisms such as social relations, shared norms, and self-enforcement to establish processes for joint action (North, 1990).

Transaction costs are at the heart of NIE. Arranging transactions among supply chain members is essential for organizing economic activities and requires complex devices both at the microlevel (modes of organizing these transfers) and at the macrolevel (institutions facilitating and enforcing these transfers) (Menard &Valceschini, 2005). The choice of organizational arrangements/modes embedded in the institutional environment are meant to minimize the cost of transactions, which are mostly determined by uncertainty surrounding these transactions and the degree of specificity of assets involved in these transactions (Klein, 2005; Williamson, 1985).

Organizational arrangements may take several forms: spot markets, vertical arrangements, and hybrid forms. In spot markets, transactions are simple and market prices are the signaling mechanism for parties' adaptation to changing circumstances. Spot markets are most effective when actors have full information and exchanges between them are not frequent and straightforward. At the other end lies full vertical integration, where one party has unified ownership and control. This governance mode appears when relationship specific assets are at stake, input markets are thin, uncertainty is high, and there is a greater need for specific investment protection, producing high transaction costs (Williamson, 1991). Hybrid forms are complex arrangements and have four distinct characteristics (Barjolle, Sylvander, & Thevenod-Mottet, 2011; Gulati & Singh, 1998; Menard, 2004; Menard & Valceschini, 2005): (1) A number of small businesses are related to each other by specific interbusiness interactions; (2) parties pool their resources and individual competences while keeping their financial autonomy and decision rights distinct; (3) sets of both formal rules (incomplete contracts) and social norms are used to govern the transactions among the parties; and (4) "coopetition" (simultaneous existence of cooperation and competition) among the parties makes the rent distribution issue problematic. Since hybrid forms can rely neither on market prices nor on commands to manage conflicts, there needs to be alternative forms of authority/regulatory bodies to govern complex arrangements among parties who maintain some distinct legal property rights.

There has been a growing demand for, and attention to, the qualities of agricultural products as a result of increased awareness of food safety and security (Goff, 2005). In particular, some of the more recent problems with food product quality (e.g., the "mad cow"/bovine spongiform encephalopathy (BSE) crisis and the H5N1 bird flu pandemic) have increased concerns for traceability and quality control issues. As Menard and Valceschini (2005) highlighted, traceability is an organizational response to the quality signaling problem. Due to concerns about the increase in genetically modified food (GMOs) and internationalization of trade, consumers are leveling up their quality control demands over processes, as well as control over products from food producers, and asking for quality certification that guarantees both health and environmental qualities. One solution to this problem may involve a choice of governance mode that minimizes the transaction cost while making credible signals, and/or the development of institutional devices that back traceability by guaranteeing adequate enforcement such as institutional guarantees. While choosing the optimal solution among the alternate solutions to quality control and traceability problems, it is important to identify their respective transaction costs and conduct a comparative assessment (Menard & Valceschini, 2005).

One way to tackle the quality control issue is the organizational approach, which focuses on the governance mode throughout the agrifood chain to ensure better quality control. GI systems, for example, adopt a hybrid governance mode where a network of interdependent members holding autonomy and distinct rights make high specific investments and develop tight coordination among themselves through a third-party private certifying organization. This effort is backed by public authorities who are responsible for approving the certifying organization and forcing it to follow strict rules with respect to the predefined quality standards, methods of production, and quality control mechanisms for guaranteeing whether the products developed conform with the quality signal. GI systems use a mix of public and private institutional devices for controlling and monitoring the control issues in the chain. Since GI product quality is certified by an independent organization whose credibility is backed by government enforcement, coordination problems in the chain are largely mitigated (Raynaud, Sauvee, & Valceschini, 2005). Given high uncertainty in the agrifood industrial environment, it is far too complex and, therefore, costly to draft and implement complete contracts ex ante. Instead, hybrid forms of arrangements are drafted as frameworks to boost the relational quality among economic actors in the chain and, therefore, make the threat of economic expulsion and social ostracism due to possible contractual hazards too costly to bear. As a result, hybrid form arrangements provide supply chain participants with confidence that the relationship is worth making a special investment in and leaves ways to ex post adjustments such that parties can still make changes in the arrangements until after it is drafted. In the next section, we discuss a GI case from Turkey, Gemlik table olives, focusing on how the hybrid mode of governance in the supply chain network is formed and its positive impact on the regional economy as well as consumer confidence.

8.4   A case study from Turkey: Gemlik table olives

As of 2014/2015, Turkey has been the third largest producer of table olives in the world [top table olives producers include the EU (e.g., Italy, Greece, Spain), Egypt, Turkey, Algeria, and Syria (IOCC, 2016)], after the EU and Egypt, with around 16% of world market share (IOCC, 2016). Eighty percent of Turkey's olive production is used for domestic consumption; the balance is exported to other countries. Though it ranks among the world's largest producers of table olives, Turkey's olive productivity is low compared to other top ranking countries. In recent years, though the number of olive trees that produce table olives has increased, the productivity rate has not met expectations (see Table 8.1). Turkey is mainly at the low-quality end of the global value chain for table olives. The issue of quality is partly attributed to dysfunctional supply chain activities, such as the lack of communication and coordination among supply chain members (e.g., farmers, traders, retailers, nongovernment organizations), and a lack of centralized management (Gurkan, 2015).

The world market for table olives has been going through significant changes, including increased global competition (e.g., table olive production increased more than 70% and exports increased 88% worldwide in the last two decades) (IOCC, 2016); progress in advanced olive production and processing technologies; new players in international markets, especially from New World producers such as the USA, Argentina, Chile, Mexico, and Peru; new conventional quality standards set mainly by the EU (e.g., PDO and PGI); and increased consumer awareness of food quality and safety (Bramley et al., 2009). All these changes combined with domestic productivity and quality issues challenge Turkish olive producers and organizations to find long-term solutions to remain competitive.

Institutions and modes of organization that should govern agricultural activities are at the heart of solving quality issues by setting quality standards. As discussed earlier, one such essential quality standard is GIs. We examine a successful PDO

Table 8.1 **Table olive production volumes and productivity levels in Turkey**

Year	Production (tons)	Average productivity (kg/tree)	Total number of olive trees
2008	512,103	15	33,599,163
2009	460,013	14	33,936,299
2010	375,000	11	35,611,525
2011	550,000	14	39,176,479
2012	480,000	12	40,252,330
2013	390,000	9	45,235,836
2014	438,000	10	45,519,208

Source: Zeytincilik Sektor Raporu (2015).

case of Turkish Gemlik table olives, focusing on how institutional arrangements are formed in the value chain and resulting improvement in performance outcomes.

The Gemlik region, the gulf of northwestern Turkey, is a significant producer in the black table olives market with Gemlik cultivar olives. These are small-to-medium sized black olives with high oil content and are processed most commonly in brine. Special microclimatic conditions (e.g., climate, topography, soil) that dominate the Gemlik region and local olive producers' treatment of the trees enable them to produce olives of unique taste, color, texture, and shape. The sign of a traditionally cured Gemlik olive is that it is meaty and the flesh comes away from the stone easily. Owing to its high oil content (29%), any fruit that cannot be used for pickling is used for olive oil production, so this cultivar is considered dual-purpose. The Gemlik region accounts for approximately 20% of overall table olive production in Turkey (GTB, 2017).

The Gemlik GI was registered as a PDO in 2003 and covers three districts: Gemlik, İznik, and Orhangazi. The PDO designation indicates that olives produced and processed in these districts are Gemlik olives. The trademark rights holder is a public organization called Gemlik Commodity Exchange (GTB). Though the PDO registration occurred in 2003, benefits from the GI system started to emerge in 2010, when GI control and monitoring mechanisms were implemented (Dokuzlu, 2016). The GTB, as a common coordinating and controlling body and public certifying organization, manages, organizes, monitors, and audits the PDO supply chain in terms of quality standards and quality measurements. The GTB auditing committee consists of several experts from the field, including farmers, academics, and NGO representatives. The auditing and certification unit is responsible for (1) quality control of the product with respect to its code of practice; (2) information and advice to the members; and (3) collective market promotion of the Gemlik olive designation of origin, R&D, arbitration among the members, as well as management of volumes and fixing of internal prices within the supply chain. The emergence of a GI system for table olives in the Gemlik region as a complex hybrid arrangement is an endogenous solution to the quality and traceability problem. The development of institutional mechanisms such as the GTB makes these solutions credible in guaranteeing adequate enforcement of the quality standard.

The supply chain network of Gemlik olives consists of olive farmers, traders, and retailers (see Fig. 8.1). Official members of the Gemlik PDO supply chain retain legal and financial autonomy and deal commercially with partners of their choice within the official group of members. If traders or retailers want to use a Gemlik olive PDO label on their product package, they must agree on terms and sign an agreement with GTB that tracks and monitors the use of the label aligned with the appropriate quality standards of the PDO (Dokuzlu, 2016).

The GI supply chain of Gemlik olives has two levels: (1) a horizontal level involving collaboration between competing olive farmers of the same level, and (2) a vertical level consisting of supply chain members at the various stages (e.g., farmers, processors, traders, retailers) (see Fig. 8.1). Members of the network are spatially proximate. GTB, as the third party and main governing institution, and the independently assigned expert auditing team (e.g., universities, NGOs located

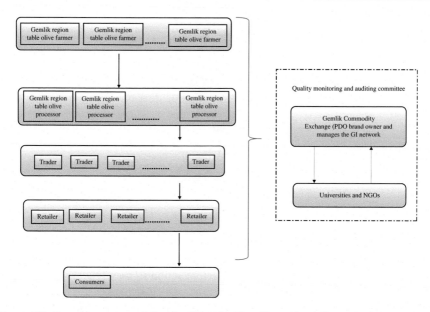

Figure 8.1 General structure of the Gemlik table olive GI supply chain network.

within the GI geographical boundaries) are responsible for inspection and monitoring of product quality. The auditing team regularly monitors production facilities and label users by taking samples from labeled products and testing for compliance with the quality standards. According to set rules and procedures, if the packaged table olives are not in compliance with the PDO quality standards, label users/traders or farmers will be penalized by the GTB for violation of agreements. The GTB has the right to pursue a legal remedy if the situation persists (Dokuzlu, 2016).

By Turkish law, the GTB is not only responsible for quality monitoring but also for overseeing tracking, marketing, sales, branding, packaging, and labeling. In contrast to producers in EU member countries, who are obliged to use PDO and PGI logos on product labels, there is no such obligation for non-EU members (Kireeva, 2011). In fact, there is no specific or standard logo for Turkish GI products. The GTB has developed its own official logo for Gemlik table olives (Fig. 8.2). GTB also closely monitors the registered farmers, their production levels, and tracks sales. When a trader/retailer applies to GTB with the receipt from a registered farmer, they earn the right to be in the system and use the Gemlik table olives GI logo on their packages. The receipt provides information on the sold quantity which is entered in a centralized database to monitor the total capacity and production levels of the farmers, and the identity of the economic actors (e.g., farmers, traders, retailers) involved in the trade. Before applicants receive the GI logo, they are required to sign an agreement with the GTB indicating that they comply with the quality standards and will face legal challenges if they do not comply with the signed agreement (GTB, 2017). GTB provides an updated list of the Gemlik table olives GI members on a timely basis on their website to inform the public and create transparency in the system.

Figure 8.2 Gemlik table olives' GI logo.

Collective action among the Gemlik region olive farmers and processors, traders/retailers, and the GTB, in conjunction with universities, reduces the transaction and coordination costs of exchanges. Collective action in the network cemented by written agreements on quality and negotiation, and a dynamic social mechanism based on trust and reciprocity, decreases the risks of opportunistic behavior and moral hazard (TEPGE, 2012). Further, due to the trust-based and collective nature of GI systems, there is a low level of direct competition between GI members of Gemlik table olive producers as well as the absence of firms with monopoly power in the region (Belletti et al., 2015). The GTB shows that institutional support can facilitate collective action by offering a negotiation framework for transaction costs, risks, and benefits. The GTB's structured organizational policies for managing membership, tracking and monitoring its members and their production levels, penalizing members who are involved in unfair practices and/or do not comply with the quality standards, and conducting marketing and branding activities on behalf of the GI network, have increased the effectiveness of the collective efforts among members (GTSO, 2016). Gemlik table olives was registered as a Turkish PDO in 2003. The GTB completed the GI project (e.g., centralized database, logo, labels, QR code creation and promotional campaigns) in 2014 (Dokuzlu, 2016). The GTB is currently working to educate farmers, traders, retailers, and consumers about the concept of GI—how it works socially, economically, and politically; how the GI logo can be an effective signaling tool for quality; and how collective efforts of the GI network can lay the foundation for virtuous economic development of the region (GTSO, 2016).

8.5 Concluding remarks

GIs are institutional constructions connecting the specific quality and reputation of a product to a specific territory (Belletti et al., 2015). GIs are collective by nature and economic actors (e.g., producers, processors, retailers, traders, nongovernmental institutions, universities) are involved in technical, social, and economic interactions. Actors in the system define what quality is and set the quality standards collectively. In this way, they shape the identity of the GI product by linking its specific quality attributes to the territory where ingredients originate and/or are produced and processed. The GI is owned by the representative organization of the GI network, which gives it the authority to manage the network (e.g., monitoring, tracking, auditing) to ensure that members of the network comply with the set GI product quality standards. GIs contribute to market regulation, reduce the information asymmetry between consumers and firms, have a positive impact on regional economies and development processes, and facilitate the protection of natural and cultural resources (Barjolle et al., 2011; Belletti et al., 2015; Bramley & Bienabe, 2012). GI protection shields domestic producers from unfair competition, but it can also discriminate against domestic producers who do not belong to the recognized consortium of product manufacturers to which GI protection has been accorded (Goff, 2005).

Based on the frequency of transactions among the economic actors, the uncertainty regarding these transactions, and the degree of asset specificity involved, GI systems adopt hybrid forms of governance to minimize transaction costs. GI networks have two levels: a horizontal level of collaboration (e.g., among farmers and producers) and a vertical level of quasiintegration among farmers, processors, traders, retailers, and a third-party representative organization. Distinct characteristics of GI systems are that economic actors share resources while keeping their financial autonomy. The main institutional mechanism is based on incomplete formal contracts, relational bonds, and social cohesiveness, and due to its hybrid and, thus, coopetitive nature, rent sharing and dispute resolution are particularly acute (Menard, 2004; Menard & Valceschini, 2005).

Drawing on the NIE literature to further our understanding of hybrid forms of governance in GI networks, we have explored a recent GI case from Turkey. The Gemlik table olives case represents the Gemlik region's need for GI to boost regional economic development and competitiveness in both domestic and international markets. Further, we have described how hybrid governance is formed and how it functions in this case, and more generally we have reviewed important legal and political institutions, such as the TPI and GTB, and how they impact the ways and costs of organizing transactions in the Gemlik table olives GI network; as well as GTB's efforts to monitor, trace, and organize transactions among actors to ensure that members of the network comply with the agreed quality standards. Though the Gemlik table olives GI registration was completed in 2003, it has taken more than a decade for GTB to show progress on the GI project regarding infrastructure building, public education, promotional campaigns, and tracking the impact of the GI system on regional competitiveness. Early reports support the positive impact of the GI system on the regional economy, as well as on reduction in moral hazard and adverse selection problems by closing the informational asymmetry between producers and consumers. More recently, the GTB was in meetings and consortiums with European Council officials to exchange information on how Gemlik table olives can be more than a Turkish regional brand and possibly expand its horizon as a strong and reputable brand in global markets (Gemliklife, 2016). Some of the initiatives the GTB executives have taken recently to solidify this objective include being engaged in partnerships with European countries to learn the general European taste/preference on olives; learning from the planting, harvesting, and cultivation processes used by European farmers; and R&D partnership projects with the European Union to enhance initiatives for olive research, education, and training (Gurkan, 2015).

References

Akerlof, G. A. (1970). The market for "lemons": Quality uncertainty and the market Mechanism. *The Quarterly Journal of Economics.*, *84*(3), 488−500.

Allaire, G. (2004). Quality in economics, a cognitive perspective. In M. Harvey, A. Mc Meekin, & A. Warde (Eds.), *Theoretical approaches to food quality.* Manchester: Manchester University Press.

Arfini, F., Mancini, M. C., & Donati, M. (2012). *Local agri-food systems in a global world: market, social and environmental challenges.* Cambridge: Cambridge Scholars Publishing.

Barjolle, D., Sylvander, B., & Thevenod-Mottet, E. (2011). Public policies and geographical indications. In E. Barham, & B. Sylvander (Eds.), *Labels of origin for food: Local development, global recognition* (pp. 92−105). Wallingford: CABI.

Belletti, G., Marescotti, A., & Touzard, J. M. (2015). Geographical indications, public goods, and sustainable development: The roles of actors' strategies and public policies. *World Development.* Available from https://doi.org/10.1016/j.worlddev.2015.05.004.

Bramley, C., & Bienabe, E. (2012). Developments and considerations around geographical indications in the developing world. *Queen Mary Journal of Intellectual Property.*, *2*(1), 14−37.

Bramley, C., Bienabe, E., & Kirsten, J. (2009). *The economics of geographical indications: Towards a conceptual framework for geographical indication research in developing countries,* (pp. 109−141). Geneva: World Intellectual Property Organization.

Cacic, J., Tratnik, M., Kljusuric, J. G., Cacic, D., & Kovacevic, D. (2011). Wine with geographical indication − Awareness of Croatian consumers. *British Food Journal.*, *113*(1), 66−77.

Dokuzlu, S. (2016). Geographical indications, implementation and traceability: Gemlik table olives. *British Food Journal.*, *118*(9), 2074−2085.

Folkeson, C. (2005). *Geographical indications and rural development in the EU.* Available at: ⟨https://lup.lub.lu.se/luur/download?func = downloadFile&recordOId = 1334511&fileOId = 1647280⟩ Accessed 4 May 2017.

Gemliklife. (2016). *Gemlik zeytini Avrupa'ya aciliyor.* Available from: ⟨http://gemliklife.net/ en-US/haberler/3365/gemlik-zeytini-avrupaya-aciliyor⟩ Accessed 4 February 2017.

Goff, P. M. (2005). It's got to be sheep's milk or nothing. In E. Hellenier, & A. Pickel (Eds.), *Economic nationalism in a globalizing world* (pp. 183−201). Ithaca: Cornell University Press.

GTB. (2017). *Gemlik Zeytini coğrafi işaret.* Available from: ⟨http://www.gemliktb.org.tr/ cografi-isaret/⟩ Accessed 17 April 2017.

GTHB. (2013). *Strategic plan of food, agriculture and husbandry*. Ministry of Turkish Republic 2013–2017. Available at: ⟨www.tarim.gov.tr/SGB/Belgeler/Stratejik%20Plan %202013-2017.pdf⟩ Accessed May 11, 2017.

GTSO. (2016). *Tarim, gida, zeytincilik sektor raporu*. Available from: ⟨http://www.gtso.org. tr/dosya/tarim-gida-zeytin-SA.pdf⟩ Accessed 3rd April 2017.

Gulati, R., & Singh, H. (1998). The architecture of cooperation: Managing coordination costs appropriation concerns in strategic alliances. *Administrative Science Quarterly.*, *43*(4), 781–814.

Gurkan, N.P. (2015). *Turkish olive oil sectoral innovation system: A functional-structural analysis*. Unpublished dissertation, Middle East Technical University.

IMF (2015). World Economic Outlook: Uneven growth-Short and long term factors. Washington, DC: IMF.

IOCC (2016). *World olive figures*. Available from: ⟨http://www.internationaloliveoil.org/esta-ticos/view/131-world-olive-oil-figures⟩ Accessed 20 March 2017.

Jena, P. R., & Grote, U. (2010). Changing institutions to protect regional heritage: A case for geographical indications in the Indian agrifood sector. *Development Policy Review.*, *28* (2), 217–236.

Josling, T. (2006). What's in a name? The economics, law and politics of geographical indications for foods and beverages. *IIIS discussion paper no. 109*.

Kireeva, I. (2011). How to register geographical indications in the European community. *World Patent Information*, *33*, 72–77.

Klein, P. G. (2005). The make or buy decisions: Lessons from empirical studies. In C. Menard, & M. Shirley (Eds.), *Handbook of new institutional economics*. (pp. 435–464). Berlin: Springer.

Koc, A.A., Asci, S., Alpas, H., Giray, F.H., & Gay, H. (2011). Food quality assurance schemes in Turkey. *JRC scientific and technical reports*.

Menard, C. (2004). The economics of hybrid organizations. *Journal of Institutional and Theoretical Economics.*, *160*, 345–376.

Menard, C., & Valceschini, V. (2005). New institutions for governing the agri-food industry. *European Review of Agricultural Economics.*, *32*(3), 421–440.

Menard, C., & Shirley, M. (2005). Introduction. In C. Menard, & M. Shirley (Eds.), *Handbook of new institutional economics*. (pp. 1–18). Berlin: Springer.

North, D. C. (1990). *Institutions, institutional change and economic performance*. Cambridge: Cambridge University Press.

O'Connor, B. (2004). *The law of geographical indications*. London: Cameron & May.

Ponte, S., & Gibbon, P. (2005). Quality standards, conventions, and the governance of global value chains. *Economy and Society.*, *34*(1), 1–31.

Raynaud, E., Sauvee, L. C., & Valceschini, E. (2005). Alignment between quality enforcement devices and governance structures in the agro-food vertical chains. *Journal of Management and Governance.*, *9*, 47–77.

Reardon, T., Codron, J. M., Busch, L., Bingen, J., & Harris, C. (2001). Global change in agrifood grades and standards: Agribusiness strategic responses in developing countries. *International Food and Agribusiness Management.*, *2*(3–4), 421–435.

TEPGE (2012). Dogu Akdeniz Bolgesinde Zeytin ve Zeytinyağı Uretimi, Pazarlaması ve Bolgede Zeytinciligi Gelistirme Olanakları. Tarımsal Ekonomi ve Politika Gelistirme Enstitusu, Ankara.

Thevenod-Mottet, E., & Marie-Vivien, D. (2011). Legal debates surrounding geographical indications. Labels of origin for food. In E. Barham, & B. Sylvander (Eds.), *Local development, global recognition*. Wallingford: CAB International.

Williamson, O. E. (1991). Comparative economic organization: The analysis of discrete structural alternatives. *Administrative Science Quarterly.*, *36*(2), 269–296.

Williamson, O. E. (1985). *The economic institutions of capitalism: Firms, markets, relational contracting*. New York: Free Press.

WIPO. (2014). *Geographical indications: What is a geographical indication?*Available from: ⟨www.wipo.int/geo_indications/en/⟩ Accessed 17 April 2017.

Zeytincilik Sektor Raporu (2015). TR63 Bölgesi: Zeytincilik Sektör Raporu 2015. Available from ⟨http://www.dogaka.gov.tr/Icerik/Dosya/www.dogaka.gov.tr_619_LZ0P55ES_ Zeytincilik-Sektor-Raporu-2015.pdf⟩ Accessed 17 May 2017.

Patanjali Ayurved Limited: Driving the ayurvedic food product market

9

Sujo Thomas[1], Abhishek[3] and Sanket Vatavwala[2]
[1]Ahmedabad University, Gujarat, India, [2]Indian Institute of Management Indore, Madhya Pradesh, India, [3]Institute of Management Technology, Ghaziabad, Uttar Pradesh, India

9.1   Introduction to the ayurvedic consumer-packaged goods industry

Indian scriptures, which are considered in India to be "Books of Wisdom," have narrated the importance of Ayurveda[1] (the medical science of ancient India) in depth. While the concept of "wellness" came into the limelight in the 1950s, its roots lie in the period 3000−1500 BC. It originated in Indian scriptures as a holistic system called Ayurveda (Global Wellness Institute, n.d.), which believes in creating harmony between soul, mind and body and offers tailor-made solutions based on an individual's bodily constitution. Since then, the tradition has been believed and practiced worldwide, though awareness of it is still limited. While a sizable section of Indians believes in the power of Ayurveda, there has not previously been enough momentum to create large scale awareness among the masses. There have been a few organizations which offered various ayurvedic products, but a huge opportunity and potential in the market remained to be seized.

The ayurvedic consumer-packaged goods industry gradually gained momentum in India and across the globe. It had a lot to offer to organizations in terms of business opportunity for consumers by offering ayurvedic products. Over the years, many players saw such opportunities and jumped into the market with various products. The ayurvedic consumer-packaged goods industry was primarily segmented into three parts: (1) Household products and personal care products such as fabric detergent, household cleaners, cosmetic products, and toiletries. This segment accounted for 50% of the sector. (2) Health care products such as over-the-counter and prescription products, which accounted for 31% of the sector. (3) Food and beverage products such as snacks, chocolate, tea/ coffee/ soft drinks, etc. This segment accounted for 19% of the sector. The consumer-packaged goods industry was inundated with numerous competitors in the marketplace, such as Procter & Gamble (P&G), Hindustan Unilever Limited (HUL), Nestlé, and many more. Patanjali, a late entrant in the market, gradually entered almost all segments of

---

[1] Appendix 1 contains a glossary of selected Hindi terms used in the case.

Case Studies in Food Retailing and Distribution. DOI: https://doi.org/10.1016/B978-0-08-102037-1.00009-8

consumer packaged goods and thus tried to compete against multiple players (IBEF, 2017).

While a plethora of companies sell food products and ayurvedic products (mostly ayurvedic medicines), Patanjali Ayurved Limited has created a revolution in the market by selling food products which have been developed on the principles of Ayurveda. These have become widely accepted by Indian consumers who believe in the benefits of the ancient medicinal system of Ayurveda but were previously limited to purchasing ayurvedic medicines only. The premise of Ayurveda rests on the belief that disease is the outcome of not living in harmony with the environment. According to Ayurveda, there are multiple components through which the human body is formed. These are five elements (ether, air, fire, water, and earth); three *doshas* (biological energies); seven *dhatus* (tissues); and numerous *srotas* (channels). Ayurveda believes that no one individual is identical to another and hence, it tries to understand the nature of the patient and disease and then suggests a remedy, which may differ from one person to the next (California College of Ayurveda, n.d.).

Patanjali, which believes in the concepts of Ayurveda and started its operations in 2006, had grown from revenues of US$68 million for the fiscal year (FY) 2011−12 to US$750 million in FY 2015−16.

(The Indian Express, 2016). Patanjali's enormous success has been majorly attributed to Baba (meaning elderly person) Ramdev, a yoga guru based in Haridwar—a small town in the northern Indian state of Uttarakhand. He founded the company along with his close disciple Acharya (meaning religious instructor) Balkrishna to spread the benefits of Ayurveda to all (Maheshwari, 2017). The *New York Times* described Baba Ramdev "a yogic fusion of Richard Simmons, Dr. Oz and Oprah Winfrey" (Polgreen, 2010). Baba Ramdev spent years studying Indian scriptures developing his knowledge of Ayurveda. In his early life, as a yoga guru, he used to teach yoga to villagers. He gradually became famous in North India. A spiritual television channel called Aastha roped in Baba Ramdev to feature him on air for a yoga program in the morning hours. A huge number of people follow him and attend his yoga camps. His followers include prominent politicians, Bollywood superstars, activists, and numerous lay people.

9.2   The growth of the consumer-packaged goods industry and the ayurvedic consumer-packaged goods industry

In recent years, the Indian consumer packaged goods sector has been in a high growth phase. As the fourth largest sector in the Indian economy, it provides jobs to three million people in the country (IANS, 2016). According to an Assocham-TechSci Research report, the sector in India was expected to grow at a Compound Annual Growth Rate (CAGR) of 20.6%, much faster in comparison to its global counterpart, which was expected to grow at CAGR of 4.4% (IANS, 2016).

The same report states that the market was expected to reach US$104 billion by 2020, from a current market size of US$49 billion in 2016 (IANS, 2016). The growth in food retailing was driven by multinational companies (MNCs) such as HUL, P&G, Nestlé, and GlaxoSmithKline. Indian companies like ITC, Dabur, Parle, and Britannia have also built extensive distribution networks and improved operational effectiveness to establish a significant market share in the Indian market. While the Indian Government's move to demonetize the country's currency in November 2016 temporarily lowered consumer spending (Chandran, 2016), industry experts expected a rise in consumption with the passage of time and expected the growth rate to remain stable. The implementation of a Goods and Services Tax, replacing central and state government taxes, in FY 2017−18 was expected to lower the incidence of taxation and help companies optimize logistics and distribution costs (Equitymaster.com, 2016). The saving was expected to be passed on to consumers, leading to lower prices and greater spending in the sector.

As per Global Industry Analysts, a US-based market research firm, the herbal market across the globe has been growing in leaps and bounds (Business Recorder, n.d.). While Europe is the largest market, Asia-Pacific was the fastest growing market for herbal products with China and India driving the growth. Ayurvedic products are part of this rapidly growing herbal industry (Balakrishnan, n.d.). Nomura estimated a high CAGR for ayurvedic products in India for the period 2010−15 and predicted that the sector would grow between 2016 and 2020 (Karnik, 2016). These key indicators showcase the potential of the ayurvedic consumer-packaged goods industry, which was not confined to India alone. There were multiple reasons for the growth of the ayurvedic industry in India, such as the presence of large, medium, and small-scale manufacturers. Many prominent individuals were proponents of Ayurveda, including spiritual gurus, ayurvedic doctors, academicians—particularly those from the fields of the arts/Sanskrit, some leading politicians, and business tycoons—who advertised their products and thus also spread awareness about the benefits of Ayurveda. Its benefits are acknowledged not just in India but internationally as well, as herbal products are considered useful for treating major to minor diseases, as well as some chronic persistent illnesses such as depression, migraines, etc. (University of New Hampshire, n.d.). Thus, the herbal industry's growth was evident and industry players were set to exploit it extensively, Patanjali being one among many.

The growth of the ayurvedic consumer-packed goods industry was expected to come from rural as well as urban markets in India. In urban markets, there has been greater awareness about hazardous ingredients. This has led to changes in consumer decisions, with greater preferences for herbal and ayurvedic products. In rural markets, consumption in the sector was driven by Indian Government initiatives such as higher minimum support prices, loan waivers, and disbursements through the National Rural Employment Guarantee Act program, which gave a lift to purchasing power among rural populations. Moreover, rural populations, who are more likely to hold beliefs in traditional medicines and practices, have been more willing to accept ayurvedic products.

9.3 Patanjali's product portfolio

Patanjali has multiple products in its portfolio, including household care, health care, personal care, and food and beverage products (see Table 9.1). The company claims that the entire range of products it sells is either natural or provide some medicinal value. For instance, *Chyawanprash* (a herbal tonic) which contained Kashmiri *kesar* (saffron) and many other ayurvedic ingredients has been highlighted as an energy increment product, which promises a revitalized life. *Ghee* (clarified butter) is focused on sharpening the brain and strengthening physique. *Amla* (gooseberry) juice and aloe vera juice are showcased as natural health drinks, focused on overall bodily heath and the immune and digestive systems. *Badampak* (a herbal supplement made from almonds) is said to encourage brain development. Honey, which has been primarily pushed in the market through a penetration pricing strategy, is presented as an energy boosting product.

While developing the ayurvedic food portfolio, Patanjali tried to ensure that it did not miss out on products that could be consumed during breakfast or as daytime snacks. As Indian lunches and dinners often involve homemade cooked food rather than packaged food, Patanjali was keen to exploit the breakfast and daytime snacks market, which could be tapped by packaged food items. It sold herbal tea, fruit and nut muesli, cornflakes, choco flakes, jam, and oats as breakfast items. Among the snacks, biscuits and noodles were two prominent categories. Patanjali offered biscuits free from trans-fat and cholesterol, claiming that consumption of such ingredients might cause conditions such as heart attack, blood pressure problems, and paralysis. The company therefore attempted to inject a "fear" factor into the mindset of consumers to boost product sales. Patanjali also sold other food categories like *namkeen* (a savory snack), spices, natural sugar, pulses, pickle, candy, *murabba* (a sweet fruit preserve) *dalia* (broken cereal), mustard oil, and rice. These were highlighted as products based on Ayurveda or as providing substantive ayurvedic benefits. Furthermore, the ayurvedic food products sold by Patanjali benefited from the positive "rub-off" of the Patanjali brand name, which the company had previously established by selling other ayurvedic goods.

The focus on natural ingredients and medicinal properties was adopted for cosmetic products and toiletries as well. The shaving cream packaging emphasized the use of *neem* (a medicinal herb), basil, turmeric, and aloe vera. All these ingredients are generally believed to have medicinal properties by the Indian masses (Lallanilla, 2015). *Shishu* (a body wash gel for children) was projected as a product which had Mother Nature's characteristics of being "extra caring" for newborns. It was a unique body lotion which had a blend of cucumber, turmeric, aloe vera, and natural oils. Similarly, in the dental care category, *Dant Kanti* toothpaste claimed to contain 26 highly valuable ayurvedic ingredients, which helped in teeth whitening, making teeth strong and maintaining healthy gums. *Dant Kanti* was presented as toothpaste which could fight tooth-related diseases as well. Some of Patanjali's other products, including aloe vera gel, beauty cream, antiaging cream, antiwrinkle cream, face packs and scrubs, face wash, moisturizer cream, body lotions, and

Household care products & personal care products			Health care products	Food & beverage products
Household care products	Personal care products			
	Cosmetic products	Toiletries		
• Hawan Samagri (items offered to fire during rituals) • Dishwash bar • Agarbatti (joss stick)	• Shaving cream • Shishu (children) care • Body care • Eye care	• Dental Care • Hair Care • Skin Care	• Digestive capsules • Chyawanprash (herbal tonic) • Ghee (clarified butter) • Health drinks • Pot • Health wellness • Badam pak (herbal supplement for brain development made from almond) • Honey • Fruit juice Ayurvedic products • Kwath (herbal supplement for strengthening liver) • Vati (herbal supplement for immune system) • Churna (mixture of powdered herbs and/or minerals) • Parpati/Ras (herbal supplement for strengthening digestive system) • Arishta (herbal supplement for treating health problems including infertility and impotence) • Sirup • Oil • Balm • Tablets • Package for diseases • Bhasma (Ash) • Guggul (gum resin) • Pishti (ayurvedic medicine used for coughs, colds, etc.) • Asava (herbal supplement for treating health problems including infertility and impotence) • Godhan ark (ayurvedic medicine from cow urine) • Lep (ointment for body pain) • Inhaler	• Biscuits and cookies • Herbal tea • Soan papdi (Sweet) • Gram flour • Cornflakes • Noodles • Namkeen (snacks) • Spices • Jam • Natural sugar • Pickle • Pulses • Oats • Bura (jaggery powder) • Candy • Murabba (sweet fruit preserve) • Broken cereal/dalia • Mustard oil • Rice • Papad (Indian food item)

Source: Compiled from https://www.patanjaliayurved.net/ [Accessed: 12 May 2017].

soaps, targeted both traditional and modern young consumers by explicitly stating the harmful effects of using chemical-based cosmetic products and highlighting the presence of natural ingredients in Patanjali products. There were also two main products sold by the company in the household care category. These were a dishwashing bar of soap and *agarbatti* (an incense stick), both of which highlighted the natural nature of their ingredients. The dishwashing bar of soap, for example, consisted of ash, *neem*, and lemon. Patanjali highlighted that these ingredients would not harm the skin while washing utensils. *Agarbattis* were available in various fragrances, such as white flower, lavender, lily, sandalwood, jasmine, etc.

9.4 Positioning and advertising

Patanjali *Chyawanprash* (a herbal tonic) was positioned as an immunity booster, with the advertisement targeting all members of the family, from granny and grandpa to kids. *Ghee* was positioned as a fitness supplement for a strong and healthy body, and the ad targeted mothers and adults. Biscuits were positioned as anticholesterol healthy cookies, and the advertisement targeted kids, mothers, and elders. Noodles were positioned as a healthy and tasty food item, and the advertisement targeted kids and mothers. *Amla* juice and aloe vera juice targeted Indian consumers by showcasing a healthy joint family (an undivided extended family where all family members stay in the same house) which consumed the natural health drinks regularly. The company roped in famous celebrities like Hema Malini (a Bollywood superstar) for some of its product advertisements, trying to capitalize on the power of celebrity endorsement.

The care taken by Patanjali in positioning and promoting its ayurvedic food products also extended to its household care products, cosmetic products, and toiletries. Patanjali's dishwashing bar of soap was positioned as an agent to fight tough grease the fastest and the ad targeted women. *Agarbattis* were positioned as pious, ritual products, and the advertisement targeted women. Shaving cream was positioned as a herbal cream and the advertisement targeted young men. *Shishu* was positioned as natural body lotion with earthly ingredients, and the advertisement targeted mothers. The *Dant Kanti* toothpaste was positioned as a *Swadeshi* product (*Swadeshi* originated in British India as an economic strategy to dethrone the British Empire and encouraged the domestic production of products and the shunning of foreign goods) and the advertisement targeted children. *Kesh Kranti* hair oil was positioned as nourishing strong and long hair, and the advertisement targeted working women in particular. *Swarnakranti* face cream was positioned as a natural beauty enhancer over chemical-containing beauty creams and the advertisement targeted college-goers.

9.5   Enabling distribution channels for ayurvedic food

Patanjali manufactures its products in its Food and Herbal Park, which is spread across 95 acres and is the largest food park in Padartha, Haridwar, India. Across India, Patanjali products were distributed through 47,000 general retail counters (Muvsi, n.d.) and 3000 distributors, and 80 super distributors (*Financial Express*, 2016) (see Fig. 9.1). Going beyond the traditional coverage of *kirana* stores (small neighborhood retail stores, usually extending to 250−1000 ft^2), the company used three main strategies to enhance its distribution of ayurvedic food products in the Indian market. First, it focused on opening either its own or franchised stores. Second, it linked up with many big retailers to increase its visibility in modern retailing. Lastly, realizing the growing preference and acceptance for e-commerce, it ensured its presence on the main e-commerce websites.

Patanjali started a network of franchises with three formats, namely, "*Arogya Kendra*" (health centers), "*Patanjali Chikitsalaya*" (clinics), and "*Swadeshi Kendra*" (centers selling nonmedical products) (see Fig. 9.2). The minimum requirement for opening a *Swadeshi Kendra* franchise was the investment ability of approximately US$1500−3000, and the availability of 300−500 ft^2 of retail space to display products. Moreover, the franchise applicant has to nominate at least 11

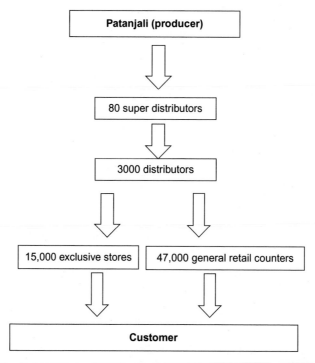

Figure 9.1 Patanjali product distribution structure.
Prepared by the authors.

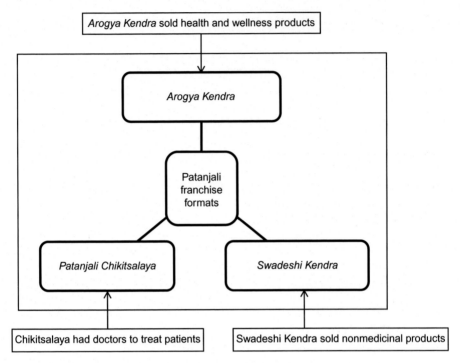

Figure 9.2 Patanjali's three franchise formats
Prepared by the authors.

members for the propagation of yoga, spirituality, cultural values, and *Swadeshi* in order to get the franchisee registration confirmed. To open a Chikitsalaya, the minimum population of the region would need to be at least 100,000. The investment required was around US$12,300−18,500, with space of 750−1000 ft^2 or more. For an *Arogya Kendra*, the region's population was required to be less than 100,000. The investment required was around US$0.6m−0.7 million with space of 350−500 ft^2 or more (Patanjali n.d.). While *Arogya Kendras* and *Swadeshi Kendras* sold health-related products, *Patanjali Chikitsalayas* were treatment centers addressing various ailments. Ayurvedic medicines and household products of the foremost quality for treatment were made available in these centers, all manufactured using indigenous procedures. These outlets also exploited all opportunities to sell ayurvedic food products. In order to provide a better consumer experience, by providing the entire range of products, the company conceptualized and started Patanjali mega marts, which was a different setup compared to the three franchise formats detailed above. These stores were spread over 3000−5000 ft^2 in selling area (Narayanan, 2016). The first Patanjali mega mart was opened in Nagpur in October 2015 and was followed by openings in other major metro cities such as Lucknow, Delhi, Bengaluru, and Kolkata.

Along with selling ayurvedic food products through its own distribution system, Patanjali also focused on modern format retail stores for better visibility. Patanjali tied up with major modern format retailers such as Big Bazaar, Spencer's, Reliance Fresh, Hyper City, and D-Mart (Kamath, 2016) to ensure that their food products were available on the shelves. Many small-scale retailers (mom-and-pop outlets), who felt a connection with the concept of *Swadeshi*, joined as distribution agents for Patanjali. Thus, Patanjali tried to reach every nook and cranny of the country. Though Patanjali focused on products with traditional roots, it identified e-commerce as an emerging channel. The company could foresee the gradual transformation in consumer psyche and the manner in which they were becoming increasingly technology savvy and were seeking ease in buying grocery and consumer-packaged goods items. Similarly, Patanjali wanted to latch on to growing internet and increased smartphone penetration, both of which were changing the consumer behavior of the Indian population. Consequently, it joined hands with major e-commerce players such as BigBasket, Amazon, and Flipkart. It also launched its own website and mobile application to sell Patanjali products. In choosing different options to ensure product availability, Patanjali tried to cater to both types of urban consumers—those who wanted to "touch and feel" goods during purchase and those who preferred to interact online (see Fig. 9.3).

9.6 Porter's generic value chain for Patanjali

Patanjali has seen increasing growth in revenue from US$68 million in the FY 2011−12 to US$750 million in the FY 2015−16, which was driven by its strong performance in the ayurvedic food category. Porter's generic value chain provides a relevant framework to understand the various primary and support activities that allowed Patanjali to create a strong brand name in the ayurvedic food sector (Porter, 2001). The value chain lists five primary activities—inbound logistics, operations, outbound logistics, marketing, sales and services as primary activities. The four support activities mentioned are the firm's infrastructure, human resource management, research and development (R&D), and procurement. Below are the key features of Patanjali's primary and support activities.

9.6.1 Primary activities

Inbound logistics: Patanjali's entry into ayurvedic food products started when a group of farmers had approached Baba Ramdev for his help in dealing with the excess production of gooseberries. Baba stopped farmers from destroying the gooseberries and instead bought them and decided to produce juice from them. Over the years, Patanjali realized the significant cost benefits of procuring raw material directly from farmers. Increasingly, the company eliminated the middleman, which eventually led to Patanjali getting better margins.

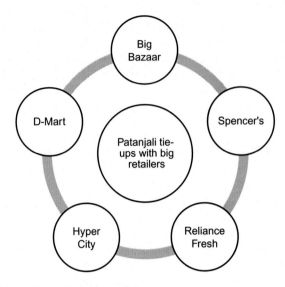

Patanjali's tie-ups with big retailers

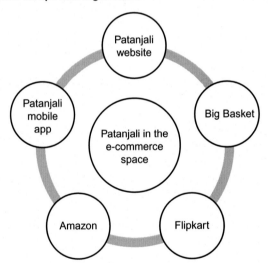

Patanjali's e-commerce activities

Figure 9.3 Patanjali market penetration strategies.
Prepared by the authors.

Operations: Almost 80% of products are manufactured in-house by Patanjali, with very limited manufacturing outsourced. This has ensured that Patanjali products are available at low cost in comparison to rivals. Patanjali has controlled production operations as it has been experiencing rapid growth and did not want to face any supply shortages in the market. While it focused on its units in Haridwar (where Baba Ramdev practiced self-discipline and meditation and has a long association with the holy city) and elsewhere in Uttarakhand, it also planned to set up plants in four other Indian states and thereby increase its production capacity and ability to reach markets quickly.

Outbound logistics: Patanjali knew that the demand for its products was growing with every passing day. While it had to focus on manufacturing operations, the company had to also ensure that its products reached all end consumers. It built a strong distribution network and supplied its products across India through retail outlets, stores, distributors, and super distributors as mentioned earlier. Patanjali opened 7000 Arogya and *Swadeshi Kendras* which were also used to sell ayurvedic consumer-packaged goods products. To gain an edge over its competitors, Patanjali joined hands with modern retailers and ensured the presence of its products online. For general trade, Patanjali had a two-stage distribution strategy. The first stage included creating a strong alternative distribution system. The primary purpose of this was to create demand and word-of-mouth advocates. Once consumers started buying from Patanjali's distribution system, Patanjali approached general trade to stock its products who were ready to join after seeing the considerable consumer base. The next stage was to pivot to general trade after generating considerable consumer base from the first stage (Gupta, Himan, & Ramdoss, 2016).

Marketing and sales: In the earlier years, Patanjali did not advertise its health and wellness products. However, with its food products, it changed its strategy and started advertising in order to reach a broader audience. It utilized media tools, often emerging as the most advertised brand. For example, Patanjali was the most advertised brand on television from 23−29 January 2016, as per data from the Broadcast Audience Research Council (Venugopal, 2016). Baba Ramdev used multiple options to market Patanjali products. For instance, he used his yoga camps to sell various products at the venues. He gave a touch of personal selling, by discussing the various benefits of Patanjali products while addressing his devotees. While other brands had the opportunity to use mass media, they did not have a captive audience for personal selling like Patanjali. The numerous devotees of Baba Ramdev could very easily be turned into Patanjali's loyal customers.

Service: Patanjali used various franchise formats and ensured that its products reached out to rural as well as urban areas. *Arogya Kendra, Patanjali Chikitsalaya*, and *Swadeshi Kendra* were three adopted formats. As mentioned earlier, the franchise formats were also used to sell ayurvedic food products. These outlets also provided an opportunity for Patanjali staff to interact with consumers and get their insights in order to improve the company's offerings. Patanjali retailers and small outlet owners' interaction with consumers helped Patanjali get feedback about its variety of products. These formats offered various services such as free consultation by certified medical practitioners, which ensured a rise in footfall. Free consultation, particularly in the area of healthcare, helped build strong faith in Patanjali, which culminated in positive word-of-mouth among friends and relatives (Gupta et al., 2016).

9.6.2 Support activities

Procurement: Patanjali explored various options to stay ahead of its competitors. It tied-up with the Maharashtra State Government, for example, to procure excessive products derived from materials available in forests. When securing raw material, it sourced directly from farmers and thus eliminated the middleman. Direct procurement from farmers helped Patanjali to increase operating profit by 20% (Prabhu, 2016).

R&D: Patanjali realized the need for technological advancement to stay ahead of its competition. It had a team of 250 scientists, who looked into product development (Sen, 2016). Scientists received input from Baba Ramdev as well and then worked on developing the final product. In order to reach tech-savvy consumers, Patanjali increased its online presence on various social media outlets like Twitter and Facebook. For instance, its Twitter account posted, "Must try and see it yourself. Save yourself from harmful chemicals … apply #Patanjali#Saundarya Body Lotion for Beautiful, soft skin in winters". Likewise, its Facebook account posted, "Herbal #WashingPowder and #Bar, which have the natural properties of *neem* and lemon that will eliminate the bacteria and give protection to the hands"; and "Aloevera juice beneficial in acidity, gas, digestion problem, joint pain". While on the one hand Patanjali linked up with e-commerce players like BigBasket, Amazon, and Flipkart, on the other hand, it started selling its products through its own website and mobile app.

Human resources: As per company recruitment regulations, Patanjali would recruit only nonsmokers, teetotalers and those who were *satvik* (which refers to a person or food which is pure and clean). Patanjali never went to top business schools to recruit well-educated professionals. On the contrary, it recruited people who would fit into the company's spiritual-cum-entrepreneurial environment. Patanjali refrained from hiring professionals from MNCs, but trained *sadhus* (religious holy men) who could manage the business for the long run. Patanjali's competitors' administrative costs were around 10% of revenue due to highly educated professional staff, whereas those of Patanjali were a mere 2%, which saved on cost (ET Bureau, 2016).

9.7 Opportunities and challenges

Patanjali Ayurved Limited, led by its brand ambassador Baba Ramdev, saw abundant opportunities in the Indian ayurvedic food market. Despite the presence of a variety of products, Indian consumers had developed an emotional bond with natural and ayurvedic products, which in a sense reflected the feeling of going back to their roots. Baba Ramdev, through the medium of the Patanjali company, was viewed as championing the cause of natural and ayurvedic products, which eventually led to greater awareness and acceptability of them. As the face of Patanjali, Baba Ramdev had a huge fan following on social media. For instance, on Facebook he had around 9.2 million followers, and on Twitter 1.4 million, by November

2017. This was an opportunity for Baba to pitch Patanjali products to his followers. He evoked a sense of belongingness in Indian consumers, by overtly employing a *Swadeshi* pitch, which intensely backed buying local products and restricting the use of foreign-made products. The halo of "*Swadeshi*-ness" had a powerful impact on consumers' minds and hearts, which led to positive consumer purchasing behavior.

In addition, retailers were also influenced by the movement and offered Patanjali products prominent shelf space without charging a premium (Bhushan, 2016). Retailers also pushed Patanjali products to consumers and tried to connect with the consumer psyche in this regard. Building upon the *Swadeshi* plank, Baba Ramdev explicitly launched a campaign against black money—income illegally obtained or not declared for tax purposes. This helped in building his aura as a patriot. The radiance around Baba Ramdev, and thus around Patanjali, was another massive opportunity to position the organization as more Indian than other competitors, thus leveraging opportunities in the name of patriotism. Baba Ramdev had very strong political connections, though they were concentrated towards a particular set of political ideologies. Thus, he had an opportunity to utilize his network to boost the prospects of Patanjali, but he was also on the radar of those who were on the other side of the political divide.

Patanjali wanted to scale up and reach out to the masses beyond its traditional followers. A report by AC Nielsen (Laugani, 2014) suggested that customers were willing to try out new products in the market, which was good news for Patanjali as it was entering in new categories with its brands at a rapid pace. The report also threw light on various aspects of Indian consumers' buying behavior. It stated that Indian consumers were aware of the prices of products they purchased regularly and that they preferred brands which would keep prices in check. Consumers viewed price rises cynically and were willing to try out new products offered to the market. Patanjali's strategy was aligned with this and involved a penetration pricing approach for its natural and ayurvedic food products, which helped it reached out to the masses. Baba Ramdev's followers were comprised mainly of middle class Indians. Patanjali kept product pricing low compared to its competitors, which visibly attracted the burgeoning Indian middle class. While the price penetration approach worked well, there were some roadblocks in pursuing this. In particular, the costs incurred in establishing a distribution structure, and heavy advertising, became a bottleneck to executing price penetration.

On the one hand, Baba Ramdev's image as that of a respected spiritual figure was helping Patanjali gain points; on the other hand, Baba Ramdev's opinionated nature sparked controversies time and again. For instance, his anti-homosexuality and anti-sex education views kept him away from well-read Indians. His views that yoga could cure AIDS and cancer were eroding the brand image of Patanjali. There were legal cases filed against Baba Ramdev for land grabbing and tax evasion, which were daunting challenges. As Patanjali grew, it became the national focus of the ayurvedic consumer-packaged goods industry. Competition with rivals intensified and so did scrutiny by the Indian regulatory authorities. There were occasions where various regulatory authorities came down heavily upon

Patanjali's claims. For instance, a sample of Patanjali noodles was found to be substandard by the Food Security and Drugs Administration (DNA, 2016). There were complaints about the presence of fungus in butter and insects in noodles (Zee News, 2015), all of which marred Patanjali's reputation. In response, Patanjali launched an attack on its competitors and claimed that they were mixing cheap palm oil with mustard oil. When asked by the regulator, the Advertising Standards Council of India (ASCI), to substantiate these charges against the competitors, Patanjali failed. Consequently, ASCI stated that Patanjali was denigrating its rivals' images by running misleading advertising campaigns (*Hindustan Times*, 2016). Patanjali was also fined around US$17,000 by a city court based at Haridwar. It was found guilty of misbranding (Section 52) and misleading advertisements (Section 53) as per the Food Safety and Standard Act, 2006. Patanjali confessed its mistake (The Times of India, 2016).

9.8   Conclusion

Though Ayurveda was considered to be a part of Vedic wisdom, and hence revered by a sizable segment of Indian society, the ayurvedic industry was not abuzz in the marketplace. The industry players were doing their business, competing with each other, chasing customers and expanding geographically. However, it was the entry of Patanjali which made the benefits of Ayurveda clear to many in the food retailing sector and beyond. Patanjali has changed the outlook of the ayurvedic industry as it created a boom in the ayurvedic food market. Its turnover for the year 2017 was up by 111% to US$1,629.6 million (The Indian Express, 2017). Though Baba Ramdev was a yoga guru, he had entrepreneurial zeal and knew that there was a huge market to be tapped. His push created a prominent space in the market for Patanjali as the company tried to reach out to consumers—rural as well as urban— to strengthen its foothold. In promoting its products, Patanjali explicitly stated the benefits of ayurvedic or natural ingredients, and Baba Ramdev also tried to persuade people to buy *Swadeshi* items. This also evoked people's emotions about nationalism and thereby associated Patanjali products with such a concept.

As suggested by various market research companies, it is evident that the ayurvedic consumer-packaged goods industry will continue to grow in leaps and bounds. This remains a positive sign for Patanjali. Despite various challenges from well-established national and MNCs, the juggernaut of Patanjali has been almost unstoppable. It is yet to be seen as to how long the wave of nationalism and *Swadeshi* will convince consumers to buy Patanjali products. This chapter has provided detailed discussion of the ayurvedic consumer-packaged goods industry, but it is yet to be seen if Patanjali can stand the test of time. The discussion provides an in-depth analysis of how a consumer packaged goods company may achieve remarkable growth in a relatively short time period. It helps new entrants understand how they can establish themselves in markets inundated by strong competitive forces through the use of prominent leaders, who tap into the spiritual capital of nations and their consumers. The ethics of such an approach are of course open to question, and this remains an area for future discussion and interrogation.

References

Balakrishnan, T. (n.d.) *Ayurvedic industry—Opportunity and challenges.* [pdf] CII. Available at: ⟨http://cii.in/WebCMS/Upload/Mr%20T%20Balakrishnan.pdf⟩ (accessed 7 March 2017).

Bhushan, R. (2016). Patanjali takes 'swadeshi' pitch to retailers; analysts say strategy likely to fail. *Economic Times.* [online] Available at: http://articles.economictimes.india-times.com/2016-03-09/news/71346539_1_swadeshi-products-swadeshi-chyawanprash [Accessed 13 November 2017].

Business Recorder (n.d.) *Global market for herbal medicines likely to reach $107bn by 2018.* [online] Available at: ⟨http://epaper.brecorder.com/2017/04/30/5-page/872897-news.html⟩ (accessed 1 April 2017).

California College of Ayurveda (n.d.). *Principles and practices of ayurveda.* [online] Available at: ⟨http://www.ayurvedacollege.com/articles/drhalpern/Principles_practices⟩ (accessed 12 November 2017).

Chandran N. (2016). *India economic growth to slow dramatically after cash crunch.* [online] CNBC. Available at: ⟨https://www.cnbc.com/2016/11/30/india-gdp-to-slow-dramati-cally-after-cash-crunch.html⟩ (accessed 1 May 2017).

DNA (2016) *After Maggi, Baba Ramdev's Patanjali Noodles found 'sub-standard', fails lab test.* DNA. [online] Available at: ⟨http://www.dnaindia.com/money/report-after-maggi-baba-ramdev-s-patanjali-noodles-found-sub-standard-fails-lab-test-2197762⟩ (accessed 5 November 2017).

Equitymaster.com, (2016). *Fast moving consumer goods sector analysis report.* [online] Available at: ⟨https://www.equitymaster.com/research-it/sector-info/consprds/Consumer-Products-Sector-Analysis-Report.asp⟩ (accessed 7 February 2017).

ET Bureau. (2016). Baba Ramdev's Patanjali and the art of levitating to the top. *Economic Times.* [online] Available at: https://brandequity.economictimes.indiatimes.com/news/business-of-brands/baba-ramdevs-patanjali-and-the-art-of-levitating-to-the-top/50899001 [Accessed 3 November 2017].

Financial Express (2016). *Patanjali may go global; needs Rs 1000 Cr investments for expansion.* [online] Available at: http://www.financialexpress.com/industry/patanjali-may-go-global-needs-rs-1000-cr-investments-for-expansion/236046/ (accessed 14 November 2017).

Global Wellness Institute (n.d.), *The history of wellness.* [online] Available at: ⟨https://www.globalwellnessinstitute.org/history-of-wellness/⟩ (accessed 25 November 2017).

Gupta, P., Himan, D., & Ramdoss, V. (2016). The secret behind Patanjali's rise and fall. *The Hindu.* [online] Available at: http://www.thehindubusinessline.com/catalyst/the-secret-behind-patanjalis-rise-and-rise/article9300591.ece [Accessed 20 November 2017].

Hindustan Times (2016). *Patanjali ads misleading, 'denigrate' rivals' products: Advertising watchdog*. [online] Available at: ⟨http://www.hindustantimes.com/business-news/patan-jali-ads-misleading-denigrate-rivals-products-advertising-watchdog/story-yanTCKmvozc iYt7rBzRRFJ.html⟩ (accessed 10 January 2018).

IANS. (2016). India's FMCG sector to reach $104 bn by 2020: Study. *Business Standard*. [online] Available at: http://www.business-standard.com/article/news-ians/india-s-fmcg-sector-to-reach-104-bn-by-2020-study-116100500613_1.html [Accessed 7 May 2017].

IBEF (2017). *Fast moving consumer goods*. [Online] Available at: ⟨https://www.ibef.org/download/FMCG-July-2017.pdf⟩ (accessed 23 November 2017).

Kamath, R. (2016). Retail chains opt for loyalty points on Patanjali items. *Business Standard*. [online] Available at: http://www.business-standard.com/article/companies/retail-chains-opt-for-loyalty-points-on-patanjali-items-116052300037_1.html [Accessed 15 February 2017].

Karnik, M. (2016). Indian consumer goods firms see a new money-spinner in an old product: Ayurveda. *Quartz India*. [online] Available at: https://qz.com/621789/indian-consumer-goods-firms-see-a-new-money-spinner-in-an-old-product-ayurveda/ (accessed 1 April 2017).

Lallanilla, M. (2015). *Ayurveda: Facts about ayurvedic medicine*. [online] Available at: ⟨https://www.livescience.com/42153-ayurveda.html⟩ (accessed 12 November 2017).

Laugani, R. (2014) *Nielsen: Keeping up with India's urban FMCG consumer*. [online] Available at: ⟨http://www.nielsen.com/in/en/insights/news/2014/keeping-up-with-indias-urban-fmcg-consumer.html⟩ (accessed 14 November 2017).

Muvsi (n.d.). *How to start Patanjali products franchise/distribution business in India*. [online] Available at: ⟨http://muvsi.in/patanjali-franchise-business/⟩ (accessed 14 November 2017).

Maheshwari, R. (2017). *The epic rise of Patanjali: Game-changer in Indian FMCG industry*. [online] Available at: ⟨https://yourstory.com/read/c5edeadc03-the-epic-rise-of-patanjali-game-changer-in-indian-fmcg-industry⟩ (accessed 7 November 2017).

Narayanan, C. (2016). From big bazaars to mega marts, how Baba Ramdev is taking Patanjali places. *The Hindu*. [online] Available at: http://www.thehindubusinessline.com/economy/from-big-bazaars-to-mega-marts-how-baba-ramdev-is-taking-patanjali-places/article7919921.ece [Accessed 2 March. 2017].

Patanjali (n.d.) *Patanjali Chikitsalaya and Arogya Kendra Form*. [online] Available at: ⟨http://patanjaliayurved.org/patanjali-chikitsalya-and-arogya-kendra-form.html⟩ (accessed 14 November 2017).

Polgreen, L. (2010). Indian who built yoga empire works on politics. *The New York Times*. [online] Available at: http://www.nytimes.com/2010/04/19/world/asia/19swami.html [Accessed 17 November 2017].

Porter, M. E. (2001). The value chain and competitive advantage. In D. Barnes (Ed.), *Understanding business: Processes* (pp. 50−66). London: Routledge.

Prabhu, U. (2016). Baba Ramdev: A pragmatic business baron. *BW Businessworld*. [online] Available at: http://indiamahesh.com/2016/01/baba-ramdev-a-pragmatic-business-baron/ [Accessed 12 November 2017].

Sen, S. (2016). How yoga guru Ramdev turned FMCG business on its head with 'upkar'. *Hindustan Times*. [online] Available at: http://www.hindustantimes.com/business/how-yoga-guru-ramdev-turned-fmcg-business-on-its-head-with-upkar/story-mGDz2Cayqt8ap Qbpwro3gM.html [Accessed 20 Nov. 2017].

The Times of India. (2016). Patanjali slapped with Rs. 11 lakh fine by Haridwar court on charges of 'misbranding'. *The Times of India*. [online] Available at: ⟨https://timesofindia.indiatimes.com/city/dehradun/patanjali-slapped-with-rs-111-fine-by-haridwar-court-on-charges-of-misbranding/articleshow/55986852.cms⟩ (accessed 20 Nov. 2017).

The Indian Express. (2016). Will double turnover to Rs 10,000 crore in 1 year: Patanjali. *The Indian Express*. [Online] Available at: http://indianexpress.com/article/business/business-others/will-double-turnover-to-rs-10k-cr-in-1-yr-patanjali-2772036/ [Accessed 7 May 2017].

The Indian Express. (2017). Patanjali's FY17 turnover up 111% to Rs 10,561 crore. *The Indian Express*. Available at: http://indianexpress.com/article/business/companies/ramdev-patanjalis-fy17-turnover-up-111-to-rs-10561-crore-4641082/ [Accessed 20 November 2017].

Venugopal, V. (2016). Is Baba Ramdev's Patanjali India's biggest FMCG advertiser? *Economic Times*. [online] Available at: https://brandequity.economictimes.indiatimes.com/news/advertising/is-baba-ramdevs-patanjali-indias-biggest-fmcg-advertiser/50874683 [Accessed 5 Nov. 2017].

University of New Hampshire (n.d.) *Office of health education promotion—Practices*. [online] Available at: ⟨https://www.unh.edu/health/ohep/complementaryalternative-health-practices/herbal-medicine⟩ (accessed 11 Nov. 2017).

Zee News. (2015) Swami Ramdev in a fix, insect found in Patanjali noodles. *Zee News India*. [online] Available at: ⟨http://zeenews.india.com/news/india/swami-ramdev-in-a-fix-insect-found-in-patanjali-noodles_1831122.html⟩ (accessed 23 Jan. 2017).

Appendix

Appendix 1 Glossary of Hindi terms used in the case

Hindi word	English translation
Ayurveda/ ayurvedic	*Ayurveda* is the medical science of ancient India. The word *ayurveda* originates from combining two words, "*ayur*" meaning life and "*veda*" meaning knowledge
Baba	*Baba* is used to refer to elderly people, paying them respect
Acharya	An instructor in religious matters
Kirana	Small neighborhood retail store, usually spread over $250-1000$ sq^2
Swadeshi	*Swadeshi* originated in British India as an economic strategy to dethrone the British Empire. It encouraged the domestic production of products and the shunning of foreign goods
Satvik	*Satvik* refers to a person/food which is pure and clean
Neem	*Neem* is a medicinal herb which contains antifungal, antibacterial, and antiviral properties

Source: Compiled by the authors.

Organic innovation: The growing importance of private label products in the United States

10

Xiaojin Wang[1], Kathryn Boys[1] and Neal H. Hooker[2]
[1]Department of Agricultural and Resource Economics, North Carolina State University, Raleigh, OH, United States, [2]John Glenn College of Public Affairs, The Ohio State University, Columbus, OH, United States

10.1 Introduction

In recent years, the burgeoning United States organic food industry has experienced double-digit growth. A new record high in total organic product sales of $47 billion [Organic products include foods, ingredients and beverages, as well as organic fibers, personal care products, pet foods, nutritional supplements, household cleaners and flowers (Organic Trade Association (OTA), 2016). See more at https://www.ota.com/about-ota.] was reached in 2016. Within this, organic food and drink sales reached $43 billion and accounted for 91% of total organic sales. This value reflects an increase of food sales of 8.4% from the previous year and far outpaced the overall food market's growth rate of 0.6% (Organic Trade Association (OTA), 2017a).

Private label (PL), also known as store brand, products have contributed significantly to this increase in organic sales. The overall PL dollar share in the consumer packaged goods (CPG) industry in the United States was approximately 16% during the period 2010−14 (IRI, 2015). PL products, which had a relatively low presence of 8% among organic products in 2003, have quickly grown and by 2008 accounted for 17% of organic products (Nielsen, 2008; Nutrition Business Journal, 2004). In 2007, about 43% of certified organic handlers manufactured PL products, which constituted approximately 19% of handlers' organic sales (Dimitri & Oberholtzer, 2009). [One example is Organic Valley, which adopted a branded, PL, and bulk products three-pronged strategy. PL output accounted for 25% of its total revenue in 2011 (Su & Cook, 2015; Organic Valley, 2013).] Increased availability, lower price points, improved quality, joint marketing of organic and health attributes, and product innovation are among the top drivers behind PL organic purchases (Jonas & Roosen, 2005; Mintel, 2015a; 2016). [Also, organic PL products are less likely to be identified as store brands and more likely to be mistaken for brand-name products. Thus, PL organic products may have more success competing with national brands (NBs) for some consumers' food dollars as a result (Mintel, 2013).] Consumers no longer view PLs simply as low-cost alternatives to national brands,

Case Studies in Food Retailing and Distribution. DOI: https://doi.org/10.1016/B978-0-08-102037-1.00010-4

or name brands, and increasingly accept them as reliable high-quality options to meet their organic needs (Hartman Group, 2014).

This chapter examines innovation through new product development in the US organic food and beverage market. In particular, innovation stemming from PL and NB product lines is explicitly considered. Due to their importance in organic market product innovation, two case studies are used to examine the processed fruit & vegetable and dairy product categories. To complement this discussion, the price premiums for organic products relative to their conventionally produced equivalents are estimated and evaluated for several food industry subsectors. We demonstrate that PL products may be leading organic innovation in certain dimensions, particularly value. Implications for PL brand strategies are considered.

This study makes use of Mintel's Global New Product Database (GNPD), which provides detailed information (including images of packaging) concerning launches of new products and is often used to monitor product innovation and retail strategies in CPG markets. The collection of the GNPD's new product launch information is primarily facilitated by a network of field associates referred to as "shoppers." The database includes product information from other sources such as press releases, trade shows, media, and company tracking (GNPD, 2016). The data provide a comprehensive resource for new product development and innovation, thus allowing comparisons of firms' new product strategies.

10.2   A premium private label—organic products

NBs, or name brands, are distributed across the United States under brand names that are owned by the producer or distributor. In contrast, PL products, also known as store brands [While the terms PL and store brands are often used interchangeably, there is some slight difference (Semeijn, Van Riel, & Ambrosini, 2004). For instance, the Private Label Manufacturers Association uses PL, whereas the Food Marketing Institute refers to 'store brands' (Stanton, Wiley, Hooker, & Salnikova, 2015)], are owned by a retailer or wholesaler. They are typically manufactured or provided by one company for offer under the retailer's brand. Retailers commonly sell a mix of NBs and PL goods to take advantage of the benefits of each. Moreover, retailers often carry a portfolio of PLs serving different quality tiers. In general, PL products can be classified into three different types: typical cheap and low quality own labels (i.e., economy PLs; "generics," or "white labels"); somewhat less expensive PLs comparable in quality to the NBs (i.e., standard PLs; "copycats," "me-toos"); and premium quality and high value added PLs (i.e., "premium PLs" and "value innovators") (Kumar & Steenkamp, 2007; Laaksonen and Reynolds,1994). Retailers often adopt a three-tiered PL approach as it can give consumers more choice and can satisfy the heterogeneous nature of the consumer market. It also offers retailers flexibility and control over a category and further creates differentiation compared to their competitors (PWC, 2011). Premium PLs usually target distinct consumer or niche segments such as healthy eating, kids, and organic foods (Hökelekli, Lamey, & Verboven, 2017). They serve not only to distinguish retailers' product lines from one another but also to place NBs and PLs more directly in price and quality competition. Organic PLs stand as an attractive alternative to branded organic goods as they undermine three main barriers preventing consumers from purchasing organic products; affordability, availability, and lack of information provided to consumers (Labajo, 2016; Martinez, 2007).

10.2.1   Overview of organic product introductions in the US market

In October 2002, the National Organic Program (NOP), part of the USDA's Agricultural Marketing Service, was established with the core mission to protect organic integrity. The NOP develops, implements, and administers national production, handling, and labeling standards for organically produced agricultural products [United States Department of Agriculture (USDA), 2017]. Prior to 2002, the term organic was unregulated and subject to various voluntary certifications by private organizations or state agencies, so there was no uniformity in standards (OTA, 2017b; Sanchez, 2014). Across food and drink, NB and PL new organic product introductions exhibited an upward trend over the period 2000−15, especially following the implementation of the NOP (see Figure 10.1). The number of new organic food and beverage products declined following the Great Recession of 2007−09, but rebounded in 2010. In 2002, there were approximately 900 new organic food and drink products launched on the US market. In 2011, the number of new organic products jumped to more than 1800. By 2015, there were 2,911 new organic product releases, which included 2,326 foods and 585 drinks and, respectively, accounted for 17% and 15% of the total food and drink introductions

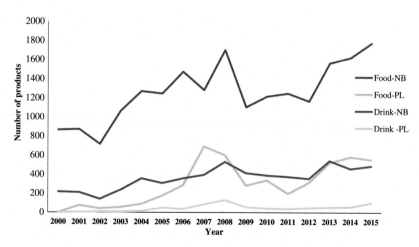

Figure 10.1 Recent trends in US NB and PL organic product innovation, 2000−15. Authors' calculations based on GNPD (2016).

(GNPD reports a total of 13,307 foods and 3,836 drinks launched in the United States in 2015). Among these, 78% of them were NBs, while PLs made up the other 22%.

Most food retailers introduced their own organic PL, and they are being developed at a fast pace. From 2001 to 2015, the average share of PLs among total new organic food and drink product introductions was 16%. The share of innovations that are PL (versus NBs) peaked in 2007 at 32%. However, its average annual growth rate during the same period is 29%, 3.6 times faster than that of NB counterparts (Fig. 10.1). In considering the type of new product launch [All GNPD records are assigned a launch type; new formulation ("New Formula," "Even Better," "Tastier," "Now Lower in Fat," "New and Improved," or "Great New Taste" on package); new packaging ("New Look," "New Packaging," or "New Size"); brand new product (representing entries of new range, line, or family of products or crossing over to a new subcategory for existing brand); new variety/range extension; and relaunch.], NBs outperformed PLs in brand new product offerings. Of the new NB organic products, 35.7% were brand new products and 32.4% were new varieties or a product line extension. These results differ from the innovation pattern of PL organic products where new varieties or product line extensions accounted for 41.7%, new packaging for 26.6%, and brand new products for 23.8% of new innovations.

It is worth also considering innovation in the amount of organic content contained in multi-ingredient food and drink products. To help consumers determine the organic content of the food they purchase, the NOP has developed labeling standards, along with production and processing requirements. As described in Appendix 1, organic products are grouped into four categories, which are largely dependent on the percentage of organic content (excluding salt and water) that a product contains: 100% organic content, 95%−99.9% organic ("organic") content, 70%−94.9% organic content, and less than 70% organic content (Calculating the Percentage of Organically Produced Ingredients, 2017). Organic content trends for PL products are more aligned with the organic content trends of all new products. Among all PL organic products, 56% were 100% organic, and an additional 31% were in the 95%−100% (organic) category. These values are lower for NB organic products where 38% were 100% organic and 27% were organic.

Product innovation is often regarded as an effective strategy for name brand firms to defend their products against PL entry. The pressure from PL sales may stimulate further product development and innovation by branded product manufacturers, thus increasing the product differentiation and quality variety for consumers. Various studies examine the impact of manufacturer brand innovation on PL consumption. For example, Pauwels and Srinivasan (2004) found that the adoption of a defensive strategy of investing in product innovation by manufacturer brands could enhance their competitive advantage and enable a sustainable price premium over PLs. Sriram, Balachander, and Kalwani (2007) showed that the introduction of new products by CPG manufacturers can improve their brand equity, thus reducing their vulnerability to the entry of PLs. Kumar and Steenkamp (2007) demonstrated that product innovation is an effective strategy to counter the success of PLs, as new products decrease the substitutability between PLs and manufacturer brands. Abril and Sanchez (2016) found that nonprice strategies, such as promotions (e.g., feature and display) and product innovation, are more effective than price strategies when competing with PLs, especially in the context of recovering consumers who have switched to PLs. Additional findings reveal that manufacturer brands' innovation intensity in a given category negatively correlates to PL share (Martos-Partal, 2010; Steenkamp & Gielens, 2003).

On the other hand, retailers are also launching new products to differentiate themselves from their competitors and to add value to their consumer franchise. PL innovation is an important market dynamic. Gielens (2012) assesses when and to what extent new products change NBs' market position and finds that products introduced by leading NBs, standard PLs, and premium PLs are more likely to increase category sales than products introduced by follower NBs or economy PLs. Abril and Martos-Partal (2013) investigated whether or not consumers accept product innovations by PLs—and if so, then to what extent. They found that PLs positively affect consumer adoption of new products, as measured by both the trial purchase level and the repurchase rate.

10.2.2   Food and beverage innovation: national brands versus private labels

Table 10.1 reports the number of organic and conventional food and drink products distinguished by brand type and category. Snacks, sauces and seasonings, and bakery products took the top three spots in both organic and conventional food categories in the number of new product introductions. When considering both organic and conventional products, NBs dominated new product introductions in almost all categories. "Processed fruit & vegetables" was the only category for which PL had a greater share than its NB counterpart in the new organic product launch. This category has increased in popularity as consumers are increasingly seeking organic forms of fresh, relatively unprocessed products, which offer convenience. (Within this category, different levels of processing are available such as prewashed, chopped, or microwave-ready bags.) Sales of organic packaged fresh produce jumped nearly 40% from 2012 to 2014, and surpassed the $200 million benchmark in January 2015, putting the category far ahead of any other organic category (Mintel, 2015b). PL had almost the half of all new organic sauces & seasonings; the fastest growing organic category in 2015 (OTA, 2016). However, except for the categories of sauces & seasoning, prepackaged fruit & vegetables, and alcoholic beverages, the share of PLs among total new organic products was lower than shares of PLs in the total new conventional products, indicating a distinct product development strategy and suggesting likely continued growth of new PL organic products in these categories.

Table 10.1 US organic and conventional food and drink new product introductions by brand type and category, 2015

Category	Organic			Conventional		
	National brand	Private label	Total	National brand	Private label	Total
Food						
Snacks	450	59	509	1230	677	1907
Sauces & seasonings	144	140	284	754	500	1254
Bakery	185	56	241	1183	565	1748
Dairy	196	40	236	705	293	998
Side dishes	78	43	121	194	167	361
Breakfast cereals	98	18	116	337	73	410
Baby food	81	27	108	37	29	66
Fruit & vegetables	37	70	107	206	132	338
Chocolate confectionery	86	16	102	408	154	562
Processed fish, meat & egg products	78	23	101	601	289	890
Meals & meal centers	80	18	98	557	193	750
Desserts & ice cream	85	7	92	388	230	618
Sweet spreads	71	17	88	140	56	196
Soup	58	7	65	95	47	142
Sweeteners & sugar	21	8	29	30	20	50
Sugar & gum confectionery	15	1	16	345	211	556
Savory spreads	12	1	13	89	46	135
Food total	**1,774**	**552**	**2,326**	**7,299**	**3,682**	**10,981**
Drink						
Juice drinks	177	27	204	256	101	357
Hot beverages	89	43	132	374	196	570
Other beverages	83	15	98	284	104	388
Ready to drink (RTD) beverages	60	6	66	113	11	124
Carbonated soft drinks	24	3	27	163	35	198
Alcoholic beverages	16	4	20	1301	23	1324
Sports & energy drinks	20	0	20	95	3	98
Water	18	0	18	112	80	192
Drink total	**487**	**98**	**585**	**2,698**	**553**	**3,251**
Total	**2,261**	**650**	**2,911**	**9,997**	**4,235**	**14,232**

Note: Categories are ordered by the total number of new organic products.
Source: Authors' calculations based on GNPD, 2016. *Global new products database: How we do it*. Mintel International Group Ltd., London, England. [Online] Available from: ⟨http://www.gnpd.com/sinatra/gnpd/&lang = uk/info/id = how⟩ (accessed: March 12, 2017).

10.2.3   The top two organic food categories—fresh fruit & vegetables and dairy products

PL penetration varies widely across product categories. According to data from the Private Label Manufacturers Association's (2016) Private Label Yearbook provided by Nielsen, the top five categories (There were more than 700 food and nonfood product categories and subcategories in US supermarkets, drug chains, mass merchandisers, the club channel, dollar store channel, and military exchanges.) in overall store brand dollar sales across all the major retail outlets were all in fresh and perishable categories in 2015: milk has the highest store brand share at 56% with $6.1 billion in sales, followed by cheese ($4.6 billion; 39%), bread & baked goods ($4.2 billion; 28.8%), fresh produce ($4 billion; 21.6%), and fresh eggs ($3 billion; 64.8%). Common features shared among these categories include minimal differentiation, low brand equity, high consumer price sensitivity, high purchase frequency, and low innovation characteristics (Batra & Sinha, 2000; Nielsen, 2014; Sethuraman & Gielens, 2014). Retailers are more likely to introduce premium PLs in categories with a higher PL share, and with greater assortment in terms of standard PLs. Thus, it is not surprising that produce and dairy are among the most common first organic PL products introduced by retailers, with the highest PL organic presence (Nielsen, 2008). They were also the top two organic food sales categories in 2015 and generated $14.4 billion and $6.0 billion in sales respectively. Together, they accounted for more than half of the total organic food sales (OTA, 2016). PL share of organic milk sold across all retail channels rose steadily and markedly through 2004−08, from about 10% to just over 50% of all organic milk sales (Jaenicke & Carlson, 2015).

Tables 10.2 and 10.3 summarize the companies responsible for the majority of US introductions of PL and NB organic processed fruit & vegetable and dairy products in 2015. All the new fruit & vegetable products were either organic or 100% organic. 70 new PL products were launched by 14 retailers while 37 new NBs were introduced by 23 companies. In contrast to PLs, the NBs were less concentrated and had a more even distribution. The top two PLs took up almost half of the total PL fruit & vegetable innovations, while the top two NBs accounted for about one-fifth of the total NB new organic products. The top PL company, Safeway, introduced more than five times more products than its closest counterpart, Earthbound Farm. Safeway's produce items, which were all sold under its brand O Organics, accounted for one-third of the total new PL fruit & vegetable products.

For dairy products, 12 retailers offered 40 new organic PLs in 2015, while 50 NBs introduced 196 new products. Both new PL and NB dairy products concentrated on the "organic" tier, as it is relatively difficult for dairy products to be 100% organic given the vitamin fortification requirements for fluid milk products (Food & Drug Administration, 2005). It is worth noticing that PL has a higher percentage of "organic" and 100% organic products (90% of introduced products) than its NB counterpart (83% of product introductions). Moreover, similar to fruit & vegetables, PL has a higher level of concentration: the top three companies introduced 68% of the total organic products compared with 30% of those introduced by the top three NB labels.

Table 10.2 Top national brand and private label companies' for new organic fruit and vegetable products, 2015

	Company	Percentage of organic content				Total
		100% Organic	Organic (95%–99%)	Made with organic ingredients (70%–94%)	Specific organic ingredients (<70%)	
NB	Earthbound Farm	4	0	0	0	4
	Hain Celestial Group	3	0	0	0	3
	Lisa's Kitchen	1	2	0	0	3
	organicgirl	3	0	0	0	3
	Enray	2	0	0	0	2
	Mann Packing	2	0	0	0	2
	Sambazon	0	2	0	0	2
	Taylor Fresh Foods	2	0	0	0	2
	Timeless Seeds	2	0	0	0	2
	Blanco & Dinapoli	1	0	0	0	1
	Subtotal	20	4	0	0	24
	Total	**30**	**7**	**0**	**0**	**37**
PL	Safeway	19	2	0	0	21
	Wegmans	10	0	0	0	10
	Supervalu	7	2	0	0	9
	Target	6	1	0	0	7
	Kroger	3	3	0	0	6
	Aldi	4	0	0	0	4
	Trader Joe's	3	0	0	0	3
	Meijer	2	0	0	0	2
	Topco Associates	2	0	0	0	2
	Whole Foods Market	2	0	0	0	2
	Subtotal	58	8	0	0	66
	Total	**62**	**8**	**0**	**0**	**70**

Source: Authors' calculations based on GNPD, 2016. *Global new products database: How we do it.* Mintel International Group Ltd., London, England. [Online] Available from: {http://

Table 10.3 Top national brand and private label companies' new organic dairy products

	Company	100% Organic	Percentage of organic content			Total
			Organic (95%–99%)	Made with organic ingredients (70%–94%)	Specific organic ingredients (<70%)	
NB	Organic Valley	2	26	0	0	28
	Stonyfield Farm	0	17	0	0	16
	Horizon Organic Dairy	1	11	3	0	15
	Forager Project	6	0	0	0	6
	Maple Hill Creamery	1	4	1	0	6
	Great Lakes Cheese	0	6	0	0	6
	Pacific Foods of Oregon	0	6	0	0	6
	So Delicious Dairy Free	0	0	2	4	6
	Rebbl	2	3	0	0	5
	Hain Celestial Group	2	0	0	3	5
	Subtotal	14	73	6	7	100
	NB Total	**35**	**127**	**7**	**27**	**196**
PL	Safeway	1	16	0	0	17
	Meijer	0	6	0	0	6
	Topco Associates	0	4	0	0	4
	Aldi	0	1	0	2	3
	Kroger	0	2	0	0	2
	Trader Joe's	0	2	0	0	2
	H-E-B	0	2	0	0	2
	Wegmans[a]	0	1	0	0	1
	Subtotal	1	33	0	2	36
	PL Total	**1**	**35**	**0**	**4**	**40**

[a]*Note:* Similarly, Whole Foods Market, Target, The Fresh Market each had one new dairy product.
Source: Authors' calculations based on GNPD, 2016. *Global new products database: How we do it.* Mintel International Group Ltd., London, England. [Online] Available from: ⟨http://www.gnpd.com/sinatra/gnpd/&lang = uk/info/id = how⟩ (accessed: March 12, 2017).

10.3 Organic product prices

When retailers launch PL organic lines, they often highlight their affordability and explicitly state their pricing goals. For example, Whole Foods Market claimed that its private brand 365 Everyday Value marked the first time that organic products have been priced at everyday low prices rather than premium prices (Whole Foods Market, 2002). When SuperValu launched Wild Harvest, the line was set to be priced approximately 15% lower than branded organic and natural products and was positioned to meet or beat competitors' private-label organic prices (SuperValu, 2008). Walmart targeted an even lower price point when launching Wild Oats, aiming to remove price premiums associated with organic products. It emphasized that customers can save 25% or more when comparing Wild Oats to NB organic products (Walmart, 2014). Appendix 2 provides additional information about retailers' expansion of their PL product portfolios through organic PL brands.

Organic PLs are often regarded as premium PLs compared with the traditional value-based PLs. Certified USDA organic PL products must hold the same standards as their NB counterparts. Nonetheless, they still have lower prices than NBs as they usually incur less research and development, image-building costs such as advertising outlays, and no slotting fees or slotting allowances—incentives paid by the food manufacturer to the retailer to encourage new products to be placed in stores (Hyman, Kopf, & Lee, 2010; Wilkie, Desrochers, & Gundlach, 2002). In addition, food retailers are able to leverage their robust sourcing and supply chain capabilities for ingredients at a lower cost, including turning to imports for procurement because they have potentially lower price points (Jaenicke, Dimitri, & Oberholtzer, 2011). Moreover, PLs are less prone to intrabrand competition. All these factors, combined, lead to higher profit margins for PLs than NBs (Ailawadi & Harlam, 2004; Hyman et al., 2010; Steiner, 2004; Tuck Communications, 2010). Supermarket profits also rise from carrying PL brands because they create loyalty to a particular supermarket chain rather than to a NB (Koschate-Fischer, Cramer, & Hoyer, 2014; Pivato, Misani, & Tencati, 2008; Seenivasan, Sudhir, & Talukdar, 2015).

At the retail level, the two top organic food sales categories, fresh fruit & vegetables and dairy products, support significant price premiums over conventional versions of these products. Using 2005 data on produce purchases for 18 fruits and 19 vegetables, Lin, Smith, and Huang (2008) found that the organic premium as a share of the corresponding conventional price was less than 30% for over two-thirds of the items examined. Using actual retail purchases from the 2006 Nielsen Homescan panel data, Smith, Huang, and Lin (2009) found that branded organic milk commanded higher premiums than PL organic milk. Organic price premiums for a half-gallon container of milk ranged from 60% to 68% for PL organic milk above the conventional counterpart and 89% to 109% for branded organic milk above the conventional counterpart. Jaenicke and Carlson (2015) estimated the organic price premiums for four diverse retail-level food products— canned soup, packaged coffee, milk, and bagged carrots, using the Nielsen consumer panel data (2004–10). In each case, they found strong organic premiums of about 30% for bagged carrots, over 40% for canned soup, over 50% for coffee, and over 70% for milk. They also found that the interaction of organic and PL was consistently negative for each of the four product categories, reflecting the fact that PL organic products were priced below their branded counterparts. Another study, also by Carlson and Jaenicke (2016), used Nielsen Homescan data and estimated the organic premiums for 17 organic products, including fresh produce, processed products, dairy, and eggs. They found that most premiums fluctuated during the study period of 2004–10. Only three products demonstrated a steady decrease in their premiums (spinach, canned beans, and coffee), while only the yogurt premium steadily increased. Eggs and milk have the highest organic premiums in 2010 at 82% and 72%, respectively, which most likely reflect high production costs compared with plant-based products. Fresh fruit and vegetables, generally recognized as the largest part of the organic market, had the widest spread of premiums (ranging from 7% for spinach to 60% for salad mixes). Processed food premiums ranged from 22% for granola to 54% for canned beans.

While the premiums for organic products have been well documented in the literature, the potential role PL organic products may have had in lowering organic price premiums for processed food has received little attention. To examine this issue, Table 10.4 presents the prices for fruit & vegetable and dairy subcategory products by brand type and unit of measurement. Data used in this analysis were drawn from GNPD and reflect average prices (by product subcategory) that shoppers paid for new products introduced to the market in 2015. We also conducted mean-comparison tests (t-tests) on the price difference between categories: organic versus conventional; organic NB/organic PL; organic NB/conventional NB; organic PL/conventional PL.

Organic premiums for all categories and subcategories were found to be significantly different from zero ($P < 1\%$) except for milk (mL), flavored milk (mL), spreads (g), and fruit (g). [Some products in the same category are available in units of grams (g)—a mass unit, or milliliters (mL)—a volume unit. No conversion between the two measures was conducted as it must take into account the density of the ingredients, which was not available.] Organic premiums for dairy products sold in g and mL were 21% and 60%, respectively. Dairy drinks and substitutes had the highest premiums (100%) among all dairy products, followed by cheese (56%) and yogurt (25%). There were no significant differences between organic and conventional prices for milk, flavored milk, or spreads. The combined organic fruit & vegetable category (g) had a 27% price premium. However, when separately examined, the category organic fruit does not command a price premium, while organic prepackaged, frozen, and canned vegetables had a 38% premium over conventional counterparts. Except for milk (mL), flavored milk (mL), and fruit (g), PL organic products for all other categories have lower prices than their NB counterparts (at the 5% significance level). In addition, the percentage differences between organic NB and organic PL prices were much larger than those between organic and conventional within each brand type. For instance, NB organic dairy products (g and mL) were priced four and two times higher than PLs, respectively. Prices of NB

Table 10.4 Prices for new organic dairy and fruit & vegetable products

Category	Subcategory	Unit	Organic (p^O)	Organic NB price (P^O_{NB})	Organic PL price (P^O_{PL})	Conventional (p^C)	Conventional NB price (P^C_{NB})	Conventional PL price (P^C_{PL})	$p^O - p^C$	$P^O_{NB} - P^O_{PL}$	$P^O_{NB} - P^C_{NB}$	$P^O_{PL} - P^C_{PL}$
Dairy	Dairy	g	0.017 (136)	0.019 (113)	0.009 (23)	0.014 (799)	0.015 (560)	0.011 (242)	0.003***	0.009***	0.004***	−0.0015
	Milk	mL	0.008 (98)	0.009 (80)	0.002 (18)	0.005 (194)	0.005 (143)	0.003 (51)	0.003***	0.007***	0.003**	−0.001
	Flavored milk	mL	0.006 (37)	0.007 (27)	0.002 (10)	0.005 (77)	0.006 (56)	0.008 (16)	0.0003	0.005*	0.0005	−0.0004
	Dairy drinks and substitutes	mL	0.003 (4)	0.003 (3)	0.004 (1)	0.003 (25)	0.003 (20)	0.003 (5)	0.0003	−0.001	0.00005	0.0008
		mL	0.009 (53)	0.011 (46)	0.002 (7)	0.005 (86)	0.005 (66)	0.004 (21)	0.005***	0.009***	0.006***	−0.002
	Yogurt	g	0.010 (69)	0.011 (58)	0.002 (13)	0.008 (275)	0.008 (252)	0.005 (26)	0.002**	0.002**	0.003	0.004***
	Cheese	g	0.029 (42)	0.031 (38)	0.013 (4)	0.018 (401)	0.023 (242)	0.012 (159)	0.010***	0.018***	0.008***	0.0007
	Spread	g	0.017 (18)	0.020 (14)	0.009 (4)	0.016 (66)	0.017 (42)	0.013 (24)	0.002	0.011**	0.002	−0.004
Fruit & vegetables	Fruit & vegetables	g	0.011 (103)	0.014 (35)	0.010 (68)	0.009 (333)	0.010 (201)	0.007 (132)	0.003***	0.004**	0.004*	0.003***
	Fruit	g	0.012 (20)	0.011 (7)	0.012 (13)	0.010 (73)	0.012 (36)	0.009 (37)	0.001	−0.0004	−0.0009	0.003**
	Vegetables	g	0.011 (83)	0.015 (28)	0.010 (55)	0.008 (260)	0.010 (165)	0.006 (95)	0.003***	0.005***	0.004**	0.004***

Notes: Numbers of product are in parentheses. Milk includes white milk, drinking yogurt & liquid cultured milk, and sweetened condensed milk. Dairy drinks and substitutes include rice/nut/grain & seed based drinks, soy based drinks and creamers. Cheese products include fresh cheese & cream cheese, hard cheese & semihard cheese, processed cheese, soft cheese & semisoft cheese, curd & quark. Spread includes butter cream margarine & other blends. Milk, flavored milk, and dairy drinks and substitutes products are measured in g. Yogurt, cheese, spread, fruit and vegetables measured in ml were omitted due to small numbers of new products. Yogurt, cheese, spread, fruit and vegetables measured in g. Single, double, and triple asterisks (*) denote statistical significance at the 10%, 5%, and 1% levels, respectively.

Source: Authors' calculations based on GNPD, 2016. *Global new products database: How we do it.* Mintel International Group Ltd., London, England. [Online] Available from: ⟨http://www.gnpd.com/sinatra/gnpd/&lang = uk/info/id = how⟩ (accessed: March 12, 2017).

organic dairy drinks and substitutes were more than five times higher than their PL counterparts. Comparing organic and conventional prices within each brand type, NB organic prices were higher for the diary category (both in g and mL), dairy drinks and substitutes (mL), cheese (g), and vegetables (g). For other subcategories, price differences were not statistically significant. PL organic prices were higher than PL conventional counterparts only for the yogurt (g), fruit (g), and vegetables (g) subcategories. We defined the organic price premium as the difference between the organic price and the nonorganic price, assuming all other factors, such as retail channel, brand, and promotional activities, are equal. In addition, the price data are prices for newly introduced products, for which promotions are highly likely. Due to the limitation of our data, we are not able to control for these product characteristics. This could explain the discrepancies between our results and previous studies. For instance, our tests do not support an organic price premium for milk, while previous studies documented a considerably high organic price premium (Carlson & Jaenicke, 2016).

10.4   Current trends and future concerns

This chapter demonstrates the important role of PL products in new organic product introductions. Although NB organic products dominated new product introductions in almost all food categories in 2015, it is likely that PL products will continue to gain market share—especially in top categories for product innovation such as snacks, sauces, and seasonings.

Moving forward, product innovation and consumer access to organic products is likely to be affected by these and other food supply-chain actors. Innovation in organic product retailing is also stemming from traditional grocers. Kroger opened a new store format called Main & Vine in Washington in 2016, which focuses on fresh produce, food preparation advice, and high-quality prepared foods. This format draws upon the company's experience in the natural and organic retail space (Tu, 2016). Publix is planning to relaunch its organic grocery store concept, Greenwise, in late 2018. And, in a similar spirit, Winn Dixie has begun remodeling many of its stores using a more contemporary, upscale, layout, and esthetic, which will focus on organic options and prepared foods. The recent entrance of German discount food retailer Lidl and growth of Aldi in the US marketplace are similarly leading to an increase in the availability of organic foods at lower prices. All of these retailers have a strong PL portfolio that includes a notable number of organic products. These and initiatives by other US retailers (e.g., Target, Wegmans) will certainly continue to increase the availability of organic foods and are likely to particularly augment demand for PL organic products.

Other important changes underway in the marketing channels through which consumers opt to purchase foods also have implications for the availability of organic products. Current trends in "channel shifting" have found consumers of organic foods are increasingly purchasing these products in traditional grocery stores. The increasing market presence of the conventional food retailers, including Kroger and Walmart, and online grocery ordering, has hampered the growth of specialty retailers. Whole Foods, Sprouts and Fresh Market, the three largest US publicly traded natural and organic supermarkets by market capitalization, saw their stock prices plummet in 2015 (Duff & Phelps, 2016). In addition, the purchase of Whole Foods by Amazon in August 2017 is expected to have significant impacts in the organic food market. In an attempt to shed its "Whole Paycheck" image (high prices) immediately following the ownership transfer, prices of some Whole Foods organic products were lowered by up to 43% (Kaplan & Boyle, 2017). Moreover, this takeover is expected to have a significant impact on the availability of organic products. Whole Foods brand products are now available for purchase through Amazon.com and Amazon's Prime platforms such as Prime Now—the same-day delivery service available in some markets (Amazon, 2017; Wattles, 2017).

While the impact of these trends on organic PL innovation is still uncertain, the growing importance of PL organic products does present a conundrum. On the one hand, the organic sector is characterized by many small manufacturers. For small firms, the production of organic PLs provide lower (entrance) costs, in particular by decreasing transaction and marketing costs (Jonas & Roosen, 2005). On the other hand, manufacturers of NB products can temper the challenge posed by PL goods as many large manufacturers are also suppliers of PLs. While PL products are increasing the availability and affordability of organic products, counterintuitively their success may decrease consumer access to a broader organic market. Competition from store PL products might be forcing out smaller organic brands. This may occur for at least two reasons. First, grocers may want to favor sales of their own PL brands. Second, organic products manufactured by smaller firms may be relatively more expensive due to their lack of economies of scale in production, distribution, and marketing. This outcome would have important, longer term implications for the US organic innovation. Overall, organic PL offerings are expected to flourish and are expected to continue to boost organic sales. However, while the number of organic products would likely continue to significantly increase, growth among smaller branded products is also likely to continue to increase but at a slower pace.

References

Abril, C., & Martos-Partal, M. (2013). Is product innovation as effective for private labels as it is for national brands? *Innovation, 15*(3), 337–349.

Abril, C., & Sanchez, J. (2016). Will they return? Getting private label consumers to come back: Price, promotion, and new product effects. *Journal of Retailing and Consumer Services, 31*, 109–116.

Ailawadi, K. L., & Harlam, B. (2004). An empirical analysis of the determinants of retail margins: The role of store-brand share. *Journal of Marketing, 68*(1), 147–165.

Amazon. (2017). Amazon and Whole Foods market announce acquisition to close this Monday, will work together to make high-quality, natural and organic food affordable for everyone". Amazon.com. Business Wire. [Online] Available from: ⟨http://phx.corporate-ir.net/phoenix.zhtml?c = 176060&p = irol-newsArticle&ID = 2295514⟩.

Barsky, R., Bergen, M., Dutta, S., & Levy, D. (2007). What can the price gap between branded and private-label products tell us about markups? *Scanner Data and Price Indexes, 64*, 165.

Batra, R., & Sinha, I. (2000). Consumer-level factors moderating the success of private label brands. *Journal of Retailing, 76*(2), 175–191.

Calculating the Percentage of Organically Produced Ingredients. (2017). Subpart D, Labeling, 7 CFR § 205.302(a).

Carlson, A., & Jaenicke, E. (2016). Changes in retail organic price premiums from 2004 to 2010, *ERR-209*. Washington, D.C: U.S. Department of Agriculture, Economic Research Service.

Dimitri, C., & Oberholtzer, L. (2009). *Marketing U.S. organic foods: Recent trends from farms to consumers*. Washington, D.C: U.S. Department of Agriculture, Economic Research Service.

Duff and Phelps. (2016). *Food retail industry insights 2016*. [Online] Available from: ⟨http://www.duffandphelps.com/assets/pdfs/publications/mergers-and-acquisitions/industry-insights/consumer/food-retail-industry-insights-2016.pdf⟩ Accessed March 6, 2017.

Food and Drug Administration. (2005). *Pasteurized milk ordinance 2005: Appendix O. Vitamin fortification of fluid milk products*. [Online] Available from: ⟨http://www.fda.gov/Food/GuidanceRegulation/GuidanceDocumentsRegulatoryInformation/Milk/ucm301969.htm⟩ Accessed: December 6, 2016.

Gielens, K. (2012). New products: The antidote to private label growth? *Journal of Marketing Research, 49*(3), 408–423.

GNPD. (2016). *Global new products database: How we do it*. London, England: Mintel International Group Ltd. [Online] Available from: ⟨http://www.gnpd.com/sinatra/gnpd/&lang = uk/info/id = how⟩ [Accessed: March 12, 2017].

Groner, A. (2016). *American heritage history of American business*. New Word City.

Hartman Group. (2014). *Organic & natural 2014 report*.

Hökelekli, G., Lamey, L., & Verboven, F. (2017). Private label line proliferation and private label tier pricing: A new dimension of competition between private labels and national brands. *Journal of Retailing and Consumer Services, 36*, 39–52.

Howard, P. H. (2016). *Concentration and power in the food system: Who controls what we eat?* Bloomsbury, New York: Bloomsbury Publishing.

Hyman, M. R., Kopf, D. A., & Lee, D. (2010). Review of literature—Future research suggestions: Private label brands: Benefits, success factors and future research. *Journal of Brand Management, 17*(5), 368–389.

IRI. (2015). *Private label & national brands: Dialing in on core shoppers*. Information Resources, Inc. [Online] Available from: ⟨https://www.foodinstitute.com/images/media/iri/TTJan2015.pdf⟩.

Jaenicke, E. C., & Carlson, A. C. (2015). Estimating and investigating organic premiums for retail-level food products. *Agribusiness, 31*(4), 453–471.

Jaenicke, E., Dimitri, C., & Oberholtzer, L. (2011). Retailer decisions about organic imports and organic private labels. *American Journal of Agricultural Economics, 93*(2), 597–603.

Jonas, A., & Roosen, J. (2005). Private labels for premium products—The example of organic food. *International Journal of Retail & Distribution Management, 33*(8), 636–653.

Kaplan, J., & Boyle, M. (2017). *Amazon cuts Whole Foods prices as much as 43% on first day*. Bloomberg.com. Available from: ⟨https://www.bloomberg.com/news/articles/2017-08-28/amazon-cuts-prices-at-whole-foods-as-much-as-50-on-first-day⟩ Accessed November 10, 2017.

Koschate-Fischer, N., Cramer, J., & Hoyer, W. D. (2014). Moderating effects of the relationship between private label share and store loyalty. *Journal of Marketing*, *78*(2), 69—82.

Kowitt, B. (2015). *Is the largest natural-foods brand even sold at Whole Foods?* Fortune. [Online] Available from: ⟨http://fortune.com/2015/10/28/kroger-natural-organic-food/⟩ Accessed: December 6, 2016.

Kumar, Nirmalya, & Steenkamp, Jan-Benedict E. M. (2007). *Private label strategy: How to meet the store brand challenge*. Boston: Harvard Business School Press.

Laaksonen, H., & Reynolds, J. (1994). Own brands in food retailing across Europe. *The Journal of Brand Management*, *2*(1), 37—47. Available from https://doi.org/10.1057/bm.1994.30.

Labajo, V. (2016). Premium and value-added private labels: The case of private labels in sustainable FMCG markets. In Gómez-Suárez M., & Martínez-Ruiz M.P. (Eds.), *Handbook of research on strategic retailing of private label products in a recovering economy* (pp. 307—332). IGI Global.

Lin, Biing-Hwan, Smith, Travis, & Huang, Chung (2008). Organic premiums of U.S. fresh produce. *Renewable Agriculture and Food Systems*, *23*(3), 208—216.

Market Force. (2016). *New market force information study finds Wegmans and Publix are America's favorite grocery retailers*. [Online] Available from: ⟨http://www.marketforce.com/wegmans-and-publix-are-america's-favorite-grocery-retailers-market-force-panel-research⟩ Accessed: March 26, 2017.

MarketLine. (2016). *Organic food in the United States*. MarketLine Industry Profile. https://store.marketline.com/report/ohmf0097-organic-food-in-the-united-states/.

Martinez, S. W. (May 2007). *The U.S. food marketing system: Recent developments, 1997—2006, economic research report no. 42*. U.S. Department of Agriculture, Economic Research Service, Washington, D.C.

Martos-Partal, M. (2010). Innovation and the market share of private labels. *Journal of Marketing Management*, *28*(5—6), 695—715.

Meijer. 2015. *Meijer expands assortment of real food products through launch of true goodness by Meijer brand*. PR Newswire. [Online] Available from: ⟨http://www.prnewswire.com/news-releases/meijer-expands-assortment-of-real-food-products-through-launch-of-true-goodness-by-meijer-brand-300143603.html⟩ Accessed: March 26, 2017.

Mintel. (2011). *Natural and organic food and beverage: The market—US*.

Mintel. (2013). *The private label food consumer*.

Mintel. (2015a). *Private label foods: What's driving purchase?—Factors considered when purchasing store brand food products*.

Mintel. (2015b). *Organic food and beverage shoppers—US*.

Mintel. (2016). *Private label food trends—US*.

Mintel. (2017). *Private label food and drink trends—US—February 2017*.

Nassauer, S. (2016). *Walmart to drop wild oats organic food brand*. [Online] Available from: ⟨http://www.wsj.com/articles/wal-mart-to-drop-wild-oats-organic-food-brand-1461605524⟩.

Nielsen. (2008). *Organics trend overview in CPG industry is the organic sales explosion over?* [Online] Available from: ⟨http://www.nielsen.com/content/dam/corporate/us/en/newswire/uploads/2008/10/organics-overview.pdf⟩ Accessed: April 6, 2017.

Nielsen. (2014). *The state of private label around the world 2014: Where it's growing, Where it's not and what the future holds*. [Online] Available from: ⟨http://www.nielsen.com/content/dam/nielsenglobal/kr/docs/global-report/2014/Nielsen%20Global%20Private%20Label%20Report%20November%202014.pdf⟩ Accessed December 6, 2016.

Nutrition Business Journal. (2004). NBJ's organic foods report 2004. Boulder, CO: New Hope Natural Media, Inc.

Organic Trade Association (OTA). (2016). *U.S. organic sales post new record of $43.3 billion in 2015.* [Online] Available from: ⟨https://www.ota.com/news/press-releases/19031#sthash.YxMUEanN.dpuf⟩ Accessed November 12, 2016.

Organic Trade Association (OTA). (2017a). *Robust organic sector stays on upward climb, posts new records in U.S. sales.* [Online] Available from: ⟨https://www.ota.com/news/press-releases/19681⟩ Accessed October 19, 2017.

Organic Trade Association (OTA). (2017b). *Organic standards.* [Online]Available from: ⟨https://www.ota.com/advocacy/organic-standards#sthash.0Dj7iiFT.dpuf⟩   Accessed November 12, 2016.

Organic Valley. (2013). *CROPP cooperative 2013 annual report.* LaFarge, WI: CROPP Cooperative.

Pauwels, Koen, & Srinivasan, Shuba (2004). Who benefits from store brand entry? *Marketing Science, 23*(3), 364−390.

Pew Research Center. (2016). *The new food fights: U.S. public divides over food science.* [Online] Available from: ⟨http://assets.pewresearch.org/wp-content/uploads/sites/14/2016/12/19170147/PS_2016.12.01_Food-Science_FINAL.pdf⟩ Accessed December 6, 2016.

Pivato, S., Misani, N., & Tencati, A. (2008). The impact of corporate social responsibility on consumer trust: the case of organic food. *Business Ethics: A European Review, 17*(1), 3−12.

Private Label Manufacturers Association. (2009). *Albertsons, Hy-Vee, Price Chopper offer Safeway brands.* [Online] Available from: ⟨http://plma.com/escanner/may2009.html⟩ Accessed December 6, 2016.

Private Label Manufacturers Association. (2016). PLMA's 2016 private label yearbook: A statistical guide to today's store brands.

Progressive Grocer. (2015). *Supervalu refreshes wild harvest free-from brand.* [Online] Available from: ⟨http://www.progressivegrocer.com/industry-news-trends/wholesalers-distributers/supervalu-refreshes-wild-harvest-free-brand⟩ Accessed December 6, 2016.

PWC. (2011). *The private labels revolution.* [Online] Available from: ⟨https://www.pwc.ru/ru/retail-consumer/assets/private-labels-eng-may2011.pdf⟩ Accessed April 10, 2017.

Sanchez, M. (2014). *Food law and regulation for non-lawyers: A US perspective.* Springer, Cham.

Seenivasan, S., Sudhir, K., & Talukdar, D. (2015). Do store brands aid store loyalty? *Management Science, 62*(3), 802−816.

Semeijn, J., Van Riel, A. C., & Ambrosini, A. B. (2004). Consumer evaluations of store brands: Effects of store image and product attributes. *Journal of Retailing and Consumer Services, 11*(4), 247−258.

Sethuraman, R., & Gielens, K. (2014). Determinants of store brand share. *Journal of Retailing, 90*(2), 141−153.

Sriram, S., Balachander, S., & Kalwani, M. U. (2007). Monitoring the dynamics of brand equity using store-level data. *Journal of Marketing, 71*, 61−67.

Smith, T. A., Huang, C. L., & Lin, B. H. (2009). Estimating organic premiums in the US fluid milk market. *Renewable Agriculture and Food systems, 24*(3), 197−204.

Stanton, J. L., Wiley, J., Hooker, N. H., & Salnikova, E. (2015). Relationship of product claims between private label and national brands: The influence of private label penetration. *International Journal of Retail and Distribution Management, 43*(9), 815−830.

Steenkamp, J. E. M., & Gielens, K. (2003). Consumer and market drivers of the trial probability of new consumer packaged goods. *Journal of Consumer Research, 30*(3), 368–384.

Steiner, R. L. (2004). The nature and benefits of national brand/private label competition. *Review of Industrial Organization, 24*(2), 105–127.

Store Brands. 2016. *Costco to offer more organic products in 2016.* [Online] Available from: ⟨http://www.storebrands.com/store-brand-insights/store-brand-news/costco-offer-more-organic-products-2016⟩ Accessed December 6, 2016.

Su, Y., & Cook, M. L. (2015). Price stability and economic sustainability—Achievable goals? A case study of organic valley. *American Journal of Agricultural Economics, 97* (2), 635–651.

SuperValu. (2008). *Supervalu launches new wild harvest brand of organic, natural foods nationwide.* [Online] Available from: ⟨http://www.supervaluinvestors.com/phoenix. zhtml?c = 93272&p = irol-newsArticle&ID = 1127452⟩ Accessed December 6, 2016.

Talevich. (2011). *The chicken, the egg and future—seeking sustainable supplies for a growing planet.* The Costco Connection (August 2011). [Online] Available from: ⟨https:// www.costco.com/sustainability-kirkland-signature.html⟩ Accessed December 6, 2016.

Target. (2017). Simply balanced—Frequently asked questions. [Online]Available from: ⟨http://static.targetimg1.com/2017/simplybalanced/pdfs/2017_SB_FAQ.pdf⟩ Accessed April 10, 2017.

Trader Joe's. (2017). *Does Trader Joe's offer organic products?* [Online] Available from: ⟨http://www.traderjoes.com/faqs/product-information⟩ Accessed April 10, 2017.

Tu, J. (2016). Kroger tries out new, green supermarket in Gig Harbor. *Seattle Times.* [Online] Available from: http://www.seattletimes.com/business/retail/cozy-green-store-tested-here-by-nations-grocery-giant/. [Accessed: April 10, 2017].

Tuck Communications. (2010). *Private-label products in the manufacturer-retailer power balance.* [Online] Available from: ⟨http://www.tuck.dartmouth.edu/news/articles/private-label-products-in-the-manufacturer-retailer-power-balance⟩ Accessed April 10, 2017.

United States Department of Agriculture (USDA). (2017). *Welcome to the National Organic Program.* [Online]Available from: ⟨http://www.ams.usda.gov/AMSv1.0/nop⟩ Accessed April 10, 2017.

Volchok, P. (2012). *Buying smart: Kirkland signature organic beef strict safety controls, higher-quality meat and Costco value.* The Costco Connection, May 2012.

Walmart. (2014). *Walmart and wild oats launch effort to drive down organic food prices.* [Online] Available from: ⟨http://corporate.walmart.com/_news_/news-archive/2014/04/ 10/walmart-and-wild-oats-launch-effort-to-drive-down-organic-food-prices⟩ Accessed April 10, 2017.

Wattles, J. (2017). *Amazon: We're lowering Whole Foods prices on Monday.* [Online] Available from: ⟨http://money.cnn.com/2017/08/24/news/companies/amazon-whole-foods/index.html⟩.

Whole Foods Market. (2002). *Whole Foods market celebrates new U.S. organic standards.* [Online]Available from: ⟨http://media.wholefoodsmarket.com/news/whole-foods-market-celebrates-new-u.s.-organic-standards⟩ Accessed April 10, 2017.

Wilkie, W. L., Desrochers, D. M., & Gundlach, G. T. (2002). Marketing research and public policy: The case of slotting fees. *Journal of Public Policy & Marketing, 21*(2), 275–288.

Winter. (2010). *Safeway's O Organics now outsells other organic brands.* New Hope Network. [Online] Available from: http://www.newhope.com/supply-news-amp-analysis/safeway-s-o-organics-now-outsells-other-organic-brands. [Accessed: April 10, 2017].

Xu, F. (2010). BioFach China 2009 and updates on organic market in Shanghai. *USDA foreign agricultural service GAIN report number: CH0804.*

Appendices

Appendix 1   USDA organic labeling standards

To help US consumers better understand the organic content of the food they purchase, the NOP has developed labeling standards, along with production and processing requirements for products to be considered organic. Labeling requirements apply to raw, fresh, and processed products that contain organic agricultural ingredients. For packaged products, organic labeling standards cover the wording allowed on both the principal display panel and the information panel. Four distinct labeling categories are used to classify organic food products by the percentage of their certified organic content (excluding salt and water):

Category	Certified organic ingredient content (%)	Product label	
		Principal display panel	**Information panel**
100% Organic	100	May include USDA organic seal and/or organic claim	Must list organic ingredients (e.g., organic whey) or indicate via an asterisk or other mark. Label must state the name of the certifying agent
Organic	95–99.9		
Made with organic__	70–94.9		
Organic ingredients	<70	Must not include USDA organic seal anywhere	May only list certified organic ingredients as organic in the ingredient list and the percentage of organic ingredients

Only products in the top two tiers, those containing a minimum of 95% organic content, can be labeled "organic" and/or bear the USDA Organic Seal. The remaining ingredients may be nonorganic agricultural products that are not commercially available in an organic form and/or nonagricultural products that are on the *National List of Allowed and Prohibited Substances* (the National List). Products that contain at least 70% certified organic ingredients may specify that they are "Made with Organic __ (insert up to three ingredients or ingredient categories)." This category of products may not include the USDA organic seal on the product or the word "organic" on the principal display panel. General statements like, "Made with organic ingredients" are also not allowed. Organic products falling within these first three categories must also state the name of their organic certifying agent on the information panel. Products that contain less than 70% organic ingredients may only list the certified organic ingredients as organic in the ingredient list and can note the percentage of organic ingredients.

Source: Federal Regulations Subpart D on Labels, Labeling and Market Information for the National Organic Program, 7 C.F.R. §205 2017. Additional information about US organic labeling requirements is available at www.ams.usda.gov/NOPOrganicLabeling.

Appendix 2 Retailers expand the availability of organic products by introducing private labels

Major retailers, including mass merchandisers such as Walmart and Target, are expanding or adding organic private label brands, making these products even more accessible to consumers. Increased availability of organic foods is predicted to continue to fuel organic food sales and consumption (Pew Research Center, 2016). In contrast to standard private label food products which can be perceived as inferior (Barsky, Bergen, Dutta, & Levy, 2007), private label organic product lines are often marketed as premium (quality or good-for-you/better-for-you) but affordable brands. These brands typically include products from the most commonly purchased food and grocery categories such as produce, dairy, bread, and breakfast cereals. Major US retailer organic private brand releases are highlighted in chronological order below.

In 1977, nearly three decades before "Organic" became a mainstream movement, Trader Joe's introduced its first private label organic item—Organic Unfiltered Apple Juice. Its organic offerings now include milk, yogurt, apples, lettuce, cereal, meat, almonds, cashews, extra virgin olive oil, beans, frozen pizza, chocolate, bread, cheese, pasta, and wine. Trader Joe's sells approximately four times more organic products than a typical grocery store (Trader Joe's, 2017).

In 2000 Topco Associates LLC, the largest food buying cooperative in the United States, introduced its natural and organic products, Full Circle. This brand was recently renamed Full Circle Market and now offers products in categories of produce, meat & seafood, dairy, grocery, beverages, frozen foods, and home care. Topco is a vertically integrated organization which produces and distributes foods to more than 50 members, including Wegman's and Meijer (Howard, 2016).

In 2002, Whole Foods launched its 365 Organic Everyday Value brand, the first national all-organic product line in the United States. In June 2003, Whole Foods Market became America's first national "Certified Organic" grocer. Among all private label natural and organic products, this brand has led new product introduction trends during the period of 2007−11 (Mintel, 2011). In total, the company offers more than 2,600 natural and organic products under its store brands: Whole Foods Market, 365 Everyday Value, and Whole Catch (MarketLine, 2016). In August 2017, the e-commerce giant Amazon acquired Whole Foods Market for $13.7 billion (Kaplan & Boyle, 2017).

In 2003, Wegmans, a chain based in New York which has stores in the mid-Atlantic and New England regions, introduced its first PL organic products under the brand name Wegmans Organic. Now more than 3,000 items are sold under the brand.

In 2004, Costco introduced its first organic item, milk, under its flagship private label line, Kirkland Signature. Now, Kirkland Signature organic products include beef, coconut oil, egg, granola cereal (by Nature's Path), honey, lemonade, nuts, quinoa, salsa, seaweed, soy milk, sugar, and pet food, among others. In order to maintain access to reliable supplies to meet a growing demand for organic products, Costco has established several programs with its suppliers. For instance, thanks to the Kirkland Signature organic ground beef program, Costco is the largest seller of organic ground beef in the United States (Volchok, 2012). Costco has supported several egg producers in converting conventional operations to organic through its regional organic egg programs (Talevich, 2011). Currently, 14% of Costco's produce, 9% of its meat items, and 20% of all its other grocery items are organic (Store Brands, 2016). Costco's annual sales of just organic products reached about $4 billion in 2014, eclipsing Whole Foods for the title of the biggest organic grocer in the United States (Kowitt, 2015).

In 2005, Safeway launched the O Organics line, which now contains over 300 products across various categories, including produce, meat, eggs, dairy, coffee, snacks, and baby food. The brand has also been carried by other retailers such as Albertsons and some regional grocers by joining the Better Living Brands Alliance formed by Safeway and sold overseas (Africa, Asia, and South America) (Private Label Manufacturers Association, 2009; Xu, 2010), which is an unusual approach for private-label goods. O Organics is among the most successful and widely known PL food brands and has strengthened Safeway's image as a leading player in the natural and organic food market (Winter, 2010).

In 2007, Meijer introduced the Meijer Organics line. In 2009, this Midwest grocer unveiled another health-focused brand, Meijer Naturals, which serves to complement the Organics line and allows Meijer to offer a wider variety of healthier alternative foods. In an effort to minimize confusion, Meijer combined the two in-house brands under the new brand True Goodness in September 2015 (Meijer, 2015).

In 2008, SuperValu, the fifth-largest food retailing company in the United States, launched a natural and organic foods brand, Wild Harvest. The initial debut featured approximately 150 items ranging from mealtime staples like milk, eggs, meat, and fresh produce to pastas and sauces, cookies, crackers, cereal, and juice. New products were available at its 10 grocery store chains: Acme, Albertsons, Bigg's, Cub Foods, Farm Fresh, Hornbacher's, Jewel-Osco, Lucky, Shaw's/Star Market, Shop 'n Save, and Shoppers Food & Pharmacy (SuperValu, 2008). In early 2015, Supervalu "refreshed" the Wild Harvest brand, with a new logo, slogan, and a "free-from" product label, as well as plans for an additional 200 new items to be added to the brand's 300 products (Progressive Grocer, 2015).

In September 2012 Kroger, the second-largest general retailer (behind Walmart; Groner, 2016), launched its Simple Truth lines of natural and organic foods. Simple Truth and Simple Truth Organic offer over 800 unique items in over 90 categories. The lines have been experiencing fast growth and exceeded $1.5 billion in annual sales in 2015. Now it is the nation's largest natural and organic brand (Kowitt, 2015).

In June 2013, Target introduced its organic and natural food line called Simply Balanced, which features over 350 products in snacks beverages, pasta, frozen fruit and vegetables, frozen seafood, dairy, and cereal categories. All products in this line are made without artificial flavors, colors, and preservatives, and many are sourced using fair trading practices from producers certified to social or environmental certifications (Target, 2017).

In January 2014, Aldi, recognized as the nation's low-price grocery leader (Market Force, 2016), launched a new brand called SimplyNature that offers products with only all-natural or organic ingredients. SimplyNature products include cereal, honey, fruit bars, pasta sauce, snacks, and apple juice. The company hopes this will improve its image of offering fresh, natural, quality products. In early 2016, Aldi announced that it was expanding its organic food brands, removing some artificial ingredients from products, and adding more gluten-free items. Additionally, it plans to remove multiple pesticides from all of its products. These changes will likely help Aldi continue to establish itself as a chain that offers incredible value and makes food and drink more affordable for everyone (Mintel, 2017).

In March 2014, the San Antonio-based grocery chain H-E-B created an organic line, H-E-B Organics. From farm fresh produce to organically raised meat, breakfast foods to salty snacks, there are about 318 products available under H-E-B Organics. The retailer also plans to add hundreds of more products to the line in the near future.

In April 2014, Walmart partnered with Wild Oats Markets to introduced nearly 100 organic products into their US stores (Walmart, 2014). However, two years later, the retail giant announced in April 2016 that it is phasing out Wild Oats products, as the line fell short of the retailer's expectations (Nassauer, 2016). In 2016, Walmart introduced 52 new organic products under the store's existing Great Value brand, with sauces & seasonings (29), dairy products (8), and fruit & vegetables (5) as the top three categories (GNPD, 2016).

Food retailing: Malaysian retailers' perception of and attitude toward organic certification

11

Muhammad Azman Ibrahim, C. Michael Hall and Paul W. Ballantine
Faculty of Commerce and Communication, Center for University of Wollongong Programs, INTI International College Subang, Malaysia

11.1 Introduction

Certification is considered to be a voluntary assurance quality scheme that is approved by a recognized accredited body (Albersmeier, Schulze, & Spiller, 2009). The purpose of standards and certification of food products is to demonstrate quality to, and obtain the trust of, consumers with whom producers do not have a direct relationship (Higgins, Dibden, & Cocklin, 2008). Aspects of the quality of food products or commodities that are sometimes regulated and referred to in certification schemes include attributes such as safety, nutritional content, labeling, production processes, and/or branding (Busch & Bain, 2004; Doherty & Campbell, 2014; Watts & Goodman, 1997). However, academic literature on organic and food certification in food retail is somewhat limited. Studies have examined the perception of producers and consumers on food certification, and in particular, determinants such as sociodemographic characteristics and willingness to pay (WTP) for food safety and quality (Probst, Houedjofonon, Ayerakwa, & Haas, 2012; Uggioni & Salay, 2014); consumers' awareness, trust, purchasing decision and WTP (Essoussi & Zahaf, 2009; Gerrard, Janssen, Smith, Hamm, & Padel, 2013); and environmental and animal welfare (Nasir & Chiew, 2010).

According to Anders, Souza-Monteiro, and Rouviere (2010), information asymmetries and uncertainty of product safety and quality are increasing in the global food retail sector. Information asymmetries occur when the processing of food products cannot be verified by the retailers or consumers of, for example, organic products. Such products are considered to be credence products (Darby & Karni, 1973; Roe & Sheldon, 2007; Voon, Sing, & Agrawal, 2011). The credibility of food certification is important to reduce food product uncertainties and the overall cost of information asymmetries between producers and retailers in the food supply chain (Anders et al., 2010; Caswell, 1998; Deaton, 2004; Manning & Baines, 2004). The credibility of food certification is related to consumers' trust in the coordination of food supply chains, which have become such a crucial element of modern international food markets (Albersmeier et al., 2009).

Case Studies in Food Retailing and Distribution. DOI: https://doi.org/10.1016/B978-0-08-102037-1.00011-6
Copyright © 2019 Elsevier Ltd. All rights reserved.

Safety and the characteristics of food product processes are becoming increasingly important in the operation of food systems (Caswell, 1998; Havinga, 2013). Quality assurance schemes are useful in food retailing as an important product and marketing attribute that offers a great opportunity to differentiate food retailers in the market and add value to their products (Botonaki, Polymeros, Tsakiridou, & Mattas, 2006; Jervell & Borgen, 2004). Hence, implementation of food certification by retailers influences consumer behavior related to food quality control and the safety of food products (Hatanaka, Bain, & Busch, 2005; Havinga, 2013).

11.2   Malaysian organic and food retailing

Food retailing in Malaysia has developed rapidly with new retail concepts emerging and competing with traditional retail formats (Chamhuri & Batt, 2013; Hassan, Sade, & Rahman, 2013; Mohd Roslin & Melewar, 2008). Chamhuri and Batt (2013) found that the development of the food retail industry has changed, together with Malaysian consumer behavior, because of several factors including personal disposable income, convenience need, high awareness of food safety and food quality, and changes in dietary habits. Mohd Roslin and Melewar (2008) agreed that retailers have to reconsider operational strategies in order to influence changes in consumer behavior patterns and increase consumption.

The Malaysian government is responsible for regulations on all food, drink, and ingredients that are locally manufactured or imported into Malaysia under the *Food Act 1983* and the *Food Regulations Act 1985*. These regulations are to ensure that food and drink are protected from any illegal ingredients that can harm people's health or safety. These regulations are implemented by the Food Safety and Quality Division of the Ministry of Health. Organic products must have obtained organic certification in order to carry the government-approved logo *Skim Organik Malaysia* (SOM) and display it on packaging (Department of Agriculture Malaysia, 2007; Stanton & Emms, 2011).

Malaysian consumers appear increasingly aware and educated about organic foods, particularly in the context of potential contributions to sustainability and wellbeing (Euromonitor, 2013). The presence of more organic specialist retail stores, as well as more space allocation to organic food products in leading hypermarkets and supermarkets, has increased consumers' awareness of organic food products (Euromonitor, 2013). According to Terano, Yahya, Mohamed, and Saimin (2014), the development of modern retail formats in Malaysia, such as hypermarkets and supermarkets, is becoming increasingly sophisticated in providing better services and products, including the introduction of organic food products. Although Malaysian consumers are increasingly aware of organic products, previous studies found that price is a major barrier in purchasing intention toward such products (Azam, Othman, Musa, AbdulFatah, & Awal, 2012; Kai et al., 2013). Other studies noted that consumers with high incomes and preferences toward the perceived benefits of organic products are likely to have the highest intention to purchase (Rezai, Teng, Mohamed, & Shamsudin, 2012; Voon et al., 2011).

There are a growing number of quality guarantee schemes at national and international levels that offer higher food welfare and food quality. For example, the Soil Association Certification, which is the biggest umbrella organization for organic farming in the United Kingdom, provides the most common logo that can be found on British organic products (Baker, Thompson, Engelken, & Huntley, 2004; Gerrard et al., 2013; Janssen & Hamm, 2012). However, in the global context, the increasing number of organic brands, certification labels, and organic stores, among other features, does not appear to have increased consumers' trust in organic products (Hamzaoui-Essoussi, Sirieix, & Zahaf, 2013). Several studies have found that international consumers are not convinced about purchasing more organic food because of skepticism and uncertainty toward organic logos and certification schemes (Aarset et al., 2004; Aertsens, Verbeke, Mondelaers, & Huylenbroeck, 2009; Hughner, Mcdonagh, Prothero, Shultz, & Stanton, 2007; Janssen & Hamm, 2012; Lea & Worsley, 2005; Padel, Röcklinsberg, & Schmid, 2009).

These views are consistent with those of Organic Monitor (2006) with respect to consumer knowledge of organic labeling in Asia, which indicated that the number of organic products imported from around the world and the accompanying plethora of organic logos was leading to confusion among Asian consumers. Dardak, Zairy, Abidin, and Ali (2009) revealed that more than 40% of the respondents in their survey did not recognize the Malaysian Organic Certification, and more than 60% had never heard about it, especially those who were from outside Kuala Lumpur, the capital. Stanton and Emms (2011) found that most Malaysian consumers tend to be confused between certified and noncertified organic food products.

However, despite research on consumer perceptions and behavior in relation to organic and food certification, the literature with respect to retailers' perceptions is extremely limited in developing countries and markets, including Malaysia. Essoussi and Zahaf (2008) emphasized that distribution, certification, and labeling are all related to consumers' confidence when consuming organic food products, reflecting the fact that consumers are concerned over trusting the certification process. However, it is inadequate to focus on the wariness of consumers over guarantees of product quality/knowledge, labeling, certification or pricing, and communication strategies. Instead, organic certification should also be observed from the supply side and retailers clearly contribute at various scales and with diverse approaches to consumers' level of knowledge of, preferences for, and trust in, organic products (Hamzaoui-Essoussi et al., 2013). Therefore, the absence of research on retailers and organic products appears to be a significant gap in knowledge of organic certification in the food system.

11.3 Hypotheses development

Three constructs with respect to organic certification have been identified and included in a series of hypotheses developments.

Product attributes

H1. Organic certification of product attributes has a positive relationship with the importance of organic certification amongst Malaysian food retailers.

H2. Organic certification of product attributes has a positive relationship with organic issue
(lack of knowledge, lack of trust, and misuse of organic certification).

Sustainability attributes

H3. Organic certification of sustainability attributes of products has a positive relationship
with importance of organic certification amongst Malaysian food retailers.

H4. Organic certification of sustainability attributes of products has a positive relationship
with organic issue (lack of knowledge, lack of trust, and misuse of organic
certification).

Organic certification issues

H5. Organic issue (lack of knowledge, lack of trust, and misuse of organic certification) has
a positive relationship with importance of organic certification in food retailing.

H6. The organic certification issue (OCI) (lack of knowledge, lack of trust, and misuse of
organic certification) mediates the relationship between the organic certification of
product attributes and the importance of organic certification in Malaysian food
retailing.

H7. The OCI (lack of knowledge, lack of trust, and misuse of organic certification) mediates
the relationship between the organic certification of sustainability attributes and the
importance of organic certification in Malaysian food retailing.

11.4   Research method

There are several major differences between the current research and previous studies on food certification in Malaysia. First, previous research that examined food
certification focused on consumers instead of retailers. For example, Dardak et al.
(2009) and Ahmad and Juhdi (2010) selected Malaysian consumers who went to
supermarkets as their sampling unit. Second, food retailers in Malaysia are considered as having a stronger presence than in other Asian markets, with an increasing
number of retail outlets such as supermarkets, hypermarkets, and convenience
stores. Running in parallel with the increase in retail outlets, it will be interesting to
know how significant food and organic certification is as a retail attribute in
Malaysian retailing. This is particularly because the number of consumers of
organic foods has significantly increased due to the presence of more organic specialist stores, as well as more shelves containing organic food in leading supermarkets and hypermarkets (Euromonitor, 2013). Third, food certification is significant
to retailers in developed and developing countries as they strengthen their structural
power in the food chain. However, previous research on food certification in

Malaysia did not cover organic certification in detail, and no research has focused specifically on Malaysian retailers' perceptions of organic certification.

A survey strategy was deployed when collecting the data in order to have a clear picture of retailers' perceptions toward organic certification. This approach allows the collection of large amounts of data from a sizeable population in a highly convenient way (Saunders, Lewis, & Thornhill, 2009) and can be carried out to collect information regarding people's knowledge, expectations, and behavior (Neuman, 2003). Therefore, this research utilized a questionnaire as the survey instrument for gathering the data.

The survey was designed to examine perceptions toward organic certification as well as other forms of food certification in the context of Malaysia. This survey used seven-point Likert scales, anchored from strongly agree to strongly disagree, in order to measure all items in the questionnaire. Using seven-point Likert scales is a suitable and most frequently used means to measure attitudes and behaviors in organizational research (Sekaran, 2006). A draft version of the questionnaire survey was also pretested with a small number of retailers before it was distributed to retailers more widely. The pretest is important to ensure the respondents understand the content of a questionnaire when answering it (Sekaran, 2006). Indeed, it helps to rectify or minimize any inadequacies or biases before administering the questionnaire.

The total population of supermarkets and hypermarkets in Malaysia is 1335 and 174 outlets, respectively (Euromonitor, 2015). The survey was conducted among modern food retailing formats (hypermarkets, supermarkets, and organic specialty stores), irrespective of whether these retailers sell organic products or not. This is in order to obtain perceptions from a range of food retailers. Due to the premium price of organic food products, it is difficult to find organic food products in any small convenience outlets. The questionnaire was distributed by mail as this is suitable for large sample sizes (Saunders et al., 2009; Sekaran, 2006). In the previous study by Ali and Suleiman (2016), which focused on small and medium enterprises regarding halal food production practices, the response rate using mail was 17%.

The distribution of the questionnaire survey was focused on food retailers located in big cities and urban areas of Malaysia; specifically the two areas of Kuala Lumpur and Selangor. Based on the list provided by the Companies Commission of Malaysia and Organic Alliance Malaysia's directory, there are 432 organic food retailers, and thus, a census approach was adopted in this study as the questionnaire was distributed by mail to all these retailers. Many food retail stores are located in the major cities, urban centers and larger towns in Malaysia. These are areas with high concentrations of middle and high income households (Euromonitor, 2014; FAS Kuala Lumpur, 2013). The questionnaire took approximately 15 minutes for respondents to complete.

Data were collected over a 4-month period from November to February 2016. A total of 432 questionnaires were delivered and administered in the two study locations. Out of the 432 questionnaires, 106 retailers responded, which is a 25% response rate. After screening of the data, four of the questionnaires were excluded due to most of the questions not being filled in. Therefore, 102 questionnaires were used for the analysis. Three types of retail format were represented in this research; supermarkets, hypermarkets, and specialty organic specialty stores. A retail store with range from 80,000 to 220,000 square feet is classed as a hypermarket, while a supermarket is usually at 20,000 square feet. An organic specialty store is a store that offers consumers a particular type of merchandize (Terano et al., 2014).

11.5   Descriptive statistics

The profile of the businesses that participated in this study was as follows: 45 (44.1%) were supermarkets followed by specialty stores (38−37.3%) and hypermarkets (19−18.6%). Given that hypermarkets accounted for less than 20% of responses, it is not surprising that the majority of retailers were small and medium size stores. Almost two-thirds of responses came from Selangor (65−63.7%) with the remainder from Kuala Lumpur (37−36.3%). Given that responses were dominated by organic specialty stores which are often relatively small in size, retailers with less than 10 employees contributed almost 60% of responses, whilst retail outlets with more than 20 employees accounted for over a quarter of respondents. Retailers that have an annual turnover of less than US$0.12 million accounted for almost 30% of the respondents. Only 12.7% of retailers had an annual turnover of more than US$0.70 million. Nearly 30% of all the retail stores surveyed had been in business for over 10 years, with 30.4% (31) established for between 6 and 10 years. This was closely followed by 29.4% of retailers that had been set up between 3 and 5 years ago. Fewer than 10% of retailers had been established for less than 3 years.

11.6   Findings

11.6.1   PLS-SEM analysis

Partial Least Squares Structural Equation Modeling (PLS-SEM) was used to analyze questions on retailers' perceptions of and attitudes toward organic certification. Many business disciplines, as well as marketing, apply PLS-SEM, and it has become a widely recognized method in recent years (Diamantopoulos, Sarstedt, Fuchs, Wilczynski, & Kaiser, 2012; Sarstedt, Ringle, & Hair, 2014a). The primary purpose of PLS-SEM in business research is to test concepts and theories (Hair, Ringle, & Sarstedt, 2011; Sarstedt, Ringle, Smith, Reams, & Hair, 2014b). PLS-SEM is a suitable technique when analyzing non-normally distributed data (Falk & Miller, 1992). Using Likert scales to measure individual perceptions will likely yield non-normally distributed responses (Aibinu & Al-Lawati, 2010). Therefore, since the survey uses Likert scales, PLS-SEM is an appropriate technique to analyze retailers' perception of and attitudes toward organic certification.

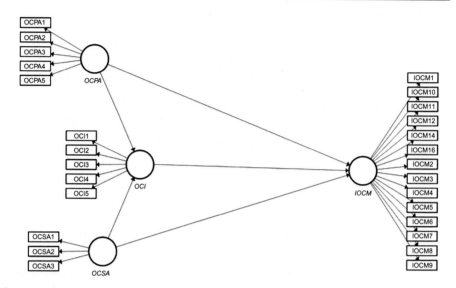

Figure 11.1 Conceptual framework of Malaysian retailers' perception of and attitude toward organic certification.

Based on the limited literature available on the organic certification process in Malaysia and elsewhere, three main themes with respect to organic certification have been identified in terms of their importance for organic certification in Malaysian food retailing: namely, product attributes, sustainability attributes, and certification issues. The conceptual framework depicted in Fig. 11.1 consists of four constructs; Organic Certification on Product Attributes (OCPA) (organic certification determines organic food product attributes based on safety, quality, taste, and appearance); Organic Certification on Sustainability Attributes (OCSA) (organic certification identifies organic food products as an important aspect of sustainability strategies with respect to environmental practices); OCI (issues that affect the credibility of organic certification such as lack of knowledge, lack of trust, and misuse of organic certification and labeling); and Importance of Organic Certification in Malaysia (IOCM) (the credibility of organic certification as a quality assurance scheme in Malaysian food retailing). Hair et al. (2011) explained that typically there is a two-step-process of PLS-SEM assessment that needs to be followed. This involves the separate assessment of the measurement models and the structural model.

11.6.2 Internal consistency reliability

Table 11.1 presents the composite reliability (CR) values. The results indicate that the measurement model has satisfactory internal consistency reliability as the CR of each construct exceeds the recommended level threshold value of 0.70. CR values are as follows: OPCA—0.919, OCSA—0.930, OCI—0.890, and IOCM—0.945.

Table 11.1 **Composite reliability**

Latent construct	Composite reliability
OCPA	0.919
OCSA	0.930
OCI	0.890
IOCM	0.945

OCPA, organic certification on product attributes; *OCSA*, organic certification on sustainability attributes; *OCI*, organic certification on issue; *IOCM*, important of organic certification in Malaysia.

11.6.3 Indicator reliability

Table 11.2 shows the indicator's outer loadings of the measurement model. There are 25 indicator outer loadings above 0.70, and two indicator values of 0.40−0.70. Five indicators were removed from the scale in order to increase the average variance extracted (AVE) above the suggested threshold value.

11.6.4 Convergent validity

All constructs for this study have AVE ranging from 0.552 to 0.817, exceeding the requirement threshold value of 0.50 (Table 11.3). Therefore, the results indicate that the measurement model in this study has adequate convergent validity.

11.6.5 Discriminant validity

Cross loadings

Cross loadings were examined, as the indicator outer loadings for each construct were greater than all of its loadings on other constructs (Table 11.4). Indeed, this method tends to be flexible when establishing discriminant validity (Hair et al., 2011). Therefore, the result of cross loading indicates that all the 27 measurement items loaded distinctly on the specified constructs as establishing the discriminant validity of the four constructs.

Fornell—Larcker criterion

Table 11.5 demonstrates that all constructs in the measurement model have met the requirement criteria, whereby the square root of AVE values is greater than the highest correlation with any other constructs.

11.7 Evaluation of the structural model

11.7.1 Collinearity assessment

Hair et al. (2011) suggested that a tolerance value of 0.20 or a variance inflation factor value of 5 and higher, respectively, indicates that the constructs probably

Table 11.2 **Outer loadings**

	IOCM	OCI	OCPA	OCSA
IOCM1	0.756			
IOCM10	0.769			
IOCM11	0.763			
IOCM12	0.766			
IOCM14	0.609			
IOCM16	0.578			
IOCM2	0.738			
IOCM3	0.750			
IOCM4	0.721			
IOCM5	0.803			
IOCM6	0.730			
IOCM7	0.838			
IOCM8	0.793			
IOCM9	0.750			
OC15		0.737		
OCI1		0.792		
OCI2		0.772		
OCI3		0.797		
OCI4		0.834		
OCPA1			0.837	
OCPA2			0.882	
OCPA3			0.921	
OCPA4			0.798	
OCPA5			0.723	
OCSA1				0.886
OCSA2				0.913
OCSA3				0.912

IOCM, important of organic certification in Malaysia; *OCI*, organic certification on issue; *OCPA*, organic certification on product attributes; *OCSA*, organic certification on sustainability attributes.

Table 11.3 **Average variance extracted**

Latent construct	AVE
OCPA	0.697
OCSA	0.817
OCI	0.619
IOCM	0.552

AVE, average variance extracted; *OCPA*, organic certification on product attributes; *OCSA*, organic certification on sustainability attributes; *OCI*, organic certification on issue; *IOCM*, important of organic certification in Malaysia.

Table 11.4 Cross loadings

	IOCM	OCI	OCPA	OCSA
IOCM1	**0.756**	0.267	0.609	0.616
IOCM10	**0.769**	0.601	0.519	0.545
IOCM11	**0.763**	0.594	0.482	0.532
IOCM12	**0.766**	0.525	0.550	0.495
IOCM14	**0.609**	0.432	0.411	0.352
IOCM16	**0.578**	0.458	0.362	0.414
IOCM2	**0.738**	0.164	0.647	0.523
IOCM3	**0.750**	0.280	0.613	0.526
IOCM4	**0.721**	0.207	0.553	0.604
IOCM5	**0.803**	0.348	0.572	0.593
IOCM6	**0.730**	0.097	0.602	0.591
IOCM7	**0.838**	0.470	0.524	0.568
IOCM8	**0.793**	0.357	0.515	0.545
IOCM9	**0.750**	0.393	0.503	0.574
OCI1	0.450	**0.792**	0.205	0.314
OCI2	0.257	**0.772**	0.117	0.163
OCI3	0.419	**0.797**	0.164	0.232
OCI4	0.411	**0.834**	0.235	0.342
OCI5	0.386	**0.737**	0.146	0.312
OCPA1	0.593	0.199	**0.837**	0.488
OCPA2	0.611	0.176	**0.882**	0.559
OCPA3	0.686	0.214	**0.921**	0.586
OCPA4	0.534	0.067	**0.798**	0.525
OCPA5	0.560	0.278	**0.723**	0.687
OCSA1	0.702	0.366	0.727	**0.886**
OCSA2	0.613	0.306	0.541	**0.913**
OCSA3	0.634	0.299	0.563	**0.912**

The results marked in bold indicate where the highest value is expected.
IOCM, important of organic certification in Malaysia; OCI, organic certification on issue; OCPA, organic certification on product attributes; OCSA, organic certification on sustainability attributes.

Table 11.5 Fornell–Lacker criterion

	IOCM	OCI	OCPA	OCSA
IOCM	0.743			
OCI	0.503	0.787		
OCPA	0.719	0.228	0.835	
OCSA	0.722	0.360	0.682	0.904

IOCM, important of organic certification in Malaysia; OCI, organic certification on issue; OCPA, organic certification on product attributes; OCSA, organic certification on sustainability attributes.

Table 11.6 **Collinearity assessment**

	IOCM	OCI
IOCM		
OCI	1.150	
OCPA	1.872	1.871
OCSA	2.039	1.871

IOCM, important of organic certification in Malaysia; *OCI*, organic certification on issue; *OCPA*, organic certification on product attributes; *OCSA*, organic certification on sustainability attributes.

Table 11.7 **Significance testing result of the structural model path coefficients**

	Path coefficients	t-Values	Significance levels	Results
OCPA → IOCM	0.434	4.888	***	0.000
OCSA → IOCM	0.322	3.610	***	0.000
OCI → IOCM	0.288	3.809	***	0.000
OCSA → OCI	0.383	2.223	**	0.026
OCPA → OCI	−0.033	0.206	NS	0.837

*$P < 0.10$, **$P < 0.05$, ***$P < 0.01$.
OCPA, organic certification on product attributes; *IOCM*, important of organic certification in Malaysia; *OCSA*, organic certification on sustainability attributes; *OCI*, organic certification on issue; *NS*, not significant.

have collinearity issues. Table 11.6 shows the collinearity assessments and reveals there were no multicollinearity issues in the structural model.

11.7.2 Structural model path coefficients

Table 11.7 shows the results of the path coefficients, t-statistics, and significance level for all hypothesized paths. Moreover, this result was used to determine the acceptance or rejection of the proposed hypotheses and is presented in Table 11.8.

Table 11.7 reveals that the relationship of OCPA and IOCM is significant with $\beta = 0.434$ and t-value = 4.888. The path coefficients between OCSA and IOCM are significant with $\beta = 0.322$ and t-value = 3.610. The relationship of OCI and IOCM is significant with $\beta = 0.288$ and t-value = 3.809. Moreover, the OCSA and OCI relationship is also significant with $\beta = 0.383$ and t-value = 2.223. In contrast, the relationship of OCPA and OCI is not significant with $\beta = -0.033$ and t-value = 0.206. Therefore, H1−H4 are supported, whilst H5 is not supported (Table 11.8).

11.7.3 Coefficient of determination (R^2 value)

The coefficient of determination (R^2 value) measures the predictive accuracy of the model and is mostly used to evaluate the structural model. The R^2 values range from 0 to 1, with higher levels indicating higher levels of predictive accuracy. In

Table 11.8 **Hypotheses testing results**

	Hypotheses statement	Result
H1	Organic certification on product attributes has a positive relationship with importance of organic certification in food retails	Supported
H2	Organic certification on sustainability attributes has a positive relationship with importance of organic certification in food retails	Supported
H3	Organic issue has a positive relationship with importance of organic certification food retails	Supported
H4	Organic certification on sustainability attributes has a positive relationship with organic issue	Supported
H5	Organic certification on product attributes has a positive relationship with organic issue	Not supported

marketing research, R^2 values of 0.75, 0.50, or 0.25 for endogenous latent variables can, as a rough rule of thumb, be respectively described as substantial, moderate, or weak (Hair et al., 2011; Hair, Sarstedt, Hopkins, & Kuppelwieser, 2014; Sarstedt et al., 2014b), while in other disciplines, R^2 value of 0.20 is considered high. In addition, the rule of thumb for acceptable R^2 values is difficult to provide because it depends on the model complexity and the research discipline. An IOCM R^2 value of 0.68 is considered moderate and OCI R^2 value of 0.11 is considered weak.

11.7.4 Effect size f^2

The f^2 effect sizes which were interpreted by following the guidelines that the f^2 values of 0.02, 0.15, and 0.35 indicate an exogenous construct's small, medium, or large effect, respectively, on an endogenous construct (Hair, Hult, Ringle, & Sarstedt, 2014; Hair et al., 2011). The result implies that the effect of OCPA has the highest effect size on the importance of organic certification in food retailing ($f^2 = 0.30$), followed by OCSA and OCI with values of $f^2 = 0.14$ and $f^2 = 0.19$, respectively. The effect of OCSA has a small to medium effect on OCIs ($f^2 = 0.08$) followed by OCPA ($f^2 = -0.009$).

11.7.5 Blindfolding and predictive relevance q^2

Hair et al. (2014) suggested that the values of 0.02, 0.15, and 0.35 indicate small, medium, or large predictive relevance of exogenous constructs for a certain endogenous construct. Consequently, the predictive value of OCPA; OCSA; and OCI for the importance of organic certification in food retailing was small ($q^2 = 0.08$, $q^2 = 0.05$, and $q^2 = 0.06$, respectively). Meanwhile, the predictive value of OCPA and sustainability attributes had a small predictive impact on OCI ($q^2 = -0.006$ and $q^2 = 0.03$).

11.7.6 The mediating effects analysis

The direct effects of OCPA and OCSA to IOCM are significant; however, the indirect effects of OCPA to IOCM are not significant compared to OCSA. This means that some of the direct effect relationships of OCSA to IOCM are absorbed by the OCI mediator.

The strength of OCI mediates the relationship between OCSA and IOCM and needs to be determined by using the variance accounted for (VAF). When the outcomes of VAF are more than 80%, it indicates a full mediation. VAF that is more than 20% and less than 80% can be characterized as partial mediation. While, VAF less than 20% can conclude that no mediation exists (Hair et al., 2014). The VAF between OCSA and IOCM via OCI is 25%. This indicates that VAF is partially mediated since it is larger than 20% but smaller than 80%. Based on these results, H6 was not supported, whilst H7 was supported (Table 11.9).

The result of the structural model path coefficients reveals that the relationships of OCPA and IOCM, OCSA and IOCM, OCI and IOCM, and OCSA and OCI were all significant, whilst the relationship between OCPA and OCI was not significant. Moreover, hypotheses testing demonstrated that H1, H2, H3, H4, and H7 were supported, and H5 and H6 were rejected (Table 11.9).

Table 11.9 Summary of hypotheses

	Hypotheses statement	Result
H1	Organic certification on product attributes has a positive relationship with importance of organic certification in food retails	Supported
H2	Organic certification on sustainability attributes has a positive relationship with importance of organic certification in food retails	Supported
H3	Organic issue has a positive relationship with importance of organic certification food retails	Supported
H4	Organic certification on sustainability attributes has a positive relationship with organic issue	Supported
H5	Organic certification on product attributes has a positive relationship with organic issue	Not supported
H6	Organic certification issue mediates the relationship between organic certification on product attributes and the importance of organic certification on food retailing in Malaysia	Not supported
H7	Organic certification issue mediates the relationship between organic certification on sustainability attribute and the importance of organic certification on food retailing in Malaysia	Supported

11.8 Discussion

Although organic certification is important in order to ensure the safety and quality of organic food products, building trust in certification, including with the inspection bodies, is also important as it helps to prevent any fraud or misuse of certification and labeling for organic food products (Munteanu, 2015). Findings from the survey showed that Malaysian food retailers perceived organic certification as a signal of trust in the related local authority approval of local and international organic food products. As highlighted previously, organic certification is a mandatory requirement for ensuring food products can be marketed as organic in Malaysia. This situation also applies to imported products that need to comply with Malaysian standards. Supporting this approach, Janssen and Hamm (2014) argue that an effective certification and labeling system is required in order to have high levels of trust and develop confidence levels among consumers and food retailers. Indeed, the level of consumers' confidence and trust is especially related to distribution, certification, and labeling concerns (Essoussi & Zahaf, 2008).

The PLS-SEM results also confirmed that product attributes were significantly associated with the importance of organic certification in Malaysian food retailing. The relationship between product attributes and the IOCM is much greater than sustainability issues. This particular finding is important for food retailers as they perceived that organic certification is a major factor in determining organic food products' attributes that are more related to the production, safety, and quality. Due to credence attributes, organic certification is an instrument that can help to verify the status of organic food products.

The relationship between organic certification of product attributes and OCIs was not significant. The reason for these insignificant relationships was because the food retailers perceived that to claim the food products to be organic they have to be certified as such by recognized certification bodies. By doing this, food retailers can prevent any misuse and mislabeling by producers or manufacturers who want to sell organic food products without any valid certification. In addition, supporting organic food products that carry organic certification will ensure that organic food products have to go through the entire certification process in order to obtain valid certification. This is particularly important for local fresh products in terms of obtaining the SOM accreditation, because this accreditation is developed by the Malaysian Department of Agriculture (DOA) which ensures all organic food products follow the Malaysian organic regulations.

The findings of this research indicated that the importance of organic certification is also related to environmental attributes, which have previously been recognized as significant in broader Malaysian society (Sinnappan & Rahman, 2011). Participants believed that organic certification indicates that organic food products are helping to preserve the environment and that this is important for consumers that are concerned about such environmental matters. In the Asian context, this finding is supported by Sirieix, Kledal, and Sulitang (2011) who similarly found that Chinese consumers perceived that organic food products make agriculture more environmental friendly. The research also confirms that the importance of organic certification in Malaysian food retailing is significantly associated with sustainability attributes. This indicates that food retailers perceive that food products that are claimed to be organic are produced by farmers or producers concerned with environmental as well as animal welfare. In addition, Malaysian food retailers' attitudes toward organic food products and certification represent their concern with sustainability issues, especially as Malaysian consumers become more aware of environmental issues as well as the benefits of consuming organic foods. Organic certification is therefore not only reassuring consumers with respect to the quality and safety of organic products but is also a symbol of sustainable agriculture and healthy living, together with process-related quality and the use of safe or natural raw materials (Midmore, Francois, & Ness, 2011).

The PLS-SEM analysis confirmed that the importance of organic certification for food retailing in Malaysia is affected by OCIs (in terms of path coefficients and statistical significance). Generally, organic food products can be found in specialist shops and supermarket chains in Malaysia and there are also traditional retail shops and wet markets that sell organic food product. However, the credibility of organic status cannot be determined without organic certification (Stanton & Emms, 2011). Aryal, Chaudhary, Pandit, and Sharma (2009) also agree that it is difficult to determine the status of organic food products without appropriate mechanisms and quality assurance schemes. This information indicates that Malaysian food retailers perceived that these issues can affect the credibility and trust of organic certification, particularly Malaysia's own certification, SOM.

11.9 Conclusions and recommendations

This research is very significant for Malaysian food retailing as there are limited academic studies on organic food certification in this context. The organic market can appear niche and this may be one of the reasons why few studies have been conducted on the subject, particularly in Malaysia and other developing countries. Nevertheless, there are many aspects that can be explored through investigating Malaysian food supply chains, and the impacts of food certification are expected to become more important in the future. For example, the study by Hamzaoui-Essoussi et al. (2013) highlighted that organic certification is an important source of consumer trust in food retailing.

Future research should focus on the elements of credibility and trust toward organic and food certification in relation to organic producers and consumers. Moreover, there is potentially great value in further comparing credibility and trust in organic certification with halal certification (Syed Marzuki, Hall, & Ballantine, 2012), as well as the potential overlap between the concepts, especially in the Malaysian food market. Finally, better insights into the equivalence of different organic certification would be advantageous, as this is an issue that has drawn attention not only in Malaysia but around the world (Willer & Lernoud, 2016). This is also likely to become even more of an issue in Malaysia given its desired positioning as an international food hub and increased Association of Southeast Asian Nations regional trade in food (Willer & Lernoud, 2016).

In conclusion, as with many types of food certifications, Malaysian food retailers perceived organic certification as an important attribute for organic food products in order for the market to grow positively for them, as well as to increase the awareness of consumers. Organic principles emphasize people's need to consume food products that can potentially benefit their personal health, but at the same time can benefit the health of the environment. It is expected that the recommendations and suggestions provided in this study will help improve the reputation of organic certification, particularly SOM, as the DOA and Malaysian food retailers want to convince consumers to support local organic food products. However, the production of organic foods in Malaysia is still limited. Malaysian food retailers believe that there are more local producers and farmers that want to produce organic food products and use SOM as a certification mark for selling these.

In order to improve Malaysian organic food retailing, it is imperative that all parties in the supply chain work together by promoting organic food and its certification. Indeed, although it would require a high commitment and effort from all the parties, food retailers should be much more active in increasing the awareness of consumers toward organic food products and organic certification. Organic food products that carry SOM or equivalent organic certification can be effective promotion tools to influence and convince consumers to purchase organic food products. Although the price of organic food products is a major barrier in organic retailing, improved cooperation between different stakeholders may be able to reduce these prices. Supporting organic producers by improving innovation processes and brand building may also help them to reduce costs. In fact, one of the initiatives the Malaysian DOA has taken is to waive the application fees when applying for SOM, and this can be a first step to reduce the price of organic food products. Such measures may also encourage greater commitment to organic certification from producers who are seeking to add value to their products. Most importantly, credible quality assurance schemes supported by food retailers will give peace of mind to consumers when purchasing organic food products.

References

Aarset, B., Beckmann, S., Bigne, E., Beveridge, M., Bjorndal, T., Bunting, J., . . . Young, J. (2004). The European consumers' understanding and perceptions of the "organic" food regime: The case of aquaculture. *British Food Journal, 106*(2), 93−105.

Aertsens, J., Verbeke, W., Mondelaers, K., & Huylenbroeck, G. Van (2009). Personal determinants of organic food consumption: A review. *British Food Journal, 111*(10), 1140−1167.

Ahmad, S. N. B., & Juhdi, N. (2010). Organic food: A study on demographic characteristics and factors influencing purchase intentions among consumers. *International Journal of Business and Management, 5*(2), 105−118.

Aibinu, A. A., & Al-Lawati, A. M. (2010). Using PLS-SEM technique to model construction organizations' willingness to participate in e-bidding. *Automation in Construction, 19* (6), 714–724.

Albersmeier, F., Schulze, H., & Spiller, A. (2009). Evaluation and reliability of the organic certification system: Perceptions by farmers in Latin America. *Sustainable Development, 17*, 311–324.

Ali, M. H., & Suleiman, N. (2016). Sustainable food production: Insights of Malaysian halal small and medium sized enterprises. *International Journal of Production Economics, 181*(Part B), 303–314.

Anders, S., Souza-Monteiro, D., & Rouviere, E. (2010). Competition and Credibility of private third-party certification in international food supply. *Journal of International Food & Agribusiness Marketing, 22*(3–4), 328–341.

Aryal, P. K., Chaudhary, P., Pandit, S., & Sharma, G. (2009). Consumers' willingness to pay for organic products. *The Journal of Agriculture and Environment, 10*, 1–59.

Azam, N.H.M., Othman, N., Musa, R., AbdulFatah, F., & Awal, A. (2012). Determinants of organic food purchase intention. In: *2012 IEEE symposium on business, engineering and industrial applications* (pp. 748–753).

Baker, S., Thompson, K. E., Engelken, J., & Huntley, K. (2004). Mapping the values driving organic food choice: Germany vs the UK. *European Journal of Marketing, 38*(8), 995–1012.

Botonaki, A., Polymeros, K., Tsakiridou, E., & Mattas, K. (2006). The role of food quality certification on consumers' food choices. *British Food Journal, 108*(2), 77–90.

Busch, L., & Bain, C. (2004). New! Improved? The transformation of the global agrifood system. *Rural Sociology, 69*(3), 321–346.

Caswell, J. A. (1998). How labeling of safety and process attributes affects markets for food. *Agricultural and Resource Economic Review, 27*(January), 152–158.

Chamhuri, N., & Batt, P. J. (2013). Exploring the factors influencing consumers' choice of retail store when purchasing fresh meat in Malaysia. *International Food and Agribusiness Management Review, 16*(3), 99–122.

Darby, M. R., & Karni, E. (1973). Free competition and the optimal amount of fraud. *The Journal of Law and Economics, 16*, 67–88.

Dardak, R. A., Zairy, A., Abidin, Z., & Ali, A. K. (2009). Consumers' perceptions, consumption and preference on organic product: Malaysian perspective. *Economic and Technology Management Review, 4*, 95–107.

Deaton, B. J. (2004). A theoretical framework for examining the role of third-party certifiers. *Food Control, 15*(8), 615–619.

Department of Agriculture Malaysia. (2007). *Standard Skim Organik Malaysia (SOM) Malaysian Organic Scheme*. Kuala Lumpur: Department of Agriculture.

Diamantopoulos, A., Sarstedt, M., Fuchs, C., Wilczynski, P., & Kaiser, S. (2012). Guidelines for choosing between multi-item and single-item scales for construct measurement: A predictive validity perspective. *Journal of the Academy of Marketing Science, 40*(3), 434–449.

Doherty, E., & Campbell, D. (2014). Demand for safety and regional certification of food: Results from Great Britain and the Republic of Ireland. *British Food Journal, 116*(4), 676–689.

Essoussi, L. H., & Zahaf, M. (2008). Decision making process of community organic food consumers: An exploratory study. *Journal of Consumer Marketing, 25*(2), 95–104.

Essoussi, L. H., & Zahaf, M. (2009). Exploring the decision-making process of Canadian organic food consumers: Motivations and trust issues. *Qualitative Market Research: An International Journal, 12*(4), 443–459.

Euromonitor. *Organic package food in Malaysia*. (2013). Retrieved from ⟨www.euromonitor. com⟩.

Euromonitor. *Retailing in Malaysia*. (2014). Retrieved from ⟨www.euromonitor.com⟩.

Euromonitor. *Retailing in Malaysia*. (2015). Retrieved from ⟨www.euromonitor.com⟩.

Falk, R. F., & Miller, N. B. (1992). *A primer for soft modeling*. Akron, OH: University of Akron Press.

FAS Kuala Lumpur. *Malaysia retail foods annual 2013*. (2013). Retrieved from ⟨http://www. fas.usda.gov/data/malaysia-retail-foods-annual-2013⟩.

Gerrard, C., Janssen, M., Smith, L., Hamm, U., & Padel, S. (2013). UK consumer reactions to organic certification logos. *British Food Journal*, *115*(5), 727−742.

Hair, J. F. J., Hult, G. T. M., Ringle, C., & Sarstedt, M. (2014). A primer on partial least squares structural equation modeling (PLS-SEM). *Long Range Planning*, *46*, 1−12.

Hair, J. F., Ringle, C. M., & Sarstedt, M. (2011). PLS-SEM: Indeed a silver bullet. *The Journal of Marketing Theory and Practice*, *19*(2), 139−152.

Hair, J. F., Sarstedt, M., Hopkins, L., & Kuppelwieser, V. G. (2014). Partial least squares structural equation modeling (PLS-SEM): An emerging tool in business research. *European Business Review*, *26*(2), 106−121.

Hamzaoui-Essoussi, L., Sirieix, L., & Zahaf, M. (2013). Trust orientations in the organic food distribution channels: A comparative study of the Canadian and French markets. *Journal of Retailing and Consumer Services*, *20*(3), 292−301.

Hassan, H., Sade, A. B., & Rahman, M. S. (2013). Malaysian hypermarket retailing development and expansion. *International Journal of Retail & Distribution Management*, *41*(8), 584−595.

Hatanaka, M., Bain, C., & Busch, L. (2005). Third-party certification in the global agrifood system. *Food Policy*, *30*(3), 354−369.

Havinga, T. (2013). *Tetty Havinga food retailers as drivers for food (2013/03 No. SSRN 2331869)*.

Higgins, V., Dibden, J., & Cocklin, C. (2008). Building alternative agri-food networks: Certification, embeddedness and agri-environmental governance. *Journal of Rural Studies*, *24*(1), 15−27.

Hughner, R. S., Mcdonagh, P., Prothero, A., Shultz, C. J., & Stanton, J. (2007). Who are organic food consumers? A compilation and review of why people purchase organic food. *Consumer Behaviour*, *6*, 94−110.

Janssen, M., & Hamm, U. (2012b). The mandatory EU logo for organic food: Consumer perceptions. *British Food Journal*, *114*(3), 335−352.

Janssen, M., & Hamm, U. (2014). Governmental and private certification labels for organic food: Consumer attitudes and preferences in Germany. *Food Policy*, *49*, 437−448.

Jervell, A. M., & Borgen, S. O. (2004). New marketing channels for food quality products in Norway. *Food Economics—Acta Agriculturae Scandinavica, Section C*, *1*, 108−118.

Kai, S. B., Chen, O. B., Chuan, C. S., Seong, L. C., Lock, L., & Kevin, T. (2013). Determinants of willingness to pay for organic products. *Middle-East Journal of Scientific Research*, *14*, 1171−1179.

Lea, E., & Worsley, T. (2005). Australians' organic food beliefs, demographics and values. *British Food Journal*, *107*(11), 855−869.

Manning, L., & Baines, R. N. (2004). Effective management of food safety and quality. *British Food Journal*, *106*(8), 598−606.

Midmore, P., Francois, M., & Ness, M. (2011). Trans-European comparison of motivations and attitudes of occasional consumers of organic products. *NJAS—Wageningen Journal of Life Sciences*, *58*(3−4), 73−78.

Mohd Roslin, R., & Melewar, T. C. (2008). Hypermarkets and the small retailers in Malaysia: Exploring retailers' competitive abilities. *Journal of Asia-Pacific Business, 9* (4), 329–343.

Munteanu, A. R. (2015). The third party certification system for organic products. *Network Intelligence Studies, III*(2), 145–151.

Nasir, M., & Chiew, E. (2010). Non-Muslims' awareness of Halal principles and related food products in Malaysia. *International Food Research Journal, 17*, 667–674.

Neuman, W. L. (2003). Social research methods: Qualitative and quantitative approaches. *Relevance of social research* (Vol. 8). New York, NY: Allyn and Bacon.

Organic Monitor (2006). The South-East Asian market for organic food & drink. Retrieved from www.ota.com.

Padel, S., Röcklinsberg, H., & Schmid, O. (2009). The implementation of organic principles and values in the European Regulation for organic food. *Food Policy, 34*(3), 245–251.

Probst, L., Houedjofonon, E., Ayerakwa, H. M., & Haas, R. (2012). Will they buy it? The potential for marketing organic vegetables in the food vending sector to strengthen vegetable safety: A choice experiment study in three West African cities. *Food Policy, 37*(3), 296–308.

Rezai, G., Teng, P. K., Mohamed, Z., & Shamsudin, M. N. (2012). Consumers' awareness and consumption intention towards green foods. *African Journal of Business Management, 6*(12), 4496.

Roe, B., & Sheldon, I. (2007). Credence good labeling: The efficiency and distributional implications of several policy approaches. *American Journal of Agricultural Economics, 89*(4), 1020–1033.

Sarstedt, M., Ringle, C. M., & Hair, J. F. (2014a). PLS-SEM: Looking back and moving forward. *Long Range Planning, 47*(3), 132–137.

Sarstedt, M., Ringle, C. M., Smith, D., Reams, R., & Hair, J. F. (2014b). Partial least squares structural equation modeling (PLS-SEM): A useful tool for family business researchers. *Journal of Family Business Strategy, 5*(1), 105–115.

Saunders, M., Lewis, P., & Thornhill, A. (2009). *Research methods for business students.* London: Pearson Education.

Sekaran, U. *Research method of business: A skill-building approach. Writing.* (2006). ⟨https://doi.org/http://www.slideshare.net/basheerahmad/research-methods-for-business-entire-ebook-by-uma-sekaran⟩.

Sinnappan, P., & Rahman, A. A. (2011). Antecedents of green purchasing behavior among Malaysian consumers. *International Business Management, 5*(3), 129–139.

Sirieix, L., Kledal, P. R., & Sulitang, T. (2011). Organic food consumers' trade-offs between local or imported, conventional or organic products: A qualitative study in Shanghai. *International Journal of Consumer Studies, 35*(6), 670–678.

Stanton, E., & Emms, S. (2011). *Malaysia's markets for functional foods, nutraceuticals and organic foods: An introduction for Canadian producers and exporters.* Retrieved from ⟨http://www.ats-sea.agr.gc.ca/ase/5842-eng.htm⟩.

Syed Marzuki, S. Z., Hall, C. M., & Ballantine, P. W. (2012). Restaurant manager's perspectives on halal certification. *Journal of Islamic Marketing, 3*(1), 47–58.

Terano, R., Yahya, R., Mohamed, Z., & Saimin, S. Bin (2014). Consumers' shopping preferences for retail format choice between modern and traditional retails in Malaysia. *Journal of Food Products Marketing, 20*(sup1), 179–192.

Uggioni, P. L., & Salay, E. (2014). Sociodemographic and knowledge influence on attitudes towards food safety certification in restaurants. *International Journal of Consumer Studies, 38*(4), 318–325.

Voon, J. P., Sing, K., & Agrawal, A. (2011). Determinants of willingness to purchase organic food: An exploratory study using structural equation modeling. *International Food and Agribusiness Management Review*, *14*(2), 103−120.

Watts, M. J., & Goodman, D. (1997). Global appetite, local metabolism: nature, culture, and industry in Fin-de-siecle agro-food systems. In D. Goodman, & M. Watts (Eds.), *Globalising food: Agrarian questions and global restructuring* (pp. 1−24). London: Routledge.

Willer, H., & Lernoud, J. (Eds.), (2016). *The world of organic agriculture 2016: Statistics and emerging trends. Frick and Bonn:* FIBL & IFOAM.

Inclusive food distribution networks in subsistence markets

12

Marcos Santos[1] and Andrés Barrios[2]
[1]Management, Faculdade Multivix, Vitoria, Brazil,
[2]Marketing, Universidad de Los Andes, Bogota, Colombia

12.1 Introduction

Traditionally, poor communities are structurally denied access to resources, capabilities, and opportunities (George, McGahan, & Prabhu, 2012). This is particularly detrimental in the food sector, where market barriers prevent low-income communities from eating healthily (Drewnowski, 2004). Since the new millennium, the private sector has been called on to develop initiatives combining market principles with social and environmental factors to help poor communities (United Nations, 2000). A business response to the previous situation has been to implement "inclusive business" strategies, whereby subsistence marketplaces are not seen simply as a segment to sell to, but rather, as a strategic partner to cooperate with (Viswanathan & Sridharan, 2009). By implementing these initiatives, both private organizations and poor communities create meaningful synergies and shared values which support inclusive growth (George et al., 2012).

The food sector has not been indifferent to this trend, and different initiatives to transform food value chains by developing inclusive food distribution networks for subsistence markets have been developed. For example, Gomez and Ricketts (2013, p. 140) suggest the implementation of what they call a "modern-to-traditional" typology, in which "domestic and multinational food manufacturers sell through the network of traditional traders and retailers." Using this approach, food organizations gain access to a wider range of consumers located in isolated markets, while low-income consumers obtain access to an increased processed product offer with low prices. However, the implementation of such initiatives for the food market is not an easy task, requiring the alignment of the region's food offer (Wrigley, Guy, & Lowe, 2002), consumers' food preferences (Ali, Kapoor, & Moorthy, 2010), and the company's logistics skill set (Baindur & Macário, 2013).

Regarding the region's food offer, poor communities are commonly found in dispersed locations with limited access to public utilities (e.g., electricity and sanitation) and restricted development of private institutions (e.g., banks) (Viswanathan, 2013). As a consequence, these poor communities are left commercially excluded, and informal networks that provide a limited food product assortment (e.g., non-perishables) at higher prices emerge (Viswanathan, Rosa, & Ruth, 2010).

Case Studies in Food Retailing and Distribution. DOI: https://doi.org/10.1016/B978-0-08-102037-1.00012-8

Consumers' food purchasing preferences are influenced by the social context (Shaw, Mathur, & Mehrotra, 1993). For example, subsistence consumers' uncertain income and lack of infrastructure often make them buy their food and grocery products from nearby marketplaces, "mom-and-pop stores," on a daily basis (Veeck and Veeck, 2000; Sabnavis, 2008). Price is considered a core factor in food choice, followed by taste and freshness (Steenhuis, Waterlander, & de Mul, 2011). In addition, consumers in subsistence marketplaces have particular exchange dynamics, based on their social capital, which are central to the development of business models that are effective and appropriate for such consumer needs (Viswanathan & Rosa, 2010).

Organizations aiming to serve subsistence marketplaces require a customer-driven approach supported by an efficient distribution and retailing network to serve these markets (Ali et al., 2010; Baindur & Macário, 2013). For example, construction (CEMEX) and technology retail (Casas Bahia) companies serving subsistence marketplaces have had to reformulate their business strategy, including developing particular products, implementing microloans schemes, and developing a new distribution system (Prahalad, 2005).

This book chapter presents an exemplar case of an "inclusive business" model for the food retailing sector. In particular, the case describes how French multinational company Danone developed a partnership with public (IDB) and social (World Vision and Aliança Empreendedora) organizations to develop a food retailing initiative—Kiteiras—with positive impacts for business, communities, and individuals.

12.2 Case description

12.2.1 The situation

In 2010, Danone Brazil launched a strategy to increase market share in the northeast of the country, involving low-income households (Lozano, Giacomini, Camara, & Borges, 2012). This region reflects the national trends of poverty and high gender inequality in the Brazilian labor market (IBGE, n.d.). With 3.7% of people below the poverty line of US$1.9 a day—representing over 7.7 million people (World Bank, n.d. a)—and a GINI index of 51.3, Brazil has a high gap in wealth distribution (World Bank, n.d. b). This situation is exacerbated when analyzed from a gender perspective. The country's unemployment rate among women (8.7%) is higher than that seen for men (5.9%), but 62% of women develop their activities in informal jobs with little income and no social benefits.[1] Some of the reasons for the unemployment gap between men and women are attributed to a lack of qualifications for vacancies due to early school dropout, linked to pregnancy, gender discrimination, and flexible schedule requirements needed to attend to family obligations (Castro & Abramovay, 2002).

[1] This refers to the population over 14 years old who are unemployed and often rely on the informal market.

12.2.2   Origin of the initiative

In 2010, the sustainability department of Danone together with Aliança Empreendedora (This is an NGO whose mission is to enable access for low-income people and communities to develop as entrepreneurs, promoting social and economic inclusion, and development.) designed the Kiteiras initiative to develop a microdistribution channel that employs only women from low-income neighborhoods to promote, sell, and distribute the company's products door-to-door in the communities in which they reside. The initiative's women-only focus responded to (1) cultural frameworks that mean women are in charge of family nutrition; (2) work barriers, described in the previous section, that women face; and (3) the community-based approach of the initiative.

At the time of the Kiteiras initiative's origin, direct sales models that included door-to-door microdistribution were very successful in the cosmetics market (e.g., Avon and Natura), but they had not been proven in the food market yet. The initiative was expected to make the company reach a new consumer base at a lower price, while providing women with a new income source, and consumers with access to a nutritious product. Danone's Sustainability Department presented the initiative to *Fonds Danone pour l'Ecosystème*, a company department that promotes intrapreneurship, and obtained an initial investment of €400,000.

12.2.3   The Kiteiras distribution system

Six processes take place in the Kiteiras distribution system. The initiative's financial model involves product markups along the distribution system. A description of the processes is now presented.

Mobilization: This process involves marketing activities to enhance neighborhood interest about the Kiteiras project and Danone products. These activities cover not only media communications, through community radio and mobile sound trucks, but also community activities such as visits to churches and bazaars.

Recruitment: This process involves the recruitment and engagement of interested women into the Kiteiras initiative. These women fill out an application form and start participating in training programs. Women identified with entrepreneurial skills, or experience in sales, can be directly selected as "godmothers." In this role, they are able to recruit and manage their selling group and act as intermediaries in the communication between the Kiteiras, which also refers to the women participating in the initiative, and the distributor in a particular location.

Development: This process involves Kiteiras' career development within the initiative across five stages: consume, sell, indicate, care, and lead. The first stage, "consume," presents Danone dairy products to the Kiteiras, detailing nutritional aspects. The second stage, "sell," presents basic sales lessons. The third stage, "indicate," offers basic training to help the new Kiteiras select appropriate dairy products for their customers' needs. The fourth stage, "care," offers basic training to monitor customers and their family. The fifth stage, "lead," offers basic leadership training to bolster women's confidence and help in sales pitches.

Along the career path, women take part in a 36-hour training program, developed by Aliança Empreendedora, involving different workshops to develop their capabilities around three main themes: entrepreneurship, nutrition, and life. Women start by being a consumer of Danone's products, and once they receive training and start to sell, they become Kiteiras.

As Kiteiras, these women are not employed by Danone or the distributor with which they work. They receive commission on sales and are referred to legally as individual microentrepreneurs by Brazilian legislation (Brazilian complementary Law 128/2008, available at: http://www.planalto.gov.br/ccivil_03/leis/LCP/Lcp128.htm). Then, those women who show leadership skills in the field develop a training program to become godmothers.

Sales: This process involves the product exchange process. Every 2 weeks, the local distributor provides product catalogs to Kiteiras. These catalogs are updated every 3 months based on shopper insights and local sales analysis. In each neighborhood, the Kiteiras, led by the godmothers, make door-to-door sales. Payment can be done either by cash or credit card. All sales orders are reported to their godmothers, who in turn inform the distributor in order to arrange product delivery.

Product delivery: This process involves transporting products to customers. The sales orders are placed each week by godmothers with the local distributor. Godmothers group the orders and send them to Danone, which then delivers the products back to the Kiteira, who in turn delivers them personally to customers.

Given that dairy products are perishable, the product delivery process needs to adhere to strict deadlines. Danone provides products to the distributor at least 20 days before the expiration date. The distributor cannot keep the product for more than 3 days in storage before making deliveries to Kiteiras. Once the products are received, they have to deliver them right away to customers. When a Kiteira has a big number of customers, she can keep part of the product in her own refrigerator for a maximum of one day. Once delivered, if the product is damaged in transit or is close to expiry, customers can return the product to Kiteiras for replacement or refund.

12.2.4   The pilot

The Kiteiras pilot started in 2011 in Salvador (in the state of Bahia), one of the biggest cities of the northeast region. At the time, Danone and Aliança Empreendedora hired Veli, a human resource consultancy, to do the mobilization and recruitment processes in Salvador's low-income neighborhoods.

By January 2013, the pilot had 210 women involved who were trained to sell around 22 tons of dairy products per month, generating an average monthly income for them of R$269 (approximately US$81). At the time, the average product markup was 55% for distributors and 25% for Kiteiras. Distributors paid an additional 3.5% as a selling commission to godmothers for their group sales.

The pilot's positive results motivated Danone to escalate the initiative from being a corporate social responsibility program from the sustainability department to a social innovation program led by the commercial department.

12.2.5   Roll out of the initiative

The escalation phase started in May 2013 with many challenges, due to difficulties in applying traditional distribution management practices among the Kiteiras. For example, the Danone distributors aimed to increase Kiteiras' sales quotas. However, the Kiteiras' customers were those embedded in their social network, making it difficult to obtain new customers and increase sales quotas in the short term. A solution was to make Kiteiras see themselves as entrepreneurs, assuming not only the sales but also storage of a small amount of products in their homes, which would take a long-term capabilities building program. As a result, the expected daily sales of the commercial department were never reached, some distributors went bankrupt, and some Kiteiras became indebted, due to them missing product expiration dates, and decided to leave the program.

In 2014, Danone signed a contract with a new distributor, Qualikits, a former beverages distributor from Salvador. The owner was from a traditional family that worked in this business and wanted to do something with a social impact. This distributor inherited a diminished salesforce of Kiteiras with only two godmothers from the previous distributor. To recruit more sellers, the sales markup scheme was changed: the local distributor markup decreased from 55% to 37% and the Kiteiras' markup increased from 25% to 30%.

In a period of 3 months, the new distributor developed a network of 350 Kiteiras and two godmothers, who were selling around 31 tons of dairy products per month. Although the initiative was economically sustainable, these results were below the commercial department breakeven point (about 450−500 Kiteiras, each selling an average of 80 kg of products per month). Danone's commercial department, which then controlled the initiative, requested its termination.

12.2.6   Escalation of the initiative

Despite the termination request, Danone's sustainability department saw an opportunity in the Multilateral Investment Fund Scala program. This program, developed by the IDB, aims to promote the economic empowerment of low-income individuals through distribution networks based on microfranchising (Fomin, n.d.) The initiative was selected by Scala and received US$2.07 million in funding and advisory services to escalate the initiative to 2000 Kiteiras nationwide. The escalation phase started at the beginning of 2016, with the implementation of several changes to the business model:

1. The sustainability department resumed the leadership of the initiative again, while the commercial department focused just on the sales processes, improving catalogs, mapping geographical distribution, and commercial intelligence operations.
2. Danone developed an alliance with World Vision Brazil, an organization that was working in the same location as the Kiteiras initiative, with the program "Groups of Local Development Opportunities" to promote collaboration among community members. World Vision recruited Kiteiras at community centers, employing people already trained by their REDES program. (The REDES program constitutes initiatives to develop income by training in economic activities such as cooking and handicraft.)

3. Aliança Empreendedora developed a career development blended program for Kiteiras, mixing classroom and virtual training via closed Facebook and WhatsApp groups.

12.2.7 Results to date

By the end of 2016, the Kiteiras initiative has promoted entrepreneurial capacity and generated income opportunities for 2100 vulnerable women that could not work otherwise due to family duties. They were selling around 144 tons of products per month, not only in Salvador but also in Fortaleza and São Paulo. (São Paulo is Danone's biggest market in the country, and Fortaleza is the second biggest market in the northeast region after Salvador.) In addition, despite the decline of the consumption of yogurt in Brazil from an annual growth of 19.7% in sales in 2012 to 3.9% in 2017 (Alvarenga, 2017), the initiative promoting yogurt consumption among low-income consumers reached 9 kg per year in 2016 (from an original 3 kg year per consumer before the program started) and introduced Danone products to new city areas where the company did not previously have a presence.

Nowadays, the Kiteiras make an average of R$500 (the minimum monthly salary in Brazil is R$788) working 20 hours a week. These women have developed their sales and entrepreneurial capacities running their own businesses, which allow them to manage their time between work and family. Moreover, due to their promotion of a healthy lifestyle and related products, these women have built and strengthened their community ties. As one expressed:

My life changed for the better, because before I had to wait for the end of the month to receive money, and working as a Kiteira I can receive every week. That makes a huge difference to me.

Others explained:

It's good because each week we get some income.

And:

[The program represents] the possibility to help my husband at home, financially.

12.3 Contribution

Over recent decades, the supply chain activities of the food industry have gained importance (Maloni & Brown, 2006). These activities play a major role in national economies, influencing aspects such as food costs, access, and quality, impacting in turn people's quality of life (Dujak, Segetlija, & Mesari, 2017). This case shows how Danone's Kiteiras initiative has created and developed a consumer centered inclusive distribution channel for subsistence marketplaces. Some of the lessons this case shows are now discussed.

12.3.1 Implementing effective cross-sector partnerships between private companies and NGOs for community development

The Kiteiras initiative demonstrates how private, social, and business organizations can complement each other to enhance their social impact, which will attract interest from peer institutions worldwide wishing to introduce similar initiatives. Developing partnerships with experienced organizations in community development and training allowed Danone to access the expertise to understand community motivations and acquire the local legitimacy needed to create the Kiteiras inclusive distribution network.

The program's legitimacy has resulted in effective recruitment processes and reduced turnover rates. The challenge in this process has been to align the different interests of each actor. For example, in the pilot phase, while Danone aimed to reach a new market with a high-quality service, Veli and distributors aimed to recruit and put into operation as many Kiteiras as possible, and Aliança Empreendedora aimed to build Kiteiras' entrepreneurial capabilities. To manage this situation, the initiative created a core team with representatives from all organizations to develop the managerial decision protocols in the field. The aim of this core team was to integrate the decision-making processes of the different companies and institutions into a single group, avoiding conflicts and improving communication.

12.3.2 Using customer knowledge as competitive advantage

With the evolution of consumers' food choices and desire for convenience, the retail market for food and grocery is changing (Ali et al., 2010). In subsistence marketplaces, companies need to develop alternative logistics systems that can adapt their operations to particular consumer preferences and infrastructure deficits (Baindur & Macário, 2013). In this context, companies' food supply chains for poor communities are striving to achieve end-customer value and satisfaction, while also improving performance outcomes for the firm (Kozlenkova, Hult, Lund, Mena, & Kekec, 2015).

The Kiteiras initiative enabled Danone to use Kiteiras' local knowledge and community access to develop new marketing strategies aligned with end-consumers' interests. Danone's distribution system normally operates outside the community. Once inside the community, goods delivery is the responsibility of Kiteiras, who have knowledge of the whereabouts and routines of their clients or can negotiate more flexible schedules to deliver products to them. Thus, Kiteiras' knowledge of the subsistence marketplace dynamics and consumer preferences has become the company's competitive advantage. As Day and Wensley (1988) affirm, competitive advantage must manifest in customer value but be based on a balance between customer and organizational perspectives.

12.3.3 Using technology to increase the escalation efficiency of inclusive distribution models

Innovations in the supply chain require a collaborative approach from the different participants and actors, as each organization is dependent on the success of their network partners (Arlbjørn & Paulraj, 2013). For this reason, any innovation in one part of the supply chain, either technological or not, must take into account the capabilities of the other parts. Danone implemented a technological innovation for serving subsistence marketplaces.

First, given the increasing number of Kiteiras who are part of the program—the goal was 2000—Danone implemented a management system that collects, analyzes, and reports information about delivery requirements, operations, and computerized cargo processing. This system provides the initiative actors with key performance indicators such as number of sales, frequency of sales, average value of sales, stock keeping units sold, turnover from the salesforce, and new Kiteiras registered.

Second, given the difficulties Kiteiras have attending the different workshops—due to their family obligations—Aliança Empreendedora developed career development blended-learning workshops using mobile apps such as YouTube, Facebook, and WhatsApp. Many of the women involved were already using these applications for their personal interactions. As a result, the company and Kiteiras obtained a new efficient and flexible communication channel with customers, which gave them information to understand their demand patterns and needs.

12.4 Conclusions

Despite intense interest in the collaborative supply chain, researchers know little regarding the collaborative process through which resources are combined and shared across supply chain networks (Fawcett, Fawcett, Watson, & Magnan, 2012). As companies initiate collaboration in the supply chain, they enter a balancing cycle that involves learning how to increase commitment and build capabilities while dealing with actors that resist collaboration and enabling factors/agents (Fawcett et al., 2012).

In the case discussed in this chapter, we hoped to shed light on how the capabilities of a multinational company, NGOs, and a local community can be merged to generate an inclusive distribution system for subsistence marketplaces—one that combines different operations at different scales with economic and social benefits. The Kiteiras business model enabled Danone to gain access to a new market of poorer consumers, thereby increasing Danone's distribution network and consumer base. In addition, the nature of the program benefits poor women in vulnerable conditions that cannot work regular hours, including them in the formal labor market and, in this way, recomposing their self-esteem and improving the quality of their children's nutrition, thus contributing to the social development of the community.

Acknowledgments

The research presented in this publication was carried out with the aid of a grant from the International Development Research Centre (IDRC), the Inter-American Development Bank (IDB), the Multilateral Investment Fund (MIF), and the City Foundation. The views expressed herein do not necessarily represent those of these institutions or foundations or their Board of Governors. For more information about the project, visit: https://observatorios-cala.uniandes.edu.co/es/.

References

Ali, J., Kapoor, S., & Moorthy, J. (2010). Buying behaviour of consumers for food products in an emerging economy. *British Food Journal, 112*(2), 109−124.

Alvarenga, D. (2017). *Cai consumo de fralda, iogurte e lâmina de barbear no Brasil; veja ranking* in Globo.com—Economia. Available from: ⟨https://g1.globo.com/economia/noticia/cai-consumo-de-fralda-iogurte-e-lamina-de-barbear-no-brasil-veja-ranking.ghtml⟩ Accessed July 4, 2017.

Arlbjørn, J. S., & Paulraj, A. (2013). Special topic forum on innovation in business networks from a supply chain perspective: Current status and opportunities for future research. *Journal of Supply Chain Management, 49*(4), 3−11.

Baindur, D., & Macário, R. M. (2013). Mumbai lunch box delivery system: A transferable benchmark in urban logistics? *Research in Transportation Economics, 38*(1), 110−121.

Castro, M. G., & Abramovay, M. (2002). Jovens em situação de pobreza, vulnerabilidades sociais e violências. *Cadernos de Pesquisa, 116*, 143−176.

Day, G. S., & Wensley, R. (1988). Assessing advantage: A framework for diagnosing competitive superiority. *Journal of Marketing, 52*(2), 1−20.

Drewnowski, A. (2004). Obesity and the food environment. *American Journal of Preventive Medicine, 27*(3), 154−162.

Dujak, D., Segetlija, Z., & Mesari, J. (2017). Efficient demand management in retailing through category management. In P. Golinska-Dawson, & A. Kolinski (Eds.), *Efficiency in sustainable supply chain* (pp. 195−216). Cham: Springer International Publishing.

Fawcett, S. E., Fawcett, A. M., Watson, B. J., & Magnan, G. M. (2012). Peeking inside the black box: Toward an understanding of supply chain collaboration dynamics. *Journal of Supply Chain Management, 48*(1), 44−72.

Fomin (n.d.) *Our latest news.* Available from: ⟨http://www.fomin.org/en-us/Home/News/PressReleases/ArtMID/3819/ArticleID/888/SCALA--An-innovative-approach-to-promote-economic-empowerment-of-the-poor-in-Latin-America-and-the-Caribbean-through-microfranchising-.aspx⟩ Accessed November 25, 2015.

George, G., McGahan, A. M., & Prabhu, J. (2012). Innovation for inclusive growth: Towards a theoretical framework and a research agenda. *Journal of Management Studies, 49*, 661−683.

Gomez, M. I., & Ricketts, K. D. (2013). Food value chain transformations in developing countries: Selected hypotheses on nutritional implications. *Food Policy, 42*(1), 139−150.

IBGE—Instituto Brasileiro de Geografia e Estatística (n.d.) *Cidades* [Online] Available from: ⟨http://cidades.ibge.gov.br/xtras/perfil.php?codmun = 292740⟩ Accessed November 20, 2016.

Kozlenkova, I. V., Hult, G. T. M., Lund, D. J., Mena, J. A., & Kekec, P. (2015). The role of marketing channels in supply chain management. *Journal of Retailing*, *91*(4), 586−609.

Lozano, M., Giacomini, E., Camara, M., & Borges, M. (2012). *Investor seminar—St.* Petersburg: Brazil: Presentation, 51 p.

Maloni, M. J., & Brown, M. E. (2006). Corporate social responsibility in the supply chain: An application in the food industry. *Journal of Business Ethics*, *68*(1), 35−52.

Prahalad, C. K. (2005). *The fortune at the bottom of the pyramid*. Upper Saddle River, NJ: Wharthon School Publishing.

Sabnavis, M. (2008), *"Why organized retail is good"*. The Hindu Business Line. [Online] Available from: ⟨www.thehindubusinessline.com/2008/05/28/stories/2008052850330800.html⟩.

Shaw, A., Mathur, P., & Mehrotra, N. N. (1993). A study of consumers' attitude towards processed food. *Indian Food Industry*, *47*, 29−41.

Steenhuis, I. H., Waterlander, W. E., & de Mul, A. (2011). Consumer food choices: The role of price and pricing strategies. *Public Health Nutrition*, *14*(12), 2220−2226.

United Nations, 2000, *Millennium Development Goals*. Washington, D.C: United Nations.

Veeck, A. and Veeck, G., 2000, Consumer segmentation and changing food purchase patterns in Nanjing, PRC, World Development 28 (3), 457−471.

Viswanathan, M. (2013). *Subsistence Marketplaces*. Urbana-Champaing, IL: Etext Inc.

Viswanathan, M., & Rosa, J. A. (2010). Understanding subsistence marketplaces: Toward sustainable consumption and commerce for a better world. *Journal of Business Research*, *63*(6), 535−537.

Viswanathan, M., Rosa, J., & Ruth, J. (2010). Exchanges in marketing systems: The case of subsistence consumer-merchants in: Chennai, India. *Journal of Marketing*, *74*, 1−17.

Viswanathan, M., & Sridharan, S. (2009). From subsistence marketplaces to sustainable marketplaces: A bottom-up perspective of the role of business in poverty alleviation. *Ivey Business Journal (On line)*, March−April.

World Bank (n.d. a). *Poverty & equity data portal—Brazil*. Available from: ⟨http://povertydata.worldbank.org/poverty/country/BRA⟩ Accessed November 12, 2016.

World Bank (n.d. b). *GINI index (World Bank estimate)—Brazil*. Available from: ⟨https://data.worldbank.org/indicator/SI.POV.GINI?locations=BR⟩ Accessed November 12, 2016.

Wrigley, N., Guy, C., & Lowe, M. (2002). Urban regeneration, social inclusion and large store development: The Seacroft Development in context. *Urban Studies*, *39*(11), 2101−2114.

Food, health, and data: Developing transformative food retailing

13

Hannu Saarijärvi[1], Leigh Sparks[2] and Sonja Lahtinen[3]
[1]Service and Retailing, Faculty of Management, University of Tampere, Finland, [2]Institute for Retail Studies, Stirling Management School, University of Stirling, United Kingdom, [3]Faculty of Management, University of Tampere, Finland

13.1 Introduction

The food retailing sector in many countries is undergoing considerable restructuring (Sparks, 2016). One of the catalysts for this has been the penetration of the hard discounters, which have established unprecedented levels of price competition. Whilst enhanced price competitiveness is one response from existing food retailers, direct head on competition with the hard discounters is not seen as a sustainable strategy. Many food retailers instead are searching for new initiatives through which they can engage themselves further into their consumers' lives, thus serving consumers in ways that go beyond traditional "simple" selling of groceries "cheaply." In this endeavor, data, digitalization in general, and mobile services play pivotal roles, as they allow retailers to integrate additional personalized resources—such as menu-planning guidance, help in tracing product origins and dietary information advice—with their consumers' everyday processes (O'Hern & Rindfleisch, 2009; Saarijärvi, Mitronen, & Yrjölä, 2013b; Saarijärvi, Kuusela, & Rintamäki, 2013c).

At the same time, it is recognized that consumers' ethical considerations and alternative dietary preferences, along with the renaissance of local food, are driving alterations in consumers' food choices. Food healthfulness has emerged as a movement influencing structures, processes, and resources in the food retailing industry. Consumer health and wellbeing are growing concerns for consumers, industries, and governments (IGD Retail Analysis, 2016). Food purchase and consumption have the potential to influence, for example, obesity rates (CDC, 2013; Cohen & Lesser, 2016), which in turn have been linked to health issues including high blood pressure, coronary heart disease, increased risk for type-2 diabetes, and several cancers (Andrews, Lin, Levy, & Lo, 2014). Food retailers thus face pressure to reconfigure their role and responsibility for health, either contributing to national health objectives or facing possible restructuring legislation, taxation, and other restrictions. Sparks and Burt (2017) provide a review of previous research in this area, focusing on potential interventions of in-store retail operations, with the aim of improving consumer choices. Their report adds a retail dimension to the often

Case Studies in Food Retailing and Distribution. DOI: https://doi.org/10.1016/B978-0-08-102037-1.00013-X

medically or health inclined systematic reviews already published (see Adam & Jensen, 2016; Afshim et al., 2017; Escaron, Meinen, Nitzke, & Marinez-Donate, 2013; Gittelsohn, Rowan, & Gadhoke, 2012; Glanz, Bader, & Iyer, 2012; Liberato, Bailie, & Brimblecombe, 2014).

This rising interest in the relationships amongst food retailing, consumption, and diet and health points to the significance of this area for health policy as well as retailers, who might increasingly feel under threat from this new emphasis. However, this emphasis also means that food retailers have an opportunity to take on a greater role and greater responsibility regarding consumers' health and wellbeing. The digital and innovative use of customer data—including point-of-sale (POS) and customer loyalty data, but possibly also data from the health services—can facilitate this paradigmatic transition, as well as developing new means to customer loyalty (Møller, 2011; Wansink, 2017). This suggests a new proactive, transformative role for retailers, arising not only from recognition of the weaknesses of current practices (in some eyes) but also from the opportunities of directly engaging more deeply with consumers' lives.

This potential new role of the food retailer necessitates an in-depth understanding of its origins, as well as its diverse implications. This chapter explores the key elements of transformative food retailing, analyses its potential, and identifies implications for consumers, companies, academics, and society. To achieve this, we discuss three key theoretical perspectives of the phenomenon, reflect on the potential opportunities and pitfalls through food retailing's four evolutionary phases, and propose a conceptual framework for transformative food retailing.

13.2 Theoretical framework

The theoretical proposition is that retailers have an opportunity to transform their role and relationship with consumers around the space of food, health, and diet. This implies the need to join together aspects of data, customer focus, and transformative service delivery. Three complementary theoretical perspectives are thus discussed. First, *transformative service research (TSR)* combines transformative consumer research and service research to identify and address the role of services in enhancing consumers' wellbeing (Mick, 2006). Second, the *reverse use of customer data* refers to the process of converting customer data into such information that directly supports customers' value creation (Saarijärvi, Grönroos, & Kuusela, 2014). Third, *customer-dominant logic (CDL)* introduces a business perspective based on the primacy of the customer (Heinonen & Strandvik, 2015).

Taken together (see Fig. 13.1), these perspectives offer a theoretical foundation for discussing, defining, and exploring transformative food retailing; i.e., they offer complementary lenses through which the evolving transition toward transformative food retailing can be viewed.

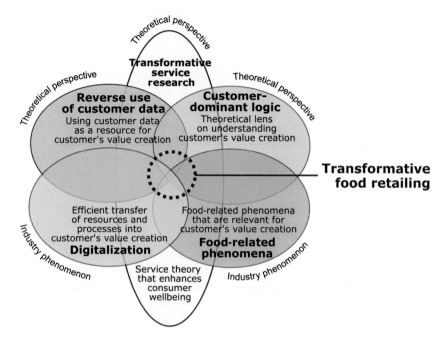

Figure 13.1 Theoretical perspectives to transformative food retailing.

13.2.1 Transformative service research

The concern for consumers' individual and collective wellbeing is increasingly gaining attention amongst service providers, as well as amongst service researchers. A specific research area, labeled TSR, is characterized by its focus on service outcomes that increase or decrease the wellbeing (physical, mental, social, and financial wellbeing) of those engaged in the service under study (Anderson & Ostrom, 2015). By definition, transformative services focus on creating "uplifting changes" to improve the wellbeing of individuals, families, communities, society, and the ecosystem at large (Anderson et al., 2013). Within TSR, transformation occurs when actors "become conscious of their roles in reproducing structures and elect to make new, imaginative choices to challenge dominant patterns" (Blocker & Barrios, 2015: 268). Current dominant patterns created and sustained by food retailers are seen to promote obesogenic environments that undermine consumers' balanced diets or overall national health (Sparks & Burt, 2017). While some food retailers are becoming aware of their role in producing these structures and experimenting in change (e.g., Adam, Jensen, Sommer, & Hansen, 2017; Wansink, 2017), being the first movers they often lack the information, examples, and guidance on what really works. We aim to elaborate the key elements that transformation in food retailing should/could entail.

Much of the current service research and practice draws on service-dominant logic (S-D logic) to highlight customers' participation in cocreating services

(Vargo & Lusch, 2008). Mende and van Doorn (2015) recognize one critical limitation in this research stream, namely that the outcomes of cocreation are predominantly linked to organizational benefits or customer responses that increase loyalty or positive word-of-mouth for the organization. To bring the wellbeing of consumers into the epicenter of value cocreation, Blocker and Barrios (2015: 268) introduce the concept of *transformative value*, which they define as "an intervention, aimed to advance greater wellbeing by facilitating enduring changes among actors." The authors point out that most of the value creation in the marketplace is *habitual* in nature, meaning that organizations focus on satisfying domain-specific, existing, everyday needs of their customers and other stakeholders. Habitual value creation sustains current patterns, practices, and structures; i.e., continues "business as usual." Transformative value creation alternatively is an option for those service providers who are willing to go beyond habitual value creation, and who can see their service system as a springboard for change in micro and macroenvironments. Transformative value not only reflects changes in beliefs, attitudes, and values, but also requires alterations in resources, processes, and practices (Blocker & Barrios, 2015). Over time, if successful, transformative value creation would become new habitual provision.

Given the centrality of customers in value cocreation (Vargo & Lusch, 2008), transformative value also requires participation from consumers. Thus, while a service provider, such as a food retailer, can facilitate transformation by creating resources, capabilities, and tools that aim to increase consumer wellbeing through its service system, the actual transformation necessitates individuals deciding to use these tools and resources to create desired outcomes (Grönroos & Voima, 2013). Mende and van Doorn (2015) investigated drivers of cocreation in transformative services by drawing from self-determination theory (SDT), developed by Ryan and Deci (2000). They point to three drivers of cocreation by providers and consumers in transformative services: service literacy (competence in SDT), involvement (autonomy in SDT), and attachment styles (relatedness in SDT). In the context of food retailing, this could imply that cocreation of transformative value may be affected by consumers' nutritional literacy, how important they see the change for a healthier diet, and the interpersonal ties with, for example, service provider, nutritional advisor, health services, and other consumers. These determinants should be considered when segmenting the customer portfolio, designing digital tools for different segments, and measuring the objective and subjective wellbeing of these service users.

13.2.2   Reverse use of customer data

Reverse use of customer data refers to the process of harnessing the potential of customer data for the benefit of the customer, not only for the benefit of the company (Saarijärvi et al., 2014). Traditionally, companies' customer data usage has focused on various customer relationship management (CRM) activities, including cross- and up-selling activities, and segmentation or identifying the most profitable and unprofitable customers. Customer data has thus been used as a resource for the company's processes: the customer has been perceived as the source of data and a passive objective of the company's diverse set of CRM activities (Saarijärvi, Karjaluoto, & Kuusela, 2013a). Loyalty schemes are a good example of this.

However, customer data holds direct potential to be used as an input resource for the customer's value-creating processes. Banking, insurance, retailing, telecommunications, or energy are examples of industries that could—and increasingly do—convert vast amounts of customer data into meaningful information; for example, related to consumers' insurance needs, real-time energy consumption, or most suitable teleoperator contracts. Customer data is thus turning into customer's data; it is converted into a resource that has potential to be used for the customer's value creation. Digitalization enables new forms of data collection, analysis, and utilization, which further exerts pressure on harnessing customer data in contemporary businesses, including food retailing. This may eventually reconfigure also the company's role: it is not only a supplier of goods or services but also serves customers through their own data.

In the context of food retailing, reverse use of customer data offers ample opportunities and space for differentiation. Food retailing is a data-driven business, but customer data is traditionally used to optimize logistics and maximize return for the business, being used as an input resource for supply chain management and retail merchandising. While these are of critical importance, they offer a narrow approach to the potential of customer data. Food retailers' customer loyalty programs and POS data, together with databases including grocery product nutrition content information and health services data, offer unique opportunities to build information that consumers could find relevant and useful for their own purposes. Moreover, information that is about the consumers (based on consumers' purchases) and for the consumers (i.e., information that supports consumers' value-creating activities) could have an impact by changing behavior—another characteristic of information. This becomes especially powerful when datasets from various domains (food retailing, health services, exercise regimes, for example), but linked by the customer, can be drawn together.

13.2.3 Customer-dominant logic

The success of transformative services is heavily dependent on in-depth understanding of the consumer's context. Recent paradigmatic initiatives in marketing and service theory, including S-D logic (e.g., Vargo & Lusch, 2004; 2016), service logic (Grönroos & Ravald, 2011, Grönroos & Voima, 2013), consumer-dominant logic (Anker, Sparks, Moutinho, & Grönroos, 2015), or service science (Maglio & Spohrer, 2008), have collectively shifted attention from a product orientation toward a customer value creation context and sphere. More recently, CDL has complemented scholarly discussion (Heinonen & Strandvik, 2015). In contrast to other established service perspectives, CDL emphasizes "how customers embed service in their processes rather than how firms provide service to customers" (Heinonen &

Strandvik, 2015: 472). As a perspective, it operates through five fundamental features.

First, *customer logic* refers to those actions, reactions, practices, preferences, and decisions accomplished by the customer to live their life. Food retailing has a key role in consumers' activities. Second, a *business perspective* refers to companies shifting focus on understanding customers' logics. In the context of food retailing, this would necessitate companies building key competence on understanding consumers' activities, experiences, and preferences, together with their goals, tasks and reasoning, and how to support them through food retailing. Third, *offering* refers to what the provider, or food retailer, offers. Service scholars (e.g., Heinonen & Strandvik, 2015) do not make any distinction between products/services or service. For example, the food retailers' offering is not only related to the product category or place of the store, but is also a combination of groceries, brand and people, offline and online services, and customized information and promotions. Fourth, *value formation* refers to the process through which value emerges. Heinonen and Strandvik (2015: 479) define it as "customers emerging behavioral and mental processes of interpreting, experiencing and integrating offerings in their everyday lives." In food retailing, this can be reflected as various dimensions of value emerging as consumers engage in different activities related to food—e.g., menu construction. Finally, *ecosystems* consist of different actors and elements that are contributing to service. In the context of food retailing, ecosystems can consist, for example, of offline and online services, suppliers, and POS data systems, but could go beyond that in their linkage to other non-retail service providers.

Table 13.1 synthesizes and summarizes the discussion thus far. These theoretical departures, i.e., TSR, reverse use of customer data, and CDL, together with their respective key concepts (consumer wellbeing, customer's data, customer logic, business perspective, offering, value formation, ecosystem), demonstrate the potential of transformative food retailing.

13.2.4 Toward a framework for transformative food retailing

Digitalization is reconfiguring the boundaries of retailing. Consumer-to-consumer e-commerce, desire for food healthfulness, and smart shopping are examples that are driving retail evolution. It can be argued that parts of food retailing are beginning to evolve toward an approach built on the insight generated from TSR, reverse use of customer data, and CDL.

The evolutionary transition can be considered through four phases that share different characteristics. During these phases, food retailing has faced varying emphases in terms of what type of customer data is being used and how, and the key challenges and strategic priorities facing food retailers and how these have contributed to shaping the food retailer's role. Table 13.2 illustrates this shift and—though not being all inclusive—illustrates the direction toward viewing food retailing as a transformative agent.

Our interest is in the fourth phase, i.e., food retailing as transformation. The key concepts of TSR, reverse use of customer data, and CDL help us explore

Table 13.1 **Implications of theoretical perspectives for transformative food retailing**

Theoretical perspective	Implications for transformative food retailing
Transformative service research	• *Shifts attention toward viewing service businesses as vehicles for enhanced consumer wellbeing* • Exerts pressure on reconfiguring food retailing's *raison d'être* and consequently, respective executive mental models • Emphasizes the understanding of the antecedents of consumers' subjective and objective wellbeing in given industries • Reduces asymmetry between consumers and service providers by bringing consumers' wellbeing to the epicenter of value cocreation in food retailing
Reverse use of customer data	• *Shifts attention toward harnessing the potential of customer data for healthier food consumption* • Provides alternative approach to customer data usage by serving customers through their own data • Exerts pressure on digital services, modernized data infrastructures, and managing consumers' willingness to share data • Emphasizes the understanding of what kind of data and how this data can support consumers' value creation
Customer-dominant logic	• *Shifts attention toward consumers embedding service in their own value-creating processes* • Provides conceptual tools to explore and analyze consumer perspective to transformative food retailing • Exerts pressure on extending focus from products and processes to people; gaining in-depth understanding of how food retailing influences consumers' lives • Emphasizes the understanding of customer logic, business perspective, offering, value formation, and ecosystem in uncovering how customers engage service in their everyday lives

systematically and analyze the potential of transformative food retailing. Table 13.3 is a synthesized conceptual framework for transformative food retailing. In order to illustrate the implications of transformative food retailing, we use a case example of nutrition code, an online-based service by a major Finnish retailer, Kesko Corporation. The service combines POS data of food grocery purchases with customer loyalty data, and the nutritional recommendations provided by the authorities (see, for example, Saarijärvi, Kuusela, Kannan, Kulkarni, & Rintamäki, 2016). This information is then offered to consumers through the online-based service. As a result, consumers are provided with information that has potential to support

Table 13.2 **Illustrating the shift toward transformative food retailing**

	Phase I Food retailing as supply	Phase II Food retailing as logistics	Phase III Food retailing as a service	Phase IV Food retailing as transformation
Theoretical orientation	Production-dominant	Goods-dominant	Service-dominant	Customer-dominant
Customer role	Passive object of generic supply	Passive object of different groceries	Active, individual consumer	Cocreator of multidimensional value
Type of customer data	POS data	POS data, loyalty program data	POS data, loyalty program data, consumer surveys, social media	POS data, loyalty program data, social media, consumer surveys, phenomenon related data (e.g., nutritive substance, CO_2 emissions, food origin, health data, locations data)
Main data usages	Ordering and delivery, inventory velocity	Store investments, category management, customer lifetime value, lean management	Consumer profiling, consumer lifestyles, personalization, customer loyalty program incentives	Supporting consumers' value-creating processes and practices, goals and desires, and consumers' wellbeing
Key challenges	Adequate level of production, balancing supply, and demand	Logistical optimization, efficient delivery system, global supply chain management, inventory management	Increasingly heterogeneous customer preferences, customer segment determination, and selection	Data infrastructure, data systems, consumers' willingness to share data, ethics, and security

Table 13.3 Conceptual framework for transformative food retailing

	Transformative food retailing	Illustrative case example: nutrition code
Consumer wellbeing	Consumers' health and wellbeing are at the epicenter of value cocreation. Wellbeing is understood both as objective wellbeing (e.g., healthfulness and nutritional balance of dietary intake) and subjective wellbeing (e.g., reduced stress and increased human agency)	Through tailored food healthfulness information, consumers have a concrete means to enhance their food dietary decisions. This influences consumers' wellbeing
Customer's data	POS data and loyalty program data combined with nutrition information of the groceries from national databases can be converted into information that supports consumers' food healthfulness. Customer data is converted into food healthfulness information; customer data becomes customer's data	Customer data is put into use for the benefit of the customer. It is converted into such information that is meaningful for consumers. Customer data becomes customer's data, the use of which they control
Customer logic	Consumers employ a multitude of food retailing related actions and practices, such as cooking, dietary practices, both utilitarian and hedonic aspects of consumption, and social eating and dining activities	Food healthfulness information is an increasingly important and useful resource that supports various of these consumers' activities
Business perspective	Food retailers place focus on understanding how they can engage in supporting consumers' wellbeing. This necessitates defining what consumer wellbeing is and offering customers vehicles for enhanced wellbeing through altered food choices (new uses of customer data, product category selections, educational activities). This alters food retailers' role: from grocery provider toward food healthfulness provider	Food retailer establishes new means for enhanced service and building customer loyalty. The more consumers concentrate their shopping to the food retailer, the more accurate food healthfulness information they receive. The food retailer's role develops into supporting consumers' wellbeing
Offering	Food retailers' offering is extended from groceries toward serving information that can be delivered to consumers through digital services. Information (resulting from innovative use of customer data) related to how consumers can enhance food-related wellbeing complements traditional offering	Food healthfulness information is a complementary element to the food retailing offering. In addition to traditional food groceries and additional services (e.g., click and collect, delivery services, etc.), consumers are serviced through meaningful information that has transformative potential for their wellbeing

(Continued)

Table 13.3 (Continued)

	Transformative food retailing	Illustrative case example: nutrition code
Value formation	Customers' value creation consists of combining various resources and processes in their everyday lives. In their endeavor toward healthier lives, food retailers can provide additional resources (information related to food healthfulness) to this process of value formation. Transformative value necessitates resource integration, where individuals are willing to use these tools to pursue better wellbeing	Customer data becomes a resource not only for the food retailer's value formation but also for the consumer's value formation. It is an additional resource to be integrated with various consumer processes centered around food consumption
Ecosystem	In addition to traditional food value chains, a food retailer's ecosystem will be characterized by various front-end and back-end data systems, food health professionals, and other service providers	Various resources and actors, e.g, POS data, customer loyalty scheme, data infrastructure, nutritional recommendations, and third-party service providers constitute an ecosystem with value and cocreative potential

various value-creating processes and consequently, enhance consumers' food healthfulness.

13.3 Discussion

Recent technological developments—together with innovative initiatives in harnessing the potential of customer data for the benefit of customers—uncover vast value potential for both consumers and food retailers. Among other things, this amplifies the transition toward transformative food retailing, i.e., reconfiguring food retailers' role from supplying groceries to supporting consumers pursuing healthier lifestyles. This evolutionary transition provides revolutionary opportunities to reconfigure the food retailer's role, not only in relation to competition but also in a societal context. This shift generates several implications for (1) consumers, (2) companies, (3) the academic community, and (4) society at large (see Table 13.4).

13.3.1 Transformative food retailing and consumers

Implications for consumers follow from the reduced information asymmetry between producers and consumers. First, this democratization of information inspires human agency, which enhances the individual's capacity to act and exert

Table 13.4 Implications of transformative food retailing

Stakeholder	Implications of transformative food retailing
Consumers	Enhancing value creation; increasing human agency; elevating identity projects; creating a sense of safety and certainty; granting access to information; assisting in informed choices; improving dietary intake; promoting healthy lifestyles
Companies	Offering a way for differentiation; establishing new markets for start-ups; motivating role reconfiguration; initiating strategic changes; inspiring new service design; reformulating service offering; engaging with different stakeholders; innovating new marketing processes; humanizing brand image; cultivating social responsibility
Academic community	Stimulating new retailing research avenues; advancing multidisciplinary collaboration; spurring new research methods; bridging the gap between scientific knowledge, business practitioners, and society; increasing research relevance and social impact
Society	Increasing the role of businesses in solving societal challenges; capturing societal potential of retailing; altering dominant social structures; stimulating social action; informing decision-making; raising public awareness of the impact of dietary choices for health; addressing socioeconomic differences; improving public health; providing new retailing-led public policy guidelines

control over their choices (Blocker & Barrios, 2015). A sense of greater agency represents a type of eudaimonic wellbeing (Bauer, McAdams, & Pals, 2008), as opposed to hedonic pleasure, and can lead to enhanced self-identity. Second, when consumers obtain the access to information that the service provider has traditionally held in their own hands, their sense of safety to operate with this provider rises. Third, since most people are unaware of the nutritional content of their dietary intake, information provided by retailers can support consumers' ability to make informed choices, holding the potential of furthering healthier eating while preserving their freedom of choice (Grunert, Bolton, & Raats, 2012). This becomes especially powerful if combined with other datasets about the consumer and their lifestyle. However, this transition requires active participation from the consumers' side; thus, consumer's competence, autonomy, and relatedness play a vital role for transformative value to be realized.

13.3.2   Transformative food retailing and companies

Implications for companies concern both traditional food retailers and new start-ups that can benefit from the transition. New start-ups that build transformative food retailing infrastructure (e.g., online services, back-end systems) can be established to target market segments that have been so far underserved. Data-driven start-ups are pioneering in acquiring, calculating, and interpreting the dietary data of groceries on behalf, and for the benefit, of consumers. Traditional food retailers are also developing their business through differentiation. Insight from transformative food retailing can provide input for initiating strategic changes, designing new service systems and offerings, engaging with different stakeholders, and innovating marketing processes. The transformative agenda can also be used as a branding tool for the retailer. By reconfiguring the role that the company plays in the community, social responsibility becomes an integral part of the company strategy and not an add-on element. However, contemporary and conservative executive mental models may hinder, or slow, the transition toward perceiving the greater role of food retailers in transforming the market. Companies may also be concerned whether the shift toward transformative food retailing is more profitable than traditional food retailing. While transformative food retailing may be a strong strategic initiative, focusing on product margins of healthy vs unhealthy products may in some cases discourage food retailers from taking an active role in facilitating behavioral changes toward healthier food consumption. However, retailers who lack in activity toward this direction could eventually meet stronger government legislation, restricting traditional retail operations and practices (Sparks & Burt, 2017).

13.3.3   Transformative food retailing and the academic community

Transformative food retailing inspires an array of new theories, concepts, methodologies, and research approaches, while encouraging multidisciplinary collaboration.

The research questions arise from real-life phenomena, addressing issues that have a great influence on individual and collective wellbeing. It is widely acknowledged that what makes research interesting and influential is its ability to challenge our current assumptions (Alvesson & Sandberg, 2011). By way of comparison, "transformative" studies in parallel fields, such as "transformative learning" or "transformative leadership," represent research streams that focus on challenging the current assumptions and altering existing patterns in their respected fields (Blocker & Barrios, 2015). Promoting wellbeing through research in the context of food retailing is a worthy endeavor and can explain, but also inspire changes in the marketplace. While retailers are being increasingly seen as part of the global health problem, transformative food retailing offers the potential to be part of the solution. Moreover, while the distance between scientific knowledge, business practitioners, society, and ordinary consumers is often described as long and wide, transformative food retailing represents context that can bring these parties closer to each other, while increasing the relevance and social impact of the research. However, transformative studies still represent a minority view, and researchers have to show courage and determination in order to further the transformation within academia.

13.3.4 Transformative food retailing and society

Implications for society are not only primarily linked to the concern of public health and wellbeing, but also to the influential role of food retailers. Despite increasing customer orientation, food retailers still have extensive power in determining what kind of groceries consumers eventually consume in respective markets. Especially in highly concentrated food retail markets, such as the Nordic countries, two or three companies may have 85%−90% of the market, which amplifies their role as major influencers of local food consumption behavior. Their in-store activities can promote unhealthy consumption; or as Sparks and Burt (2017: 3) articulate: "consumer 'desire' for unhealthy products has been encouraged and manipulated by the in-store and retail environment." However, in a similar way, in these highly concentrated markets, a single food retailer initiative, in terms of food healthfulness, can have a major societal impact on national health. Through transformative food retailing, society as a whole can benefit from addressing health problems linked to unhealthy food consumption, such as high blood pressure, increased type-2 diabetes, and several cancers. Furthermore, this has direct positive effects on public costs, work wellbeing, and productivity, further characterizing the societal potential of transformative food retailing. Finally, we referred earlier to the energized human agency at the individual level, which can aggregate to a collective moral agency, reflecting the proactive responsibility to "do good" (Bandura, 2002). Thus, we suggest that at the societal level, transformative value can cultivate uplifting social change.

13.4 Conclusion

Food retailing is at a crossroads. It is viewed increasingly as part of the problem of health and diet in society. It need not, however, be like this. There is a choice to be made between a battle over regulation or an embrace of digitalization, data availability, and health concerns. Transformative food retailing offers new venues for value creation, both for firms and customers, and points toward potential areas of strategic differentiation. New uses of customer data play a pivotal role and provide a new means for building customer loyalty. Changes in consumer behavior toward healthier food consumption at the individual level contribute to potentially major impacts at the societal level. Taken together, transformative food retailing has multiple value potentials and can extend the food retailers' role from supplying products and services toward facilitating consumers' personal and societal transformation toward healthier lives. This redefines and concretizes the importance of food retailing in contemporary society.

References

Adam, A., & Jensen, J. D. (2016). What is the effectiveness of obesity related interventions at retail grocery stores and supermarkets? − A systematic review. *BMC Public Health*, *16*(1247).

Adam, A., Jensen, J. D., Sommer, I., & Hansen, G. L. (2017). Does shelf space management intervention have an effect on calorie turnover at supermarkets? *Journal of Retailing and Consumer Services*, *34*, 311−318.

Afshim, A., Penalvo, J. L., Del Gobbo, L., Silva, J., Michaelson, M., O'Flaherty, M., ... Mozaffarian, D. (2017). The prospective impact of food pricing on improving dietary consumption: A systematic review and meta-analysis. *PLoS ONE*, *12*(3).

Alvesson, M., & Sandberg, J. (2011). Generating research questions through problematization. *Academy of Management Review*, *36*(2), 247−271.

Anderson, L., & Ostrom, A. L. (2015). Transformative service research: Advancing our knowledge about service and well-being. *Journal of Service Research*, *18*(3), 243−249.

Anderson, L., Ostrom, A. L., Corus, C., Fisk, R. P., Gallan, A. S., Giraldo, M., ... Williams, J. D. (2013). Transformative service research: An agenda for the future. *Journal of Business Research*, *66*(8), 1203−1210.

Andrews, J. C., Lin, C.-T. J., Levy, A. S., & Lo, S. (2014). Consumer research needs from the food and drug administration on front-of-package nutritional labeling. *Journal of Public Policy & Marketing*, *33*(1), 10−16.

Anker, T., Sparks, L., Moutinho, L., & Grönroos, C. (2015). Consumer dominant value creation. *European Journal of Marketing*, *49*(3), 532−560.

Bandura, A. (2002). Social cognitive theory in cultural context. *Applied Psychology*, *51*(2), 269−290.

Bauer, J. J., McAdams, D. P., & Pals, J. L. (2008). Narrative identity and eudaimonic well-being. *Journal of Happiness Studies*, *9*(1), 81−104.

Blocker, C. P., & Barrios, A. (2015). The transformative value of a service experience. *Journal of Service Research*, *18*(3), 265−283.

CDC (Centers for Disease Control and Prevention). (2013). *National health and nutrition examination survey: 2009−2010 data documentation, codebook, and frequencies, consumer behavior phone follow-up module-adults (CBQBFA_F)*. Available at: ⟨https://wwwn.cdc.gov/Nchs/Nhanes/2009-2010/CBQPFA_F.htm⟩.

Cohen, D. A., & Lesser, L. I. (2016). Obesity prevention at the point of purchase. *Obesity Reviews*, *17*, 389−396.

Escaron, A. L., Meinen, A. M., Nitzke, S. A., & Marinez-Donate, A. P. (2013). Supermarket and grocery store interventions to promote healthful food choices and eating practices: A systematic review. *Preventing Chronic Disease*, *10*, 120156. Available from https://doi.org/10.5888/pcd10.120156.

Gittelsohn, J., Rowan, M., & Gadhoke, P. (2012). Interventions in small food stores to change the food environment, improve diet and reduce risk of chronic disease. *Preventing Chronic Disease*, *9*, 110015. Available from https://doi.org/10.5888/pcd9.110015.

Glanz, K., Bader, M. D. M., & Iyer, S. (2012). Retail grocery store marketing strategies and obesity: An integrative review. *American Journal of Preventative Medicine*, *42*(5), 503−513.

Grunert, K. G., Bolton, L. E., & Raats, M. M. (2012). Processing and acting on nutrition labeling on food. In D. G. Mick, S. Pettigrew, C. Pechmann, & J. L. Ozanne (Eds.), *Transformative consumer research: For personal and collective well-being*. New York, NY: Routledge.

Grönroos, C., & Ravald, A. (2011). Service as a business logic: Implications for value creation and marketing. *Journal of Service Management*, *22*(1), 5−22.

Grönroos, C., & Voima, P. (2013). Critical service logic: Making sense of value creation and co-creation. *Journal of the Academy of Marketing Science*, *41*(2), 133−150.

Heinonen, K., & Strandvik, T. (2015). Customer-dominant logic: Foundations and implications. *Journal of Services Marketing*, *29*(6/7), 472−484.

Liberato, S. C., Bailie, R., & Brimblecombe, J. (2014). Nutrition interventions at point-of-sale to encourage healthier food purchasing: A systematic review. *BMC Public Health*, *14*, 919. Available at: http://www.biomedcentral.com/1571-2458/14/919.

IGD Retail Analysis. (2016). *Top five global grocery trends 2016*. Available at: ⟨retailanalysis.igd.com/⟩.

Maglio, P. P., & Spohrer, J. (2008). Fundamentals of service science. *Journal of the Academy of Marketing Science*, *36*(1), 18−20.

Mende, M., & van Doorn, J. (2015). Coproduction of transformative services as a pathway to improved consumer well-being: Findings from a longitudinal study on financial counseling. *Journal of Service Research*, *18*(3), 351−368.

Mick, D. G. (2006). Presidential address: Meaning and mattering through transformative consumer research. *Advances in Consumer Research*, *33*, 1−4.

Møller, J. J. (2011). Consumer loyalty on the grocery product market: An empirical application of Dick and Basu's framework. *Journal of Consumer Marketing*, *28*(5), 333−343.

O'Hern, M., & Rindfleisch, A. (2009). Customer co-creation: A typology and research agenda. *Review of Marketing Research*, *6*, 84−106.

Ryan, R. M., & Deci, E. L. (2000). Self-determination theory and the facilitation of intrinsic motivation, social development, and well-being. *American Psychologist*, *55*(1), 68−78.

Saarijärvi, H., Grönroos, C., & Kuusela, H. (2014). Reverse use of customer data: Implications for service-based business models. *Journal of Services Marketing*, *28*(7), 529−537.

Saarijärvi, H., Karjaluoto, H., & Kuusela, H. (2013a). Customer relationship management: The evolving role of customer data. *Marketing Intelligence & Planning, 31*(6), 584−600.
Saarijärvi, H., Mitronen, L., & Yrjölä, M. (2013b). From selling to supporting − Leveraging mobile services in the context of food retailing. *Journal of Retailing and Consumer Services, 21*(1), 26−36.
Saarijärvi, H., Kuusela, H., & Rintamäki, T. (2013c). Facilitating customers' post-purchase food retail experiences. *British Food Journal, 115*(5), 653−665.
Saarijärvi, H., Kuusela, H., Kannan, P. K., Kulkarni, G., & Rintamäki, T. (2016). Unlocking the transformative potential of customer data in retailing. *The International Review of Retail, Distribution and Consumer Research, 26*(3), 225−241.
Sparks, L. (2016). Spatial-structural change in food retailing in the UK. In A. Lindgreen, M. Hingley, R. Angell, J. Memery, & J. Vanhamme (Eds.), *A stakeholder approach to managing food*. Routledge.
Sparks, L., & Burt, S. (2017). *Identifying and understanding the factors that can transform the retail environment to enable healthier purchasing by consumers. Report for food standards Scotland (FSS 2016 013)*. University of Stirling. Available at: www.stirlingretail.com.
Vargo, S. L., & Lusch, R. F. (2004). Evolving to a new dominant logic for marketing. *Journal of Marketing, 68*(1), 1−17.
Vargo, S. L., & Lusch, R. F. (2008). Service-dominant logic: Continuing the evolution. *Journal of Academy of Marketing Science, 36*(1), 1−10.
Vargo, S. L., & Lusch, R. F. (2016). Institutions and axioms: An extension and update of service-dominant logic. *Journal of the Academy of Marketing Science, 44*(1), 5−23.
Wansink, B. (2017). Healthy profits: An interdisciplinary retail framework that increases the sales of healthy foods. *Journal of Retailing*. Available from https://doi.org/10.1016/j.retai.2016.12.007. (in press).

Building consumer trust and satisfaction through sustainable business practices with organic supermarkets: The case of Alnatura

14

Adrienne Steffen[1] and Susanne Doppler[2]
[1]Department of Consumer Behavior and Market Research, Hochschule Fresenius Heidelberg, Germany, [2]Department of Event Management, Hochschule Fresenius Heidelberg, Germany

14.1   Introduction to sustainable marketing practices

Today, the term sustainability is visible in all forms of communication such as magazines, newspapers, on company homepages, as well as on product packaging. The idea is, however, not new. Belz and Peattie (2012, pp. 23–31) review the evolution of marketing and point out that green marketing or environmental marketing has its origins in the 1970s and has further emerged in the late 1980s in Western Europe and North America (Belz & Peattie, 2012, p. 27).

A common understanding has emerged that sustainability has three interdependent dimensions: environmental, social, and economic. The aim of "environmental sustainability is the ongoing preservation of essential ecosystems and their function" (Martin & Schouten, 2014, p. 12). Economic sustainability is defined in its broader sense as "the ongoing ability of an economic system to provide for all human needs," and social sustainability refers to "the ongoing ability of communities to provide for the well-being of their members" (Martin & Schouten, 2014, p. 12).

Businesses have adapted since the 1970s and have undergone a major shift toward environmental and social sustainability. Marketing as one of many business functions has also changed toward more sustainable goals, starting to identify consumer needs and values and adapting its communication strategy accordingly. Sustainable marketing is the "process of creating, communicating, and delivering value to customers in such a way that both natural and human capital are preserved or enhanced throughout" (Martin & Schouten, 2014, p. 18). In the view of Belz and Peattie (2012, p. 28), sustainable marketing fosters institutional change toward sustainable development and is thus a macromarketing concept which takes on the idea that a change in behavior of virtually everyone is needed; both consumer and producer (Belz & Peattie, 2009, p. 30).

Case Studies in Food Retailing and Distribution. DOI: https://doi.org/10.1016/B978-0-08-102037-1.00014-1

Overall, there is a lack of literature on how consumers view sustainable products and services and even fewer studies which link the whole supply chain to consumer choice (Bask, Halme, Kallio, & Kuula, 2013). This chapter reviews the current literature on sustainable consumption in the context of food retailing. Food safety is of consumer concern, and the media has repeatedly reported on food scandals or food scares where firms have removed products from their shelves (Harvey, 2011). Major food scandals like the mad cow disease, which re-emerged around 2000, and also the Enterohaemorrhagic *Escherichia coli* (EHEC) (*Escherichia coli—e. coli*) vegetable crisis in Germany in 2011 have made consumers more aware of food quality, not only for meat but also for vegetables (Rampl, Eberhardt, Schütte, & Kenning, 2012). Besides official warnings around the incidence of these diseases, there also exists a government-administered website (www.lebensmittelwarnung.de) that collects firm-specific food recalls in Germany and makes them available to the general public (BVL, n.d).

Rather than looking into organic products and satisfaction on a general level, this case study takes a more holistic approach and investigates how companies can build consumer trust and satisfaction through sustainable business practices. The case of Alnatura is chosen, due to it being one of the biggest organic supermarkets in Germany in terms of turnover (Statista, 2014, p. 6), and one that has managed to gain a positive reputation through sustainable business practices.

The chapter objectives are

- To outline the sustainable consumption process.
- To identify sustainable food retailing trends.
- To analyze the role trust and satisfaction play in sustainable business practices.
- To evaluate the role of consumer trust and satisfaction in the case of Alnatura.

To understand the role trust and satisfaction play in sustainable business practices, the chapter first describes the sustainable consumption process and the conflicting attitudes consumers have about sustainable consumption. Then sustainable food trends are identified, and the role of trust and satisfaction for the firm—customer relationship in the supply chain is discussed. Interviews with consumers were conducted to evaluate the role of trust and satisfaction for Alnatura. Finally, the results of these consumer interviews are presented, and managerial implications are discussed.

14.2   The sustainable consumption process

At first glance, the terms "sustainable" and "consumption" seem to be a contradiction (Emery, 2012, p. 90), and there is not even any consensus about their definition as a combined term, only that they are competing positions. Jones, Hillier, and Comfort (2011, p. 938) refer to Jackson's (2006) review and note that there is a consensus that sustainable consumption goes hand in hand with a change in consumer behavior and lifestyle, thereby raising the question whether people should consume more efficiently, more responsibly, or whether they should consume less. Sustainable consumption is also identified as a behavior which "meets people's needs without compromising the ability of other people to meet their needs, either now or in the future" (Martin & Schouten, 2014, p. 72). From this definition, it is evident that sustainable consumption is complex. Indeed, Jones, Hillier, and Comfort (2014) review different definitions and note that the term is defined in various ways. The European Commission (2012, p. 32) even proposes that some people see sustainable consumption as a step back in life quality. Emery (2012, p. 91) argues that in comparison to regular nonsustainable consumption decisions, sustainable product attributes and the benefits of sustainable products have added complexity to the consumer decision-making process, whereas the basic decision process with its five steps (need recognition, information search, evaluation of alternatives, purchase decision process, and postpurchase behavior) is the same, the sustainability dimension is able to influence all five consumer decision-making steps (Martin & Schouten, 2014, p. 77). For example, at the information search stage, consumers might not drive long distances to get information in-store for environmental reasons, or information may be gathered about employee working conditions as a brand choice criterion.

In the past, the green marketer has only seen green consumers as a subsegment or specialist niche of the larger consumer base (Emery, 2012, p. 70). The new challenge of adapting sustainability marketing is now to target everyone, since sustainable consumption is no longer a niche market. Yet, of course, there are some consumers who do not have a sustainable attitude at all and thus exhibit contradictory behavior (Emery, 2012, p. 70—71). It can be assumed that some consumers are perhaps overwhelmed by this additional complexity: e.g., for understanding how to assess organic products, which ecolabels are officially certified, and how to find shops which sell sustainable products that can be trusted.

When both marketers and consumers have understood the concept of sustainable consumption and have been educated about the concept, consumers can be guided toward more sustainable consumption practices (Emery, 2012, p. 90). Here retailers have the potential to greatly influence consumers' sustainable consumption patterns (Jones et al., 2011) "through their own actions, through partnerships with suppliers and through their daily interaction with consumers" (Jones et al., 2014, p. 703). One way of doing this is to treat consumers as "co-marketers" instead of sales targets and to engage them in the marketing process meaningfully—e.g., with word-of-mouth or word-of-mouse activities (Martin & Schouten, 2014, p. 73).

Without this further re-education or active collaboration, consumers often have difficulties in understanding sustainability and do not adapt to sustainable practices because they associate sustainable consumption with "sacrifice, inconvenience, higher prices or poorer quality" (Emery, 2012, p. 90), or because it is a "frustrating sacrifice of our present, tangible needs and desires in the name of uncertain promotion for a better future" (European Commission, 2012, p. 32). If this is the consumer's understanding of sustainable consumption, such a misperception can lead to a lack of consumer trust (Emery, 2012, p. 91). Examples from the mobile phone industry show that if consumers have understood the concept of sustainable consumption, some are willing to pay a premium for phones with sustainable features (Bask et al., 2013). The same authors even conclude that many consumers are now interested "in the social and environmental impact of the entire supply chain of a firm" (Bask et al., 2013, p. 381).

14.2.1 Sustainable food retailing and consumption trends

Many consumers are concerned about the environmental and social impact of their consumption (Steffen & Günther, 2013) rather than just satisfying their own consumption needs (Belz & Peattie, 2009, p. 82−84). One quarter of the German population buys organic food regularly, and demand for organic eggs, fruit, and vegetables is especially large (Infas, 2016). The increasing number of food scandals Germany has faced in the years from 2011 onward (e.g., the EHEC vegetable scandal or the egg dioxin scandal) has left consumers with uncertainty about their food choices.

Due to these food scandals, there is a lot of pressure on retailers from the European Food Safety Authority, and from their own customers, to promote sustainable consumption (Rampl et al., 2012). Companies have responded by doing business in a sustainable and ethical way to fulfill the attitude of these ethical consumers and to achieve a competitive advantage in the long term (West, Ford, & Ibrahim, 2010, p. 446). However, retailers do not usually engage with all aspects of the sustainable consumption discourse and "focus on a subset of possible focus areas" (Lehner, 2014, p. 398). Consumers prefer to buy in supermarkets, but more than half of the population also buys in specialist stores selling organic products (Infas, 2016). They do that perhaps because consumers are tired of checking product information in traditional retail stores which use untrustworthy sustainability logos whose credentials are frequently questioned in the media (Schröder & McEachern, 2004).

14.2.2 The role of sustainability in creating consumer trust and satisfaction

Although consumers express concerns about environmental aspects of production, and despite the growing appearance of "green products" in retail outlets, the consumption of sustainable products becoming mainstream is still a challenging goal (Zhang, Liu, Sayogo, Picazo-Vela, & Luna-Reyes, 2016). This is partly because consumers are less willing to pay for them (Godelnik, 2012; Carvalho, de, Salgueiro, de, & Rita, 2016). Hence, besides price, the main barriers when it comes to consumers buying sustainable products or not are seen in the availability of sustainable products (Carvalho, et al., 2016), in consumers' expertise concerning sustainable products, and in consumer trust (reviewed by Zhang et al., 2016).

For the purpose of this study, we refer to the literature review and research on trust in food retailers by Rampl et al. (2012), who use Morgan and Hunt's (1994) definition of trust in buyer−seller relationships, namely, "the confidence of one party (the buyer) in the reliability and integrity of the exchange partner (the seller —i.e., retailer). Trust is founded on the anticipated capacity of the company to regularly satisfy consumer expectations and to avoid anything that might harm the consumer" (Rampl et al., 2012, p. 256−257). In defining satisfaction, we refer to the review of Loureiro, Dias Sardinha, and Reijnders (2012, p. 173), who define satisfaction as "an overall evaluation of a consumer's total purchasing and consumption experience of goods or services over time."

14.2.2.1 Factors influencing consumer trust

Bonn, Cronin, and Cho (2016) identified two determining factors of trust: consumer perceptions of sustainable practices of suppliers (producers and retailers) and consumer perceptions of product attributes (i.e., environmental attributes, sensory attributes, product health, and price). Based on their review, Bonn et al. (2016) suggest that consumers' sustainability perceptions and their future behavioral intentions are mediated by trust and commitment, rather than by satisfaction. Also, in developing and maintaining positive relationships between consumers and suppliers, trust is seen as one of the most influencing factors (Bonn et al., 2016; Daugbjerg, Smed, Andersen, & Schvartzman, 2014). Both studies provide evidence that consumers' trust in the sustainable practices of producers strongly influences buying decisions. In addition, Bonn et al. (2016) found that the sustainable practices of retailers have no significant impact on consumers' purchase intentions and that trust has the power to completely reverse the impact of higher prices from negative to positive. Referring to the framework of Mayer, Davis, and Schoorman (1995), Rampl et al. (2012) provide evidence that "integrity" is an important driver in building consumers' trust in food retailers: "the retailer must consistently follow a set of principles, e.g. support for environmentally friendly products and services … these principles must be accepted by the customer" (p. 258). As a second driver, these same authors identify "ability" or the "expertise and competence of the retailer, [and] the variety of offered goods and services" as an important driver of consumers' trust in food retailers (Rampl et al. 2012, p. 259). Furthermore, peer information, e.g., word-of-mouth and internet comments, have a great influence in building consumer trust (Carvalho et al., 2016).

Consumer perceptions of product attributes can be influenced by ecolabels. These transmit accessible messages to consumers, in a sense that a particular product has environmental or healthy attributes distinguishing it from non-labeled alternatives (Daugbjerg et al., 2014). As consumers usually purchase food through a retailer, they do not have the information needed to assess production standards at the point of sale. Therefore, consumers "have to rely on the eco-label for reassurance that the products do actually contain the desired [sustainable] attributes" (Daugbjerg et al., 2014, p. 564). A study by Samant and Seo (2016) provides empirical support for this argument. In addition, Castaldo, Perrini, Misani, and Tencati (2009) suggest that customers tend to trust retailers that have a reputation for sustainable conduct more than other retailers that do not.

Nevertheless, sustainable as well as organic labels are facing the problem that the production process cannot easily be verified and controlled by laboratory analysis through institutions or the customer. Therefore, beyond the level of sustainability labels, various product-certifying systems have been developed to provide quality assurance via certificated labels (Zhang et al., 2016). Currently, the "Ecolabel Index" (Ecolabel Index, n.d.) is tracking 465 sustainable labels in 199 countries, of which 59 match the search term "food," such as the labels "Demeter," "Fairtrade," or "Bioland". The Ecolabel Index was established in 2007 as an advisory service that helps all stakeholders of ecolabels to pass through the complex and diverse international certified ecolabel landscape, "associated with sustainable practices with varying standards and certification processes [which make it] difficult for consumers to assess the meaning and robustness of the certification criteria, and thus, trustworthiness of a label" (Zhang et al., 2016, p. 554).

Furthermore, ratings and awards seek to measure, compare, or reward corporate sustainability performance (SustainAbility, 2010). Gebhardt (2016a) analyzed the most common 24 German sustainability awards and found that these are a well-known and popular marketing communication tool for retailers, but that consumers are not aware of them. Rather, consumers place trust in sustainable and organic labels like "Fairtrade," "Demeter," or "Bioland" (Gebhardt, 2016b).

The determinants of consumer trust in the food retail context are shown in Fig. 14.1. All determinants besides certificates and awards are expected to have a positive influence in building consumer trust.

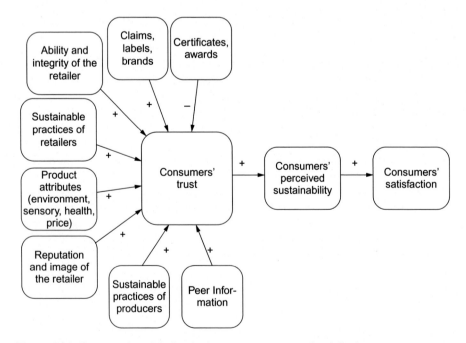

Figure 14.1 Conceptual model: developing consumer trust and satisfaction. Developed by the authors.

14.3 The case of Alnatura: How Alnatura builds trust and satisfaction through sustainable business practices

14.3.1 Company background

Alnatura was founded in 1984 by Götz Rehn in Mannheim, Germany, where the first supermarket for organically produced food was established in 1987. In 2015/16, 2760 employees generated revenues of €762 m. Götz Rehn supports the philosophy of anthroposophy based on Rudolf Steiner, which is closely linked to the organic production philosophy of Demeter. Based on the ideas of Rudolf Steiner, a special feature is the use of so-called biodynamic preparations for soil and plant treatment. German farmers started to apply those biodynamic treatments in 1924 and founded the association "Demeter," which is the oldest organic label in Germany (Demeter, n.d.). Over the past 30 years, the company has opened 119 organic supermarkets—branded as "Super Nature Market"—in 56 cities in Germany, an example of which can be seen in Fig. 14.2. The average sales area of each shop is 550 sq. m, and the assortment consists of approximately 6000 products. Most of the products are sourced regionally, and natural cosmetic and children's textiles are made of organic cotton (Alnatura, 2017).

Alnatura is classified as a producer and a retailer. The company has 14,700 distribution and trade partners in 17 countries. In Germany there are, for example, numerous national retailers like Edeka or Globus, who sell Alnatura products (Alnatura, 2017). Other European partnerships exist in Austria and in Switzerland (Alnatura, 2017). Distribution takes place centrally. Alnatura's distribution center is located in Lorsch, Germany and is the largest high rack warehouse of wooden construction in the world. It has photovoltaic cells, an air—water heat pump, and uses

Figure 14.2 Interior of Alnatura organic supermarket.
The authors.

geothermal energy (Alnatura, 2017; Materialfluss, 2014). Currently, the most important distribution partner, "dm," which is a national drugstore, delisted Alnatura products in their stores while discontinuing the partnership. This delisting caused a tremendous decline in Alnatura's revenue. As a consequence, Alnatura had to reorganize its partner network and is now associated with new partners such as "Edeka," "Globus," "Müller," and "Rossmann," which are major supermarkets or drugstores in Germany, as well as "Migros" in Switzerland (Eschbacher, 2017).

Sustainability is the key concept of Alnatura's business (Alnatura, n.d. a). Alnatura's (n.d. a); understanding of sustainability is based on its corporate vision to act sensibly for both humankind and the earth (see also Fig. 14.3). In terms of the economic and social aspects of sustainability, long-lasting partnerships with suppliers and farmers are aimed for, as well as a peaceful, fair, and family-friendly management (Alnatura, n.d. b, n.d. c). Since 2014, Alnatura, (n.d. b) has been obliging its manufacturers to comply with the "Alnatura Policy of Social Standards" to implement rules for working and living conditions along the Alnatura product supply chain. In the field of ecology, the responsible treatment of energy, water, soil, and air resources, as well as farm practices with the treatment of plants and animals, are addressed. According to Alnatura's understanding, certified organic farming principles like "Bioland," "Demeter," and "Naturland" are best suited to these ideas. Consequently, all agricultural ingredients of the more than 1000 Alnatura products come from organic farming (Alnatura, n.d. d). Alnatura even designs its store layout sustainably. The shelves are made from regional wood, the tiles are made of natural stone, recycled material is employed, and ecological, nonnuclear

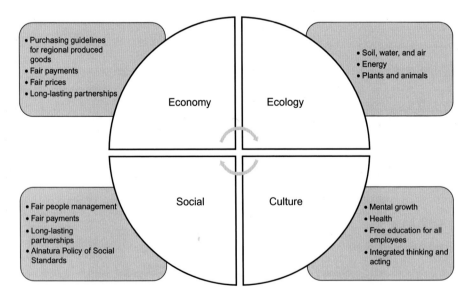

Figure 14.3 Alnatura's understanding of sustainability.
Alnatura (n.d. a) (translated by the authors).

power from the electrical power station located in the city of Schönau is used (Alnatura, n.d. e).

Besides the social and ecological dimensions of sustainability, Alnatura (n.d. a) attaches importance to "culture, mental growth, and health" in a sense of free education for all employees, supporting integrated thinking and acting. Due to this mission statement, employees are invited to special workshops to gain knowledge in the field of sustainability. Moreover, theatre workshops and dialogue events for employees and customers enhance Alnatura's cultural dimensions (Alnatura, n.d. e).

In 2016, Alnatura was awarded the "Deutscher Nachhaltigkeitspreis" (German Sustainability Award), being the most sustainable company in the category of "medium-sized companies" (Alnatura, n.d. f) and the "Corporate Social Responsibility (CSR) Award" in the "maintenance of worldwide biodiversity" category (Alnatura, n.d. g). The CSR Award recognizes companies and institutions that take on social responsibility and are characterized by sustainable management (Alnatura, n.d. h).

14.3.2   Methodology and research design

A case study approach was chosen to investigate how Alnatura builds trust and satisfaction through sustainable business practices. It is a single, retrospective case using mainly an explanatory but also a theory-building approach (as described in Thomas, 2016, pp. 114−116). Three main methods were used to gain insights for the case study. First, a literature review on sustainable business practices and the role of satisfaction and trust in consumption was conducted. Second, a review of the brand's media coverage was carried out. Third, 10 semistructured personal interviews with Alnatura customers were undertaken. A convenience sample of 10 Alnatura customers was chosen by both researchers, since no access to Alnatura's customer base was possible. Participants were selected based on whether they had purchased Alnatura products in the past. Thus, a prescreening which included this question took place before the interviews. A gender and age mix took place as far as possible, considering the brand purchase criteria.

The interview guidelines were developed based on the findings of the literature review, in particular, the determinants and influencers depicted in the conceptual framework (see Fig. 14.1). The interview first asked about the general meaning of sustainable consumption for the consumer and his or her sustainable consumption habits. Then, product-related sustainability factors were explored before a shift toward discussion of Alnatura took place. Here, purchase behavior, attitude, and the role of trust and satisfaction were investigated. The interview ended with a discussion of the chances and opportunities Alnatura has concerning trust, satisfaction, and sustainability. The duration of interviews varied from 30 to 40 minutes.

The interviews were recorded after the interviewee signed a participant consent form. All interviews were transcribed, and the content was organized in an Excel file. The data analysis took place in the form of coding, which is the starting point for most qualitative data analysis (Bryman & Bell, 2011, p. 584). Relevant text passages were translated from German to English and paraphrased. Codes were assigned and reviewed again in a second round to find bigger themes in the text, as proposed by Bryman and Bell (2011, p. 586). Computer-assisted qualitative data analysis was not seen as necessary with a relatively small sample size.

14.3.3 Sample characteristics

The sample consisted of six females and four male participants (see Table 14.1). Their ages ranged from 22 to 66 years. Most participants used several purchase channels and bought organic products frequently. Supermarkets and organic supermarkets are popular organic purchase channels amongst consumers. Convenience and store accessibility played the biggest role in deciding where to shop. Several participants noted that the store has to be closed to their normal travel routes, e.g., on their way to work or elsewhere, because the thought of driving somewhere to buy organic products was against the principles of sustainable consumption.

14.4 Building trust and customer satisfaction at Alnatura

14.4.1 Spontaneous reaction to the term sustainable consumption

When asking for spontaneous ideas about the term "sustainable consumption" several themes emerged. These included the use of sustainable, respectively, organic labels and claims and product attributes like environment and health. Some comments included

> I solely check if the product is produced organically. (Female, 50).

> [I trust the] Fairtrade claim, because they often correspond with organic production. (Male, 22).

On a more abstract level, the determinants "sustainable practices of producers" as well as the "integrity of retailers" appeared to be relevant when it comes to consuming in a sustainable way:

> Adequate animal housing; furthermore, I distinguish larger from smaller agricultural plants, e.g. for milk production; this is an important criterion. (Male, 60).

> The retailers' context. I don't trust organically produced food offered at discounters like Aldi or Lidl. (Female, 54).

14.4.2 Determinants of trust and satisfaction in sustainable consumption

In the second step, the drivers of trust and satisfaction from Fig. 14.1 were explored individually.

Table 14.1 Sample characteristics

	Profession	Age	Gender	Organic product consumption	Frequency of organic product purchase location	Biopurchase channels	Criteria for organic product purchase location
1	Pensioner (self-employed before)	66	Female	Buys organic products e.g., vegetables and fruits, muesli, meat, coffee	Frequently but not every purchase	Supermarket, drugstore, organic supermarket, health food store, farmer's market, farm shop	Convenience, shops where she currently is close to
2	Physiotherapist	44	Female	Buys organic products occasionally when there is a lack of alternatives, mainly vegetables, cereal, oat milk, muesli	Every week or two	Discounter, supermarket, organic supermarket, drug store	Convenience, shops where she can get everything, needs to be on her way
3	Pensioner (salesperson and nanny before)	65	Female	Often buys organic products, mainly milk, cheese, and meat products, wine	Frequently, but not every purchase	Supermarket, organic supermarket, drugstore, discounter, farmer's market, farm shop	Accessibility, needs to be on her way

(Continued)

Table 14.1 (Continued)

	Profession	Age	Gender	Organic product consumption	Frequency of organic product purchase location	Biopurchase channels	Criteria for organic product purchase location
4	Student, part-time	22	Male	Buys organic products often if it is not too expensive, mainly groceries and clothes	Every day (goes shopping daily)	Avoids organic supermarkets, supermarket, farmer's market, farm shop, internet	Convenience, the supermarket is close to where I live
5	Chemical laboratory worker (quality control)	51	Male	Buys organic fruit and vegetables, especially bananas	Occasionally	Supermarket, discounter	Convenience, needs to be on his way, distance is key
6	Geologist	60	Male	Family buys 90%–100% organic products for everything they buy (only son uses conventional products for body hygiene)	Once a week	Organic supermarket (farm shop now closed)	Accessibility and availability of parking
7	Nonmedical practitioner	50	Female	Buys organic groceries, cosmetics, washing detergent, and vegan products	Almost all products are bio, two to three times a week	Organic supermarket, health food store, farmer's market	Accessibility and product quality

8	Chemical engineer	58	Female	Buys fresh products like fruits and vegetables, fresh tea (no bags), oats, oil, tofu products, pulse, spices, bread	Every 2 weeks	Organic supermarket, farm shop, supermarket, discounter (buys only bioproducts at discounters)	Distance and accessibility
9	Yoga teacher and nonmedical practitioner	54	Female	Buys vegetables, fruits, and meat organic products if she buys them. Milk, oil, tea, juice, and clothes if they are not too expensive	Three times a month in bulk	Organic supermarket, supermarket, farm shop	Distance and accessibility
10	Police officer	59	Male	Buys organic products like muesli, vinegar, oil, frozen fruits, sweats, vegetables, fruits, sometimes chicken, nuts, yoghurt, milk products, fish, and eggs	A few times a week	Supermarket, Organic supermarket, health food store	Product assortment and quality

14.4.2.1 Trust

Concerning the use of sustainable and organic claims, on the one hand, participants attest that they trust in certified labels like Demeter, Bioland, and Fairtrade. In building trust, transparency about the label seems to be a critical issue. Furthermore, consumers are suspicious of discounters' labels and a lack of comprehensibility:

> *Yes, it plays a big role, because I think that the farmers are checked to get the labels. I would trust that label. (Female, 60).*

> *I trust in big claims that are used extensively. Bioland, Fairtrade and eco. But I do not trust any claim that the companies lend themselves, like discounters' organic claims. (Male, 22).*

Nevertheless, organic labels like Demeter and Bioland seem to be more trustworthy than the more socially responsible label Fairtrade:

> *At Demeter, I know that it is very high-quality; that creates more trust than Bioland and Fairtrade. (Male, 60).*

> *That is a question of conscience, e.g., Fairtrade in the case of coffee: [for me] the claim isn't comprehensible and reliable, therefore the trustworthiness is questionable and I wonder if it's true [that the coffee is produced and traded under fair and organic conditions], because I can't understand [if the coffee is produced and traded under fair conditions]. (Female, 66).*

On the other side, interviewees stated that claims do not play a role when it comes to building trust:

> *I do not buy because of the claim. (Female, 44).*

> *Not necessarily. They are known, but I do not pay attention to them. (Male, 50).*

To differentiate between labels and brands, a second question was introduced, where the role of brands was explored. The statements confirm a positive influence of brands and reflect that the role of the brand relates to the availability of them at the preferred point of sale and is clearly linked to individual experiences:

> *I buy Alnatura products, which I assess as good but not outstanding quality, at the supermarket. Others like Rapunzel or Demeter I assess to be of higher quality. (Male, 59).*

> *[The brand plays a role] only when I experienced the brand individually or when I read about the brand in the press. (Female, 50).*

The role of "sustainable practices of producers" is confirmed by the statements provided above. Environmental aspects, as well as social aspects, are seen to be critical aspects:

This plays the biggest role for me. That's why I prefer Fairtrade rather than organic labels. My internal ranking is first to look at people who are currently living. Secondly, I look for future generations and how their wellbeing will be influenced when I buy the product; and thirdly how I feel with the product concerning my health. (Male, 22).

At the same time, consumers state a lack of knowledge about the sustainable practices of producers:

All in all, it is clear to me that organic farming is different to conventional farming practices, but for the individual organic labels I can't distinguish between farming practices and standards. (Female, 54).

The role of sustainable practices for the retailer is perceived to be fair people management and payment:

If the salespeople or employees are treated well, then yes, or if they are paid fair. (Female, 65).

I'd be shocked hearing about food scandals at Alnatura. (Female, 58).

The assortment and service offer of the retailers play a role by means of the combination of organic assortment and service, in the sense of friendliness and atmosphere. However, the assortment appears to be seen as more important than service:

[The assortment and services are] a main criterion for choosing my retailer. (Female, 54).

The assortment of goods is important; service only if I look for something special. (Female, 58).

The expertise of retailers, expressed in the knowledge of employees, is discussed critically. The study participants do not experience, and do not generally expect specialist knowledge, of supermarket employees. Although some of the participants would appreciate knowledge, none of the interviewees claimed it as a critical factor:

In the food sector, I do not need any specialist knowledge. In other matters, e.g. technology or clothes, it would play a bigger role. (Female, 44).

According to my experience, there is a lack of knowledge; therefore, I prefer to inform myself. But I would appreciate it if I'd have the opportunity to ask employees. (Female, 50).

The reputation, integrity, and image of retailers seems to be a weaker determinant, as interviewees do not feel influenced by a special image and tend to build their own personal view about the retailers:

> *I build up my own opinion about the retailer. (Male, 59).*

> *Probably [I get influenced by image], but I don't notice. (Female, 50).*

Word-of-mouth and peer information given by friends and family make experiences more concrete. Therefore, they are assessed to be more relevant than "promotion", which does not appear to play a role in building trust:

> *Word-of-mouth given by friends and family makes experiences easier to grasp. It's more important than promotion. (Female, 50).*

> *I do not really look at advertising. I am more interested in documentations about sustainably produced products, organically produced food, or about health. (Male, 59).*

> *Promotion has little influence. (Female, 58).*

All interviewees confirmed that certificates and awards do not have an impact on their trust in products and companies:

> *Certificates don't play a role at all; I'm not aware of certificates. (Male, 59).*

> *I don't trust in certificates; they are a question of money if you pay for it − you'll get it. I trust more in regional products. (Female, 44).*

To conclude, in food retailing, the use of big, well-known claims, labels, and brands like Bioland, Demeter, Fairtrade, and Alnatura, as well as the sustainable practices of producers and retailers, are very strong drivers for building trust in sustainable consumption; followed by the ability of retailers in relation to assortment size and quality, service, and employee knowledge. Peer information has a stronger impact than retailer reputation and image. Promotion and advertising have a rather weak and unconscious impact. Certificates and awards seem to be unsuitable instruments to develop trust.

14.4.2.2 Sustainable consumption and satisfaction

Overall, satisfaction seems to play an important role in sustainable consumption decisions:

> *Dissatisfaction would lead to a change in buying behavior. (Male, 50).*

In detail, products that address environmental protection tend to lead to higher enjoyment and positive feelings through the idea of protecting the environment:

Yes, buying an organically produced product makes me feel happier and maybe I enjoy it more than not organically produced alternatives. (Male, 59).

Definitely, because I'm fascinated by ideas and innovations in offering products and protecting the environment at the same time. This causes enthusiasm and higher expectations and a more positive feeling. (Male, 22).

Yes, because I can influence the environment with my organic consumption. This positive environmental influence creates satisfaction in me. (Male, 60).

However, interviewees also stated that environmental topics do not always serve individual needs:

No, I'm rather selfish. I want to have the product, that's all. I am not the organic type. This is really a bad attitude, but that's how it is. But on the other side, I wouldn't buy a peeled cucumber or an egg in a plastic package. (Female, 44).

Others emphasize that, in their opinion, it is not enough to address solely organic production and environmental protection. Thus, while for most of the interviewees aspects of environmental protection and nature conservation seem to have an impact on satisfaction, they are not more important than questions of quality and taste:

Well, for me quality is the most important criteria. If I care for managed grassland with apple trees but I don't like their taste then it is not satisfying. I tried organic coffee, but I didn't find one that suited me and therefore I don't buy it. (Male, 59).

Besides environmental protection and nature conservation, social aspects of production seem to be the strongest determinant confirmed by all interviewees, as it touches on consumers' sense of conscience, and consequently, consumers are more satisfied with themselves:

Yes, one has better peace of consciousness. (Female, 44).

Yes, that would strongly influence me. But in my opinion, we have no influence on it, e.g. child labor in less developed countries. (Female, 54).

Yes, I'm more satisfied with myself, because I behaved better. If I feel pleased with myself it leads to satisfaction with the product. (Male, 22).

When being asked for the meaning of trust and satisfaction, all interviewees confirmed the importance of trust and satisfaction when it comes to purchasing organically produced food, whereas satisfaction seems to have a stronger influence than trust:

Trust is important, but nevertheless I always have to think about products myself. (Female, 54).

Purchasing behavior would suffer in case of dissatisfaction. (Male, 50).

14.4.3 Alnatura, trust and satisfaction

When asked for activities which Alnatura applies to develop customers' trust, interviewees confirmed that these included product-focused promotions, friendly salespeople, a good servicescape like bakeries, and visible signs of sustainability like the huge wooden high rack warehouse:

> Nearly every Alnatura has a bakery where freshly baked products are sold;
> Alnatura tries to connect the salesperson and customer. They try to improve their
> sustainable image through the large wooden high rack warehouse, for example.
> Secondly, they offer their products at Edeka [a popular supermarket in Germany] and
> other markets to reach customers who otherwise wouldn't buy organic products.
> (Male, 22).

The interviewees argue that a lack of bad experiences with Alnatura products, an absence of invasive promotion, good quality and taste of products, and the brand Alnatura itself, lead them to trust the retailer:

> For me, their recall actions are trustworthy. So far, I have not had any bad
> experiences. The brand doesn't make unnecessary promotion or internet
> advertising, from a visual point of view everything looks smart and is thereby
> trusting. (Female, 44).

> For me, the brand Alnatura stands for sustainability, organic and competence.
> (Male, 50).

Counter to this, one reason for potential mistrust in Alnatura is seen in the similarity to conventional supermarket strategies:

> I trust the brand Alnatura with restrictions. I have heard that the founder
> comes out of a rather conventional business area and when I look at the products,
> the procedure (convenience products, packed products) is very similar to that
> of conventional supermarkets. For sure, Alnatura is one of the driving forces to
> develop organic supermarkets towards an extremely wide range of
> offerings, like refrigerated products, packaged meat, sausage and cheese.
> (Female, 60).

All in all, it can be said that Alnatura's communication covers all determinants of trust and corresponds with the statements given by interviewees (Fig. 14.4). For most interviewees, satisfaction seems to be linked to product attributes like quality and taste, as well as the absence of negative public affairs. Previous marketing efforts by Alnatura in building consumer trust and satisfaction seem to have been successful. Because Alnatura has built strong trust and satisfaction in the past, the removal of Alnatura products at the German drugstore "dm" has not affected consumers' trust and satisfaction. This fits the conclusion that activities in the field of promotion, advertisement, image, and reputation were not noticed or perceived to be negative by customers.

Alnatura's determinants of trust:	
The use of organic labels and claims	✓
The sustainable practices of producers	✓
The sustainable practices of retailers	✓
The ability of retailers	✓
Reputation, integrity, and image of retailers	✓
Peer information/word of mouth	✓
Role of certificates, awards	⊘

Figure 14.4 Alnatura's determinants of trust.
The authors.

14.5   Implications for consumers and organizational strategy

The results should be interpreted with caution and by taking into consideration the limitations of the study. The sample consisted mainly of people above the age of 40 and contained more females than males. The studies of Infas (2016, p. 6) and of Niva, Mäkelä, Kahma, and Kjærnes (2014) showed that older consumers, particularly women, engage more in sustainable food consumption than younger consumers and males in general. Other studies show that there generally seems to be an attitude−behavior gap for sustainable consumption where attitude and behavior are not consistent (Vermeir & Verbeke, 2006). The results of Niva et al. (2014, p. 479) also indicate that many people have a positive attitude about sustainable food consumption but do not practice it themselves. Because in our study all participants buy organic products, this hesitant behavior can only be partially confirmed; in that, taste and store location or convenience appear more important than buying organic for some participants. The findings also provide insights into consumers who buy organic products and do not need to be convinced by retailers about sustainable consumption practices. Overall, the results suggest that consumers generally trust Alnatura's claims and that they are very satisfied with Alnatura's organic supermarket concept. Yet only people who have previously purchased at Alnatura were interviewed, so perhaps these results are not surprising. One might assume that these customers are biased toward Alnatura, because they are trying to justify their previous store selection decision.

Research results have shown that retailers' sustainability agendas are influenced by their own communication goals. A firm's focus might be on balancing the three sustainability pillars (economic, social, and environmental) rather than "reducing demands on finite natural resources" (Jones et al. 2014, p. 712). One could thus critique Alnatura's very heavy growth strategy, which is most closely linked to the economic pillar. The firm is expanding quickly, and new organic supermarkets are opening across Germany. Thus, it is questionable which of the defined values they in fact follow. It can be suspected that the economic sustainability pillar is more prevalent than the social or environmental at Alnatura, or at least more important than the firm admits openly. A few respondents, especially those who are well-informed, seem to critique Alnatura for this economic focus. They raised a critical voice about Alnatura because they put "economic drivers in the foreground" (Male, 60). This respondent has previously worked in a store which sells organic products and evaluates Alnatura's business practices as being too similar to conventional retailers. He notes that Alnatura's "boss is from conventional retail" and is shocked because the "practice with ready-made food and prepacking mirrors conventional retailing" too much (Male, 60). He further critiques the availability of fruit and vegetables year-round from overseas. Further critique was raised by a younger interviewee (Male, 22) who was also very knowledgeable about sustainable food as a vegan consumer and sees himself as an opinion leader in terms of organic food. He has the feeling that Alnatura has a bad image, does not treat its employees well, and notes that the brand is too expensive in comparison to other organic brands.

The interview extracts suggest that Alnatura as an organic brand seems to appeal to mainstream consumers. Those consumers with special dietary needs, e.g., vegan consumers or those who are extremely knowledgeable because they buy only organic products, have a more critical attitude toward sustainable consumption. Here, perhaps the definition of sustainable consumption is key to understanding the deviation. Very knowledgeable and rather extreme organic purchasers perhaps have another understanding of sustainability. Thus, it probably depends on how the firm defines sustainability itself and how this definition is the perceived by others, particularly potential customers. For Alnatura, sustainability is a balance of four main factors (see Fig. 14.3), and this is clearly communicated. Yet the examples of the two critical consumers who have another understanding of sustainability show that communication, especially about how sustainability is defined by the firm, could be a key factor in creating consumer trust and satisfaction.

Clearly, retailers should work on providing information about their supply chain at the point-of-sale. Research from Jones et al. (2011, p. 935) showed that the top 10 UK grocery retailers had little in-store information available that leads customers to engage with sustainable consumption and instead were heavily encouraging mass consumption. The authors concluded that these 10 retailers were only following the weak model of sustainable consumption (Jones et al., 2011, p. 935). A follow-up study by the same authors (Jones et al, 2014, p. 712) showed that customers in the UK only had limited information about sustainable production processes in-store and were bombarded with dominant marketing messages instead.

The Alnatura case shows that it is not so important to really have in-store information on every product available, but rather, to build a brand which consumers trust. Some consumers who are inexperienced in buying organic products perhaps need more in-store guidance for their sustainable shopping behavior, but more experienced consumers trust the brand and are less likely to require such information. The homepage can be a central hub for all communication activities, and it should contain all relevant information about the firm's sustainable activities. The firm's understanding of sustainability and all company activities and processes concerning sustainability need to be published to build consumer trust. Any additional communication activities which the firm undertakes should link to the homepage. With the help of Quick Response (QR) codes on product packaging, consumers are able to access additional information about the production location, transportation, or other sustainability-related issues on their mobile devices at the time of purchase. In addition, QR codes or weblinks in customer magazines or on posters can link to the firm's homepage. Further, communication in customer magazines, e-mail newsletters, on social media, and on YouTube (via videos which document all steps in the sustainable value chain) are all important communication activities which help to distribute information on sustainability.

The developed conceptual model in Fig. 14.1 shows that sustainable practices of the producer and the retailer influence consumer trust positively. In addition, the ability, integrity, reputation, and image of the retailer have a positive influence on consumer trust. Peer information also seems to drive consumer trust. Although the product attributes, claims, labels and brands have a positive effect on consumer trust at a product level, non-accredited certificates and sustainability awards have no influence on consumer trust. Thus, one could conclude that the firm should provide as much product information as possible. Yet, the effort needed to provide this product information needs to be seen in relation to the generated benefit and the generated return. The verification of such information needs to be manageable in a constantly changing business environment with a large number of suppliers (Jones et al., 2011, p. 944). As an alternative to providing information for every product, Jones et al. (2011, p. 944) propose that food retailers should practice "choice editing" which means not stocking environmentally damaging products in the first place. When retailers have gained consumer trust, like Alnatura appear to have, they can concentrate their efforts on monitoring that sustainability is driven throughout the supply chain.

14.6   Conclusion

The case study of Alnatura highlights the importance of communication in building consumer trust and consumer satisfaction. A conceptual model was derived from the literature to explain the drivers of consumer trust and satisfaction. Here the determinants of trust were peer information, sustainable practices of producers, reputation and image of the retailer, product attributes, sustainable practices of the retailers, ability and integrity of the retailer, claims, labels and brands, and certificates and awards. All of these factors are shown to have a positive influence on trust, besides certificates and awards, which influence consumer trust negatively. All of these mentioned factors influence consumers' perceived sustainability and lastly form consumer satisfaction.

Alnatura was a pioneer and has developed into one of the biggest organic supermarkets in Germany. The firm has a holistic understanding of sustainability by adding culture as a dimension to the three pillars of economy, ecology, and society which are traditionally cited in the literature. Interviews with Alnatura customers show that Alnatura generally does a very good job in creating consumer trust and satisfaction by communicating thoroughly. Only two very knowledgeable and critical customers raised negative remarks about the economic orientation of the firm, perhaps because they have a different understanding of the term "sustainability." Yet the results need to be evaluated critically, because only those Alnatura customers were interviewed who like the brand and have shopped there in the past. They are not necessarily objective because they want to justify their past shopping behavior.

This case shows that communication is necessary to explain a firm's understanding of sustainability. Only when a firm's understanding of sustainability is clearly communicated, can the customer decide whether there is a match with his or her own understanding. In turn, this is helpful for consumers in deciding how much they trust the firm's sustainability claims. Communication is the key to building the right consumer expectations, e.g., concerning the supply chain but also concerning human resources. The homepage, the firm's information brochures, customer magazines, social media, QR codes with links to the producers or their own YouTube channel, help consumers to trust in the retailer's sustainability claims.

References

Alnatura. (n.d. a). Wie wir Nachhaltigkeit verstehen [How we understand sustainability]. < https://www.alnatura.de/de-de/ueber-uns/nachhaltigkeit/wie-wir-nach-haltigkeit-verstehen > Accessed 26.06.17.
Alnatura. (n.d. b). Dimension Soziales [Social dimensions]. < https://www.alnatura.de/de-de/ueber-uns/nachhaltigkeit/wie-wir-nachhaltigkeit-verstehen > Accessed 26.06.17.
Alnatura. (n.d. c). Dimension Wirtschaft [Economic dimensions]. < https://www.alnatura.de/de-de/ueber-uns/nachhaltigkeit/wie-wir-nachhaltigkeit-verstehen/dimension-wirtschaft > Accessed 30.06.17.
Alnatura. (n.d. d). Was ist die Alnatura Bio-Qualität [What is the Alnatura organic quality?]. < https://www.alnatura.de/de-de/ueber-uns/qualit%c3%a4t > Accessed 07.11.17.
Alnatura. (n.d. e). Dimension Kultur [Cultural dimension]. < https://www.alnatura.de/de-de/ueber-uns/nachhaltigkeit/wie-wir-nachhaltigkeit-verstehen/dimension-kultur > Accessed 26.06.17.
Alnatura. (n.d. f). Alnatura ist Deutschlands nachhaltigstes Unternehmen [Alnatura is Germany's most sustainable company]. < https://www.alnatura.de/de-de/magazin/alnatura-aktuell/archiv-2016/gewinner-alnatura-gewinnt-deuschen-nachhaltigkeitspreis-2016 > Accessed 26.06.17.
Alnatura. (n.d. g). Alnatura erhält Deutschen CSR-Preis [Alnatura receives German CSR award]. < https://www.alnatura.de/de-de/magazin/alnatura-aktuell/archiv-2016/csr-preis-2016-gewonnen > Accessed 26.06.17.
Alnatura. (n.d. h). Alnatura löst Coca-Cola als beliebteste Lebensmittelmarke der Deutschen ab [Alnatura replaces Coca-Cola as the most popular food brand in Germany]. < https://www.alnatura.de/de-de/magazin/alnatura-aktuell/archiv-2014/alnatura-beliebt-este-lebensmittelmarke > Accessed 26.06.17.
Alnatura. (2017). Daten und Fakten [Dates and facts]. < https://www.alnatura.de/de-de/ueber-uns/presse/daten-und-fakten > Accessed 28.06.17.

Bask, Anu, Halme, Merja, Kallio, Markku, & Kuula, Markku (2013). Consumer preferences for sustainability and their impact on supply chain management, In *International Journal of Physical Distribution & Logistics Management*, *43*(5/6), 380–406.

Belz, F. M., & Peattie, K. (2009). *Sustainability marketing: A global perspective*. Chichester: Wiley.

Belz, F. M., & Peattie, K. (2012). *Sustainability marketing: A global perspective*. Chichester: Wiley.

Bonn, M. A., Cronin, J. J., & Cho, M. (2016). Do environmental sustainable practices of organic wine suppliers affect consumers' behavioral intentions? The moderating role of trust. *Cornell Hospitality Quarterly*, *57*(1), 21–37.

Bryman, A., & Bell, E. (2011). *Business research methods* (3rd ed.). Oxford University Press, Oxford

BVL. (n.d.). Bundesamt für Verbraucherschutz und Lebensmittelsicherheit [Federal Office of Consumer Protection and Food Safety], Lebensmittelwarnungen, < http://www.lebensmittelwarnung.de/bvl-lmw-de/app/process/warnung/start/bvllmwde.p_oeffentlicher_bereich.ss_aktuelle_warnungen > Accessed 07.04.17.

Carvalho, B. L., de, Salgueiro, M., de, F., & Rita, P. (2016). Accessibility and trust: The two dimensions of consumers' perception on sustainable purchase intention. *International Journal of Sustainable Development & World Ecology*, *23*(2), 203–209.

Castaldo, S., Perrini, F., Misani, N., & Tencati, A. (2009). The missing link between corporate social responsibility and consumer trust: The case of fair trade products. *Journal of Business Ethics.*, *84*, 1–15.

Daugbjerg, C., Smed, S., Andersen, L. M., & Schvartzman, Y. (2014). Improving eco-labelling as an environmental policy instrument: Knowledge, trust and organic consumption. *Journal of Environmental Policy & Planning*, *16*(4), 559–575.

Demeter. (n.d. i). Demeter – Pioniere der Bio-Branche [Demeter – pioneers in the organic sector]. < https://www.demeter.de/organisation > Accessed 07.11.17.

Ecolabel Index. (n.d.). Ecolabel Index | Who's deciding what's green? Available at: < http://www.ecolabelindex.com/ > Accessed 08.06.17.

Emery, B. (2012). *Sustainable marketing*. Harlow: Pearson.

Eschbacher, Bettina. (2017). 'Es ist wie seelisches Turnen'. Interview mit Alnatura-Chef Götz Rehn, Schwetzinger Zeitung, 2 Mai 2017, Wirtschaft, S. 4.

European Commission. (2012). *Policies to encourage sustainable consumption*. Available at: < http://ec.europa.eu/environment/eussd/pdf/report_22082012.pdf > Accessed 30 June 2017.

Gebhardt, B. (2016a). Ausgezeichnet.pdf, Nachhaltiges Management Studien zur nachhaltigen Unternehmensführung, Band 27. Verlag Dr. Kovac, Hamburg.

Gebhardt, B. (2016b). 'Nachhaltigkeitspreise: Ernährungsbranche schöpft Potenzial noch nicht aus [Sustainability awards: food industry is not exploiting potential yet]. Available at: < https://www.uni-hohenheim.de/pressemitteilung?tx_ttnews%5Btt_news%5D = 31902& cHash = 36d76e864118bb704fb58525e84 > Accessed 07.11.17.

Godelnik, R. (2012). *Consumers increasingly sceptical of green products? Maybe not.* Available at: < http://www.triplepundit.com/2012/10/consumers-pay-premium-green-products-study-data/ > Accessed 08.06.17.

Harvey, J. (2011). A ratings-based stakeholder analysis for a food production company, including trust and risk implications. *Business Strategy Series*, *12*(2), 98–104.

Infas. (2016). *Infas Institut für angewandte Sozialwissenschaft GmbH Ökobarometer im Auftrag des Bundesministeriums für Ernährung und Landwirtschaft (BMEL)*. < https://de.statista. com/statistik/studie/id/32670/dokument/studie-zum-einkauf-von-biolebensmitteln-2016/ > Accessed 15.11.17.

Jackson, T. (2006). Readings in sustainable consumption. In T. Jackson (Ed.), *The Earthscan Reader in Sustainable Consumption*. London: Earthscan.

Jones, P., Hillier, D., & Comfort, D. (2011). Shopping for tomorrow: Promoting sustainable consumption within food stores. *British Food Journal, 113*(7), 935−948.

Jones, P., Hillier, D., & Comfort, D. (2014). Sustainable consumption and the UK's leading retailers. *Social Responsibility Journal, 10*(4), 702−715.

Lehner, M. (2014). Translating sustainability: The role of the retail store. *International Journal of Retail & Distribution Management, 43*(4), 386−402.

Loureiro, S. M. C., Dias Sardinha, I. M., & Reijnders, L. (2012). The effect of corporate social responsibility on consumer satisfaction and perceived value: The case of the automobile industry sector in Portugal. *Journal of Cleaner Production, 37*, 172−178.

Martin, D., & Schouten, J. (2014). *Sustainable marketing*. Harlow: Pearson.

Materialfluss. (2014). *Hochregallager von Alnatura komplett aus Holz [High-bay warehouse made entirely of wood by Alnatura]*. < http://www.materialfluss.de/lager-und-kommissioniertechnik/hochregallager-von-alnatura-komplett-aus-holz/ > Accessed 15 November 2016.

Mayer, R. C., Davis, J. H., & Schoorman, F. D. (1995). An integrative model of organizational trust. *The Academy of Management Review, 20*(3), 709−734.

Morgan, R. M., & Hunt, S. (1994). The commitment-trust theory of relationship marketing. *Journal of Marketing, 58*(3), 20−38.

Niva, M., Mäkelä, J., Kahma, N., & Kjærnes, U. (2014). Eating sustainably? Practices and background factors of ecological food consumption in four Nordic countries. *Journal of Consumer Policy, 37*, 465−484.

Rampl, L. V., Eberhardt, T., Schütte, R., & Kenning, P. (2012). Consumer trust in food retailers: conceptual framework and empirical evidence. *International Journal of Retail & Distribution Management, 40*(4), 254−272.

Samant, S., & Seo, H. (2016). Quality perception and acceptability of chicken breast meat labeled with sustainability claims vary as a function of consumers' label-understanding level. *Food. Quality and Preference, 49*, 151−160.

Schröder, M., & McEachern, M. (2004). Consumer value conflicts surrounding ethical food purchase decisions: a focus on animal welfare. *International Journal of Consumer Studies, 28*(2), 168−177.

Statista. (2014). *Dossier: Bio-Supermärkte in Deutschland [Organic supermarkets in Germany]*. < https://de.statista.com/statistik/studie/id/25061/dokument/bio-supermaerkte-in-deutschland-statista-dossier/ > Accessed 03.07.17.

Steffen, A., & Günther, S. (2013). Success factors of cause-related marketing—What developing countries can learn from a German sweets campaign. *The Mena Journal of Business Case Studies*.

SustainAbility. (2010). *Rate the raters. Phase one. Look back and current state*. < http://www.sustainability.com/library/rate-the-raters-phase-one# > Accessed 28.11.17.

Thomas, G. (2016). *How to do your case study* (2nd ed). London: Sage.

Vermeir, I., & Verbeke, W. (2006). Sustainable food consumption: Exploring the consumer attitude−behavioral intention gap. *Journal of Agricultural and Environmental Ethics, 19*(2), 169−194.

West, D., Ford, J., & Ibrahim, E. (2010). *Strategic marketing* (2nd ed). Oxford: Oxford University Press.

Zhang, J., Liu, H., Sayogo, D. S., Picazo-Vela, S., & Luna-Reyes, L. (2016). Strengthening institutional-based trust for sustainable consumption: Lessons for smart disclosure. *Government Information Quarterly, 33*(3), 552−561.

Spices of the future: Forecasting the future of food retailing and distribution with patent analysis techniques

15

Daniel Boller[1,2] and Johanna F. Gollnhofer[1,3]
[1]Institute for Customer Insight, University of St. Gallen, Switzerland, [2]Stanford Graduate School of Business, Stanford, CA, United States, [3]Institute for Marketing and Management (Consumption, Culture and Commerce), University of Southern Denmark, Odense, Denmark

15.1 Introduction

Food retailing is a vibrant and highly competitive industry (Colla, 2004). Besides consumer-driven initiatives (for instance "slow food" (Sebastiani, Montagnini, & Dalli, 2012; Tencati & Zsolnai, 2012; Chaudhury & Albinsson, 2015)) or community-supported agriculture (Thompson & Coskuner-Balli, 2007), technological advances challenge the food retailing industry. For instance, Amazon is about to revolutionize the retail sector with automatic checkout ("Amazon Go") in grocery stores, providing Amazon with a competitive edge. These technological advances do not literally fall from the sky, but are developed over the years by companies, research institutions, and other stakeholders in the marketplace. Patents materialize those technological advances on paper and can be used to study these technological advances. For instance, already in 2012, Amazon had applied for patents which substantiate the basis for Amazon Go (Dwarakanath, Blakey, Casteel, & Cooper, 2012).

In 2015 alone, 42,991 patents, representing major technological advances (such as bioprinting for food products; Gatenholm, Backdahl, Tzavaras, Davalos, & Sano, 2010) with specific regard to food retailing and distribution were registered. Understanding such technological advances has been shown to be vital for the long-term success of businesses (Fabry, Ernst, Langholz, & Köster, 2006; Rust & Espinoza, 2006). However, anticipating and forecasting such technological advances is difficult since they are characterized by a high degree of uncertainty and complexity (Lee, Yoon, Lee, & Park, 2009; Lee, Yoon, & Park, 2009) and often lack common understanding and comprehension among practitioners and researchers. Building on this, the current research provides not only a holistic overview of the most relevant techniques in analyzing patent data, but, most importantly, shows how patent analysis techniques can be applied to allow business practitioners to assess future trends in the food retail market.

Case Studies in Food Retailing and Distribution. DOI: https://doi.org/10.1016/B978-0-08-102037-1.00015-3

Patents, patent applications, and utility models (hereafter patents) offer crucial insights into how technologies in a specific field will develop and advance by mapping out available and future technologies (i.e., current R&D activities). Patents can be seen as a representation of R&D development in a specific field of interest as they officially record and institutionalize technological advances (Danneels, 2004). In other words, patents outline the development of (new) technologies, which in turn are likely to affect whole industries and related business practices. Research that relies on patent analysis techniques provides crucial insights into future developments: For instance, Daim, Rueda, Martin, and Gerdsri (2006) analyzed patents in the field of "fuel cell", "food safety" and "optical storage" technologies in order to forecast emerging technologies. Tseng, Lin, and Lin (2007) analyzed patents of the National Science Council Taiwan (i.e., the government agency that sponsors research activities in Taiwan) to detect the core areas of research and development across different sectors and industries. Lee et al. (2009) examined "RFID" (radio-frequency identification) patents yielding to technology-driven strategy assessments for companies.

Recent methodological developments offer innovative avenues to explore technological advancements via patent analysis. The application of these technological advancements in the food retailing industry lies at the core of this book chapter. The chapter proceeds as follows: first, we provide a brief overview of the most recent patent analysis techniques in order to identify technological advances that will drive the future of food retailing. Second, two case studies in the food retailing industry illuminate how patent analysis techniques can be applied in order to detect future technological trends in the food retailing industry. Finally, we summarize the results, outline future potential for the food market based on the applied methods, and stress the implications for practitioners in their strategic decisions.

15.2   Patent analysis techniques

This section explains in colloquial terms trend analysis techniques for patent data with a focus on (1) trend detection and analysis and (2) content analysis. We outline the basic mechanisms to allow business analysts in the food retailing industry to study, investigate, and analyze patent data in order to uncover future trends. The goal of this section is to provide a basic understanding of the most relevant approaches.

15.2.1   Trend detection and analysis

There are two concepts to trend detection and analysis, namely internal and external. The concept of internal trend detection derives relevant trends solely on a specific patent database. Three core concepts can be distinguished. First, patents are grouped and categorized based on topic modeling techniques and are modeled as time series (based on the publication date of the patent) in order to identify promising and emerging trends within the set of patents in an explorative way. Second, patents are grouped and categorized by patent applicants or inventors and modeled as time series to identify promising and emerging technology leaders within the set of patents. Third, patents are weighted by their relative importance (according to their citations and references by other inventions) in order to identify which patents lie at the nexus of R&D activities and drive technological development in a specific field of interest.

The concept of external trend detection builds on an existing understanding in the area of interest, whereby two core concepts can be distinguished. First, publicly available information (e.g., within newspaper articles, blog posts, etc.) are collected and analyzed with respect to a specific field of interest. The resulting potentially relevant topics are linked to available patent data in order to validate their relevance against actual technological developments. Second, a panel of experts is interrogated with respect to their perception of relevant trends in a specific field of interest, resulting in a set of potentially relevant topics. These potentially relevant topics are linked to available patent data in order to assess their relevance against actual technological developments.

15.2.2 Content analysis

Content analysis techniques do not allow for predictions, but display the status quo of a specific technology. Content analysis techniques provide an initial understanding of patents and, thus, may serve as the basis for the above-outlined concepts of external and internal trend detection (Wanner et al., 2008). Two concepts of content analysis can be distinguished. First, frequent term analysis offers crucial initial insights into patents within a specific field of technologies by showing which terms (or phrases) occur frequently in those patents (see, for example, Schecter & Mortinger, 2013; Tseng et al., 2007). Second, sentiment analysis (based on a predefined set of words or phrases; e.g., Akhondi et al., 2016) or machine learning methods (e.g., Kim, Suh, & Park, 2008) allow for patent data to be structured based on textual information and for higher order categories to be derived, which finally can be used to assess the most relevant topics.

The presented techniques constitute the basis of patent analysis approaches. Depending on the research questions and the study objectives, the different patent analysis techniques can be applied simultaneously, consecutively, and in combination. The following two case studies serve as hands-on illustrations of patent analysis techniques with the objective to provide novel insights into the future of food retailing.

15.3 Case studies

15.3.1 Case study 1

The aim of case study 1 is to show how future trends in the field of food retailing can be derived based on external information and patent data (i.e., external trend detection). In particular, we first identified a set of potentially relevant trend indications based on external information (i.e., newspaper and website articles), and, second, linked those trend indications to patent data. This process assures the validity of the trend indications.

In the first step, we obtained 768 newspaper and website articles (published in the period from January 2016 to December 2016) via LexisNexis, with specific regard to future trends in food retailing (search term: "future food retailing"). The collected newspaper and website articles included the name of the publishing house, publication date, headline, publication type, publication language, and publication texts. Based on term frequency analyses of the publication texts, we identified and defined a set of nine trend indications (see Fig. 15.1). The abscissa indicates the percentage of how often a trend indication was mentioned across all the collected newspaper and website articles.

The derived set of trend indications set the stage for the following patent analysis, which was based on 75,075 patents, patent applications, and utility models dealing with "retailing" (published in the period between January 1899 and October 2016; requested via the European Patent Office). The collected patent data included the publication number, publication date, title, abstract, description texts, International Patent Classification (IPC), applicants, and inventors for each of the patents. The subsequent analysis relied on the abstracts of each patent given that the patents' abstracts reflect the most relevant aspects of the invention.

First, building on the corpus of 75,075 patents, we extracted those patents which explicitly referred to the previously derived set of nine trend indications (based on key term extraction procedures for the respective nine trend indications; e.g., smartphone: patents including the phrase/term "mobile device," "mobile phone," "cellphone," or "smartphone") and that were published between January 1990 and December 2015, finally resulting in a set of 24,066 patents. Second, we aggregated

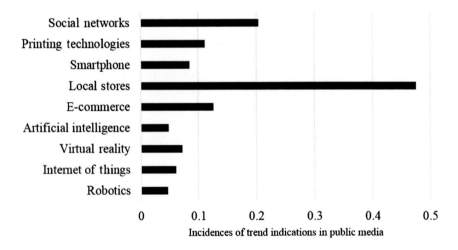

Figure 15.1 Incidences of trend indications in public media.

the number of patents per trend indication and month (based on the publication date of the patents) to control for divergent dates of disclosure among the patent and trademark offices. This procedure resulted in nine-time series (i.e., the number of monthly issued patents for each trend indication during the entire time period). Third, we applied the unobserved components model (UCM) (Harvey, 1989) to forecast the future development of the previously identified trend indications (forecasting period: 4 years; January 2016−December 2019). Finally, we calculated the growth factor for each of the nine trend indications (i.e., (monthly) average number of patents that are expected to be published within the next 4 years divided by the (monthly) average number of patents that were published during the last 4 years). The results show us which trend indications are supposed to be relevant in the future of food retailing and merit explicit attention from a technology perspective (compare Fig. 15.2).

Patent analysis techniques offer valuable insights into the future of an industry and offer guidance to top executives. By juxtaposing those results with the current trend indicators (compare Fig. 15.1), we argue that managers need to make a mental shift from current topics to future topics. Future topics and trends might not be so prominent in public media right now, but our patent analysis indicates a significant shift in some areas. According to our results, the "internet of things," "robotics," "e-commerce," and "printing technologies" (i.e., food printing technologies) are the four biggest trends that will shape the retailing industry in the future. However, local stores and social media seem to lose traction. This does not mean that they become irrelevant but only that innovation pace loses traction in those domains.

Next, we will turn to those domains where we expect—according to our results—major changes, challenges, and opportunities:

- The "Internet of Things"—referring to objects exchanging data with other objects—will provide the technology for developing decentralized, integrated, networked, and adaptive

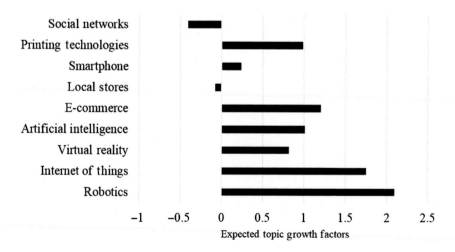

Figure 15.2 Expected topic growth factors.

order options for customers. For instance, Walmart (the US retailer) developed a technology, which aligns different channels (i.e., omnichannel management). This technology puts the consumer in the center of interest and organizes the different channels around the consumer (Natarajan & High, 2015).

• "Robotics" will improve food storage by replacing established inventory systems with intelligent and integrated robotic systems (a network of supply and demand resources). Moreover, robot technologies are expected to revolutionize distribution channels through autonomous agents, such as vehicles and drones. For example, Zume Pizza (American pizza baker) developed a technology, which allows pizza delivery by drones and, thus, has replaced traditional delivery channels (Garden, 2013).

• E-commerce technologies remain a relevant topic in the food retailing industry since e-commerce technologies lie at the nexus of retailer−consumer interactions in our increasingly digital world. Those technologies allow for improving pricing strategies (e.g., Mohapatra, Sahu, Chowdhury, & Panda, 2012) and for seamlessly aligning online and offline distribution channels (e.g., Chelly, Levi, Dekel, & Menipaz, 2013).

• Finally, in our analysis, we identified printing technologies as a driver in the retail industry. Printing technologies for food, including, but not limited to, 3D techniques or composition of ingredients, will change food retailing as well as the food consumption habits of consumers. For example, Natural Machines (a Spanish technology company) developed a 3D printer, which allows the consumer to create food items (e.g., cookies) at home (Kuo et al., 2013).

Understanding and developing these four technologies in-depth is expected to be crucial for companies and managers as it may influence competitive advantage and consolidation within the industry. It is noteworthy to mention that not all topics that receive a lot of "buzz" in our current times are expected to grow in the future (for instance, social media). Such insights might allow managers to reschedule their priorities.

15.3.2   Case study 2

The objective of case study 2 was to gain a better understanding of future consumer-centric retail technologies without any preexisting knowledge (internal trend detection). The analysis was based on 412,078 patents, patent applications, and utility models dealing with "food" (published in the period between January 1893 and November 2016; requested via the European Patent Office). The collected patent data included the publication number, publication date, title, abstract, description texts, IPC, applicants, and the name of inventors for each of the patents. The subsequent analysis relied on the abstracts of each patent given that the patents' abstracts reflect the most relevant aspects of the invention. We narrowed our dataset by focusing on those patents that explicitly referred to consumer-centric retail technologies (based on key term extraction: "consumer," "buyer," and "shopper"). In order to capture the most recent status of the field, we grounded the analysis in patents that were published between January 2010 and December 2015 (based on the publication date), leaving us with a set of 463 patents for further analysis. We applied topic modeling techniques to the patents in order to obtain a set of topics (i.e., terms or phrases which occur in thematically or temporally related contexts and can be grouped into higher order topics). The latent Dirichlet allocation (LDA) algorithm (i.e., three-level hierarchical Bayesian model; Blei, Ng, & Jordan, 2003) allowed us to detect the main topics within the patents (i.e., the patents' abstracts). In LDA algorithms "each item of a collection is modeled as a finite mixture over an underlying set of latent topics. Each observed word originates from a topic not directly observable. Each topic is, in turn, modeled as an infinite mixture over an underlying set of topic probabilities." (Blei et al., 2003, p. 993).

The number of topics (10) was determined by maximizing the log likelihood values for a prespecified number of $1-100$ topics (Griffiths & Steyvers, 2004; Hornik & Grün, 2011). Following this, the LDA algorithm computed term to topic ratios for each of the 10 topics. We opted for a two-step approach in order to summarize the terms that were assigned to each topic. In the first step (Step I), 29 participants ($M_{Age} = 33.38$, $SD_{Age} = 10.32$, and female $= 51.72\%$; recruited from an online panel) were presented with the derived terms assigned to each topic and asked to propose a name for each of the 10 topics. In the second step (Step II), 32 participants ($M_{Age} = 32.84$, $SD_{Age} = 8.91$, and female $= 25.00\%$; recruited from an online panel) rated which of the 29 proposed names from Step I best characterizes the terms assigned to each of the 10 topics. We took the proposed names with the highest approval rate to name the 10 topics.

Finally, we applied the UCM method to forecast the future development of the previously derived trends (see Table 15.1) for the following 4 years (January 2016 until December 2019). The UCM method allows us to predict the number of patents for each of the 10 topics. As in case study 1, we calculated the growth factor for the 10 topics (i.e., (monthly) average number of patents that are expected to be published within the next 4 years divided by the (monthly) average number of patents that were published during the last 4 years). Fig. 15.3 displays the trends that are supposed to shape the future of retailing and related R&D activities.

Our results indicate the future trends of customer-centric retail technologies. As such, they offer first indications to managers in terms of allocation of human, monetary, and material resources. The most important trends should be addressed. First,

Table 15.1 Topics and example terms of topics

Topic	Example terms
Customer services	Cloud, self-service, and wireless
Food production	Fresh, instant, and vegetable
Storage technologies	Contain, chamber, and portion
Human physical comfort	Need, health, and diet
Order/Information services	Phone, menu, and mobile
Organization/Inventory	Reorder, inventory, and reusable
Marketing/Information issues	Develop, advertise, and marketing
Delivery technologies	Transport, configure, and move
Hardware technologies	Device, automat, and machine
Kitchen technologies	Cook, cabinet, and box

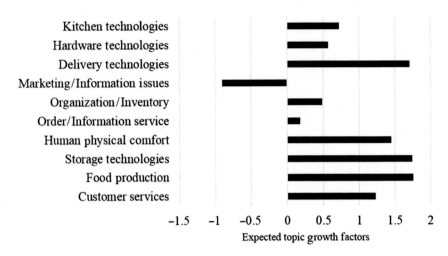

Figure 15.3 Expected topic growth factors.

food production technologies are found to be a driver for future food retailing, including, for example, in-store and local as well as fresh production technologies. For example, Restaurant Technologies (a US company) developed an RFID-based technology that automatically monitors food items (e.g., in a restaurant) and creates automatic orders or even produces the food itself (Schackmuth & Sus, 2006). Second, storage technologies, like intelligent food portioning and packaging systems are suspected to play a major role in future food retailing. For example, Kraft Foods (the US food production company) developed an intelligent food packing systems, which allows consumers to cook instant food items without unpacking them (Raymond et al., 2006). Third, food delivery, including local or at home production systems and transport infrastructures (e.g., drone delivery) constitute, as already outlined in case study 1, a major pillar in current and future R&D activities, and are suspected to shape the way in which retailers interact with customers and food producers.

15.4 Discussion and future research

In this chapter, we have demonstrated how patent analysis techniques allow managers and researchers to shift their focus from contemporary topics (such as social media) to future trends that are supposed to shape significantly the food retailing industry. Patent analysis techniques offer a competitive edge in highly competitive and vibrant industries as they allow managers not only to make judgments and decisions based on their gut feeling, but based on rigorous analysis. According to our analysis, topics that are highly relevant this year in food retailing, such as social media, will lose relevance, whereas other topics, such as robotics, will be of paramount importance for the food retailing industry (compare case study 1).

Moreover, the production, storage, and delivery of food products are shaped by major technological advances in the near future (compare case study 2). Our analyses offer first valuable insights into the trends in the retailing industry. These trends could constitute the basis for further research that might focus on one specific trend and explore this trend in-depth (through qualitative as well as quantitative approaches).

Besides this, the chapter also offers insights into patent analysis techniques by outlining relevant methodological approaches in the area of patent analysis and by applying those in two case studies in the food retailing industry. In case study 1, we illustrated the concept of external trend detection, which draws on public media coverage. We link these insights with patents in order to identify relevant technology trends in the food retail industry. In case study 2, we employed the concept of internal trend detection that identifies a set of topics within a given patent database (through topic modeling techniques). Both the external and internal trend detection approaches allow us to forecast the future developments of the identified topics in the food retailing industry.

We argue that patent analysis techniques constitute a major competitive advantage in the food retailing industry as they allow practitioners to evaluate future business opportunities based on a technology perspective. Both internal and external trend detection represent essential approaches for business analysts in order to identify future trends in their defined field of practice. Once the understanding for future trends is created, subsequent strategic decisions regarding the company's future can be considered and implemented. This understanding does not only allow for an assessment of promising future market technologies, but also constitutes a major advantage in the rapidly consolidating food retailing industry as potential merger and acquisition (M&A) activities can be evaluated against the identified trends or with regard to relevant technology leaders.

The presentation of the applied patent analysis approaches in this research work can be seen as an introduction for business analysts who are interested in future perspectives and trends in the food retailing industry. Thus, this book chapter not only provides insights with regard to future perspectives and trends in the food retailing industry, but also introduces a powerful tool that opens up more advanced and nuanced ways of investigation into the future compared to common market research and trend analysis techniques.

The following steps provide guidance for implementing and applying patent analysis in the food retailing industry:

1. *Definition of Purpose.* The main power of patent analysis is to draw meaningful conclusions (e.g., definition of future market trends or segments) from a technology perspective for a defined topic of interest. Thus, it is important to develop an initial scope and goal for the envisioned patent analysis. These goals might refer, for example, to considerations of novel market segments, product developments, M&A activities, or food distribution channels. Based on the defined goal, practitioners should assess the different possible patent analysis approaches, that is, external trend detection or internal trend detection. Following this, a relevant cluster of technologies, industries, patent applicants, or regions for the analysis purposes has to be selected.

2. *Data*. Patent data constitute an integral part of the analysis. Patent data can be retrieved from public patent and trademark authorities (e.g., United States Patent and Trademark Office, European Patent and Trademark Office, or World Intellectual Property Organization), although it should be noted that those platforms work with different search parameters and modalities. Finally, the obtained patent data should be verified with regard to data consistency and correctness.

3. *Analysis and Modeling*.
   a. Data Preparation. The most relevant variables are represented by the applicant/inventor/assignee, publication/filing/priority date, patent text (title, full text, and abstract of the patent), citations, publication/patent number, and the responsible patent and trademark authority. The textual information has to be prepared for further analysis by applying text preparation procedures.
   b. Data Analysis. The prepared data can be analyzed using longitudinal or cross-sectional analysis procedures. The relevant variables, models, and model parameters have to be set in accordance with the general analysis purpose and available variables.

4. *Verification*. The results of the analysis have to be checked in order to ensure the results' overall validity. The verification includes, for example, the modification of sensitive model parameters and scenario-based model applications.

5. *Interpretation and Preparation*. Finally, the results of the analysis should be interpreted and presented in a practical way to make the results accessible and visible to the targeted audience (i.e., to make findings applicable to a defined business strategy).

Patent analysis techniques are an emerging methodological toolbox that promise to deliver intriguing results about future developments; thus, they are a perfect supplement to established market research activities for companies operating in the retail industry.

References

Akhondi, S. A., Pons, E., Afzal, Z., van Haagen, H., Becker, B. F. H., Hettne, K. M., ... Kors, J. A. (2016). Chemical entity recognition in patents by combining dictionary-based and statistical approaches. *Database*, *2016*, 1−8.

Blei, D. M., Ng, A. Y., & Jordan, M. I. (2003). Latent Dirichlet allocation. *Journal of Machine Learning Research*, *3*(1), 993−1022.

Chaudhury, S. R., & Albinsson, P. A. (2015). Citizen-consumer oriented practices in naturalistic foodways: The case of the slow food movement. *Journal of Macromarketing*, *35*(1), 36−52.

Chelly, Y., Levi, O. S., Dekel, T., & Menipaz, A. R. (2013). *Online and offline ecommerce connections*. U.S. Patent No. US20140279013 A1. Washington, DC: U.S. Patent and Trademark Office.

Colla, E. (2004). The outlook for European grocery retailing: Competition and format development. *The International Review of Retail, Distribution and Consumer Research*, *14*(1), 47−69.

Daim, T. U., Rueda, G., Martin, H., & Gerdsri, P. (2006). Forecasting emerging technologies: Use of bibliometrics and patent analysis. *Technological Forecasting and Social Change*, *73*(8), 981−1012.

Danneels, E. (2004). Disruptive technology reconsidered: A critique and research agenda. *Journal of Product Innovation Management*, *21*(4), 246−258.

Dwarakanath, S. H., Blakey, S. W., Casteel, B. R., & Cooper, B. J. (2012). *Inventory system with climate-controlled inventory*. U.S. Patent No. US9185998 B1. Washington, DC: U. S. Patent and Trademark Office.

Fabry, B., Ernst, H., Langholz, J., & Köster, M. (2006). Patent portfolio analysis as a useful tool for identifying R&D and business opportunities—An empirical application in the nutrition and health industry. *World Patent Information, 28*(3), 215–225.

Garden, A. J. (2013). *Systems and methods of preparing food products*. U.S. Patent No. US9292889 B2. Washington, DC: U.S. Patent and Trademark Office.

Gatenholm, P., Backdahl, H., Tzavaras, T. J., Davalos, R. V., & Sano, M. B. (2010). *Three-dimensional bioprinting of biosynthetic cellulose (BC) implants and scaffolds for tissue engineering*. U.S. Patent No. US8691974 B2. Washington, DC: U.S. Patent and Trademark Office.

Griffiths, T. L., & Steyvers, M. (2004). Finding scientific topics. *Proceedings of the National Academy of Sciences of the United States of America, 101*(1), 5228–5235.

Harvey, A. (1989). *Forecasting, structural time series models and the Kalman filter*. Cambridge, NY, USA: Cambridge University Press.

Hornik, K., & Grün, B. (2011). Topicmodels: An R package for fitting topic models. *Journal of Statistical Software, 40*(13), 1–30.

Kim, Y. G., Suh, J. H., & Park, S. C. (2008). Visualization of patent analysis for emerging technology. *Expert Systems with Applications, 34*(3), 1804–1812.

Kuo, C. J., Huang, S. H., Hsu, T. H., Rodriguez, L., Olivé, X., Mao, C. Y., ... & Delgado, V. (2013). *Apparatus, method and system for manufacturing food using additive manufacturing 3D printing technology*. U.S. Patent No. US20160135493 A1. Washington, DC: U.S. Patent and Trademark Office.

Lee, S., Yoon, B., Lee, C., & Park, J. (2009). Business planning based on technological capabilities: Patent analysis for technology-driven roadmapping. *Technological Forecasting and Social Change, 76*(6), 769–786.

Lee, S., Yoon, B., & Park, Y. (2009). An approach to discovering new technology opportunities: Keyword-based patent map approach. *Technovation, 29*(6), 481–497.

Mohapatra, M. P., Sahu, D., Chowdhury, S., & Panda, D. (2012). *Product pricing in E-commerce*. Patent No. EP2626827 A1. Munich, Germany: European Patent and Trademark Office.

Natarajan, C., & High, D. R. (2015). Retail subscription in internet of things environment. U.S. Patent No. US20170124633 A1. Washington, DC: U.S. Patent and Trademark Office.

Raymond, M. N., Teasdale, A. C., Kwiat, C. L., Forneck, K. D., Pai, S., & Gan, R. (2006). *Packaging system for storage and microwave heating of food products*. U.S. Patent No. US20080063760 A1. Washington, DC: U.S. Patent and Trademark Office.

Rust, R. T., & Espinoza, F. (2006). How technology advances influence business research and marketing strategy. *Journal of Business Research, 59*(10), 1072–1078.

Schackmuth, G., & Sus, G. (2006). *RFID food production, inventory and delivery management method for a restaurant*. U.S. Patent No. US20070254080 A1. Washington, DC: U.S. Patent and Trademark Office.

Schecter, M., & Mortinger, A. (2013). *Using analytics to generate glossaries in patent applications*. Available from ⟨https://www.uspto.gov/sites/default/files/patents/init_events/swglossary_e_ibm_2013oct03att.pdf⟩ Accessed 09.01.17.

Sebastiani, R., Montagnini, F., & Dalli, D. (2012). Ethical consumption and new business models in the food industry. Evidence from the Eataly case. *Journal of Business Ethics, 114*(3), 473–488.

Tencati, A., & Zsolnai, L. (2012). Collaborative enterprise and sustainability: The case of slow food. *Journal of Business Ethics*, *110*(3), 345–354.

Thompson, C. J., & Coskuner-Balli, G. (2007). Countervailing market responses to corporate co-optation and the ideological recruitment of consumption communities. *Journal of Consumer Research*, *34*(2), 135–152.

Tseng, Y. H., Lin, C. J., & Lin, Y. I. (2007). Text mining techniques for patent analysis. *Information Processing & Management*, *43*(5), 1216–1247.

Wanner, L., Baeza-Yatesa, R., Brügmann, S., Codina, J., Diallo, B., Escorsa, E., ... Zervakig, V. (2008). Towards content-oriented patent document processing. *World Patent Information*, *30*(1), 21–33.

(No) time to cook: Promoting meal-kits to the time-poor consumer

16

Beverley Hill[1] and Sarah Maddock[2]
[1]Swansea University, Swansea, United Kingdom, [2]Royal Agricultural University, Cirencester, United Kingdom

16.1   Introduction

Convenience foods have become very much a part of our daily lives, lauded for saving time and reducing the physical and mental effort required for food preparation, consumption, and clean-up (Brunner, van der Horst, & Siegrist, 2010). A growing, but still relatively new, entrant to the convenience food market is "meal-kit" products. These range from supermarket packs of measured ingredients to meal-kit box schemes, which deliver premeasured ingredients and full recipe instructions to the doorstep, ready to be cooked at home. These latter, often subscription-based, meal-kit options differ from what we would typically term "convenience foods," in that they are often more nutritious, more environmentally friendly (less food waste and packaging), and have the added benefit of avoiding the negative associations linked with convenience foods (Hertz & Halkier, 2017). Furthermore, just as the combination of freezer and microwave technology changed food-related routines by allowing us to store food longer and to cook and reheat a meal swiftly (Shove & Southerton, 2000), meal-kit box schemes, ordered online and delivered to the door, have the potential to change consumption patterns and daily food habits. As such, recognizing that meal-kits are more closely associated with proper, healthy food rather than junk or highly processed *convenience* foods, it is useful to refer to the concept of "*convenient* food provision, cooking and eating" (Halkier, 2017, p. 136). This also shifts the focus from particular food items to the processes and practices of shopping, cooking, and eating.

Historically, purveyors of convenience foods have promised the consumer speed, domestic efficiency, and freedom (Smith, 2001), and this is no less the case with meal-kit services. Providers such as Gousto and Freshly, with statements such as "Gousto gives you back some 'you' time" and "Great tasting healthy food should take time—just not yours" rely on communicating the time-saving benefits of speed and convenience to consumers who may perceive themselves as being "time-poor" (Brewis & Jack, 2005). Such texts can also be perceived as offering "new combinations and configurations of doing" (Shove, 2009, p. 26), which change the relationship between the food consumer and providers. Tacit rules (Benwell & Stokoe,

Case Studies in Food Retailing and Distribution. DOI: https://doi.org/10.1016/B978-0-08-102037-1.00016-5

2006) which support these "new ways of doing" can be discerned in meal-kit promotional texts which provide an alternative food script (Block et al., 2011). These promotional messages produce subject positions for consumers (Hall, 2000; Serazio & Szarek, 2012), which may have negative long-term consequences for consumer relationships with food providers and the food system more widely. Although messages such as "Let us take care of the meal planning and the grocery shopping" (Hello Fresh) may resonate with the time-pressed consumer, they have potential to further disconnect food consumption from food production, distancing the consumer from important aspects of food preparation and potentially reducing consumer food knowledge and involvement while increasing the power and control of the provider. This is a concern at a time when ease of access to food, the extent of food education, and a lack of cooking skills are being discussed across Europe (European Commission, 2007).

In light of these concerns, this chapter provides a critical textual analysis of meal-kit promotional messages, exploring the subject positions they offer to consumers and the "rules" of consumption they infer. The aim is to outline the means by which meal-kit promotional messages draw the time-poor consumer into a consumption relationship based on time-saving promises which imply tacit new "rules" (Benwell & Stokoe, 2006) for food planning, shopping, and cooking—mobilizing the concept of time to encourage purchase and to influence existing food-related practices. Drawing on promotional text from meal-kit websites, we ask:

1. How are time-saving promises articulated in promotional messages (and what metadiscourses about time in today's society do they imply)?
2. How are consumers interpellated or "hailed" by the text?
3. What tacit new "rules" are implied for food planning, shopping, and cooking?

Our analysis is informed by similar studies (see Benwell & Stokoe, 2006; Serazio & Szarek, 2012), which view promotional texts as ideological documents that imply specific understandings of society and culture and which "hail" (to use Althusser's term) the consumer as a subject, encouraging them to adopt particular subject positions in order to understand the preferred meaning of the text as everyday common sense. To commence, we review the meal-kit sector and discuss the temporal aspects of consumer food-related behaviors, before exploring the implicit rules of meal-kit promotion and their potential consequences for the consumer.

16.2   The meal-kit market

Variously referred to as recipe boxes or meal boxes, meal-kits are boxes of premeasured food items, with a recipe, ordered from supplier websites and delivered to the door, on a subscription or on-demand basis. They are available in many forms, from the complete meal boxes discussed in this chapter to spice kits, all designed to add convenience to food provisioning.

The meal-kit sector, at 1%, is a very small part of the UK's online grocery market (Weinbren, 2016). Subscription-based meal-kit providers Gousto and Hello

Fresh dominate the UK market, while Blue Apron is the US market leader. Nonsubscription based services include Plated and Chef'd in the United States, while Marley Spoon, originally a one-off demand service, has also recently switched to a subscription-based service. Simply Cook, The Spicery, and Tastesmiths offer more limited kits, consisting of recipe cards and ingredients such as cooking pastes and spices, which consumers need to supplement with fresh ingredients (Mintel, 2016a). Existing UK organic veg box suppliers, such as Abel & Cole and Riverford Organics, now provide portioned recipe boxes. Mindful Chef (United Kingdom) and Purple Carrot (United States) offer specialist food options, while retailers Waitrose and Tesco have added their own meal-kits to their online grocery offering. Additionally, big US brands such as Quaker and the Campbell's Soup company are also entering the meal-kit market (Martino, 2017).

Removing the need to shop for ingredients, meal-kits are aimed at consumers who "aspire" to cook from scratch, but lack the time or skill to do so (Mintel, 2016a). They appeal to those seeking "life-hacks" or "tools to help them improve their productivity and save time" (Mintel, 2016b)—most likely the "affluent achiever" and the "rising prosperity" groups (Weinbren, 2016). Hello Fresh MD Ian Marsh believes meal-kits can have a relatively mass appeal, pointing to Sweden where 2%—3% of the population purchase meal-kits, while 4% of British consumers claim to have tried an online food specialist, rising to 7% amongst the urban population (Weinbren, 2016).

Meal-kits primarily adopt a convenience positioning, promoting themselves as a bridge between convenience and scratch cooking, but are also marketed as healthy due to their fresh ingredients (Mintel, 2016b). While the UK grocery market has seen an increasing focus on "ease of use" claims (e.g., "easy to cook," "ready to cook," and "simply") (Mintel, 2016a), Blue Apron in the United States has recently dropped its convenience positioning in favor of emphasizing its close working relationship with farms and small suppliers, leading some to suggest that perhaps convenience is no longer sufficient as a basis for a profitable business (Martino, 2017).

Technological innovations such as the introduction of meal-kits are disrupting the food sector, reconfiguring food provisioning. However, there are some doubts about the long-term viability of meal-kit services. While they may be enjoyable occasional experiences, it is much more difficult to embed such services into the everyday habits of consumers (BMI, 2016). Nevertheless, market analysts suggest that the rise in meal-kit provision, the increase in mobile apps to enable ordering, and a growth in third-party delivery services, is indicative of a speeding up of consumer food procurement as food service becomes an "on-demand" industry (Henkes, 2017).

16.3 Time and food provisioning

Food has always demanded an investment of time and effort (Walker, 2016). Time, however, has become a highly valued resource and its perceived scarcity has an influence on food-related consumer behaviors and routines (Herrington & Capella, 1995). When it comes to food provisioning, time is of the essence (Brewis & Jack, 2005). Indeed Warde (1999) attributes the widespread use of convenience foods more to time pressures than food preference, stating "Many people are constrained, in the face of more pressing social obligations, to eat convenience foods as a provisional response to intransigent problems of scheduling in a de-routinised society" (p. 525). Time-use studies in the United Kingdom, United States, and the Netherlands, as well as the expansion of the fast-food industry and an ever-increasing range of time-saving meal options, all point to a decrease in food-related time (Cheng, Olsen, Southerton, & Warde, 2007; Mandemakers & Roeters, 2015; Tashiro & Lo, 2012). We are now less concerned about whether there is *enough* food than "the manner of its delivery and provision" (Warde, Cheng, Olsen, & Southerton, 2007, p. 378) and, as many of us seem unwilling to let food take away from other pursuits (Walker, 2016), the time we take to choose, prepare, and cook food is under pressure.

Feeling pressed for time leads consumers to adopt a number of coping strategies, depending on their degree of confidence and skill in meal planning and cooking (Alm & Olsen, 2017). This may result in spending less time preparing food at home and depending more on "outsourcing" (Mandemakers & Roeters, 2015) via choosing to consume fast-foods, convenience and ready-prepared foods (Jabs & Devine, 2006). Although some consumers find time pressure impedes healthy eating (Welch, McNaughton, Hunter, Hume, & Crawford, 2008), Kalenkoski and Hamrick (2013) suggest that time-poor consumers are not necessarily turning to unhealthy convenience foods. Instead, they are likely to source alternative prepreparel foods; that is, other "convenient" food options such as meal-kits (Halkier, 2017; Hertz & Halkier, 2017).

In the discussion of meal-kit promotion below we explore further the relationship between food provisioning and time. First, we suggest how time is offered as a commodity to the busy consumer to invest in selected parts of the food provisioning process. Next, we show how the consumer is encouraged to become less involved in the parts of the process that meal-kit providers wish to annex for themselves.

16.4   The time-poor consumer and the commodification of food-related time

All convenience foods rely on the promise that they can expedite the process of consumer food provisioning in some way, from reducing the need to shop to simplifying the cooking process. Meal-kit promotions draw on existing discourses of time scarcity in modern society, mobilizing the concept of time as a valuable commodity to attract the "time-poor" consumer. Of course, the consumer needs to be willing to accept the presupposition that time is a scarce commodity in order to make sense of any text which promises to return "time" to them. To that end, as we show in the examples below, meal-kit promotion confirms to consumers that time is scarce before identifying certain aspects of food provisioning as problematic. Meal-kit services are presented as offering a solution by removing the hard work of planning and shopping, so that consumers can concentrate just on cooking.

16.4.1 Busy as a bee: appealing to busy lifestyles

Meal-kit promotional text echoes the cultural refrain of time scarcity, reflecting back to consumers their perceptions of being time-pressed. "Busy as a bee?" enquires Abel & Cole, while Chef'd tells students "You're swamped. This meal plan will allow you to spend more time on more important things-like passing classes." Gousto and Mindful Chef, below, attempt to simulate intimacy with consumers through suggesting a shared common experience of modern life:

> Everyone is so darn busy being busy these days. *The result is that when* we *finally flop down on the couch, we end up wondering just where the time went. Gousto gives* you back some 'you time', *enabling you to enjoy one of the oldest and most relaxing activities in the world as it was meant to be. Enjoy your time, savour every moment with Gousto! (Gousto FAQs). [Authors' emphasis in italics throughout].*

> Time matters: We know a modern lifestyle is busy *and so all recipes are 20—30 minutes and have less than 8 steps. Our boxes are designed to slot into your lifestyle,* not get in the way of it *(Mindful Chef)*.

The text suggests that food-related activities "get in the way of" consumers being able to get on with their busy lives. This has the effect of reducing the relative priority of food-related time in everyday life. Indeed, for US company Plated, "everything else" is more important than taking time for food-related activities:

> *Deciding what to do for dinner shouldn't be a struggle. Skip the "What's for dinner?" debate so* you can focus on everything else you have going on. *In less than 5 minutes you can choose recipes, set your delivery day, and find out how delightfully simple meal planning with Plated can be (Plated).*

Through routinely categorizing aspects of the process as negative, certain parts of food provisioning are problematized in the promotional text; either as being time-consuming or being as hard work. Cognitive behaviors of thinking, planning, and deciding what to eat are negatively indicated through language choice. For example, Hello Fresh tells us that "*thinking* about a week's worth of meals is *time consuming*," "Take the *hassle* out of weekly meal *planning*" suggests Chef'd, while Plated, in the extract above, reassures us that "*Deciding* what to do for dinner shouldn't be a *struggle*." In addition to these cognitive processes, some (although not all, as we later discuss) routine physical behaviors are negatively categorized. Customer reviews testify that it is a chore to check the store-cupboard, find ingredients, and to go to the grocery store. Existing customers enthuse "it's such a relief not to have to check I have everything in the cupboards" (Abel & Cole) because meal-kits remove "the stress and thinking out of shopping/planning/prepping" (Mindful Chef). The most negatively categorized physical behavior is grocery store shopping. Hello Fresh vows "you don't have to go to three different supermarkets to get everything you need for one recipe ever again." Chef'd promises "no grocery shopping, no stress." And Plated suggests "Skip the shopping." In the example below, Gousto contrasts the hard work of the routine shopping trip with the simplicity of their weekly box offering:

> With Gousto you don't need that epic *shopping list,* soul-crushing *queue at the checkout or to take part in the* heavyweight *grocery bag Olympics. All you need is your weekly box (Gousto).*

Such messages express the hard work required for some aspects of food provisioning, specifically meal-planning and grocery shopping, i.e., the parts of the process that meal-kit providers seek to change. Cooking, however, particularly cooking facilitated by meal-kits, as we show below, is much more positively portrayed.

16.4.2   (No) time to cook? Meal-kits make cooking simple and speedy

How we choose to spend our time indicates what is important to us—what we value. The underlying assumption in the messages above is that spending time planning and shopping for food is not a good use of valuable time. Cooking, however, is always positively presented. Typically cooking appears as the final stage in a three-part process, based around a common construction with very little variation across providers; "*We create* delicious plant-based recipes each week. *We deliver* fresh, perfectly portioned ingredients. *You cook* power-packed meals that are easy to make and outrageously good to eat" (Purple Carrot). Or in the case of Blue Apron, keen to emphasize their work directly with farmers, "Nature creates. We deliver. You cook." In the meal-kit vision, cooking is always "fun" and "a pleasure":

> Marley Spoon is making it easier *to cook the food you want to eat. We bring together seasonal ingredients,* simple, *delicious recipes, and make it* easy *so* you can enjoy the fun part – cooking. *(Marley Spoon)*

Cooking can be positively described because meal-kit providers have removed the "hard work," essentially taking "the planning and trekking out of home cooking, leaving you with the fun bit" (Gousto). Cooking, preceded by simple preparation, is associated with positive language choices linked to temporal and convenience aspects of food provisioning:

> You cook & enjoy – *Healthy evening meals with* simple *and* speedy *prep (Mindful Chef).*

> *We deliver every ingredient − All you need to do is* cook, eat & feel great! *(Mindful Chef).*

> Simply *cook, serve and* enjoy *your culinary creations! (Gousto).*

Cooking meal-kit meals is portrayed as both enjoyable and easy, enabling the consumer to "Relax and Enjoy" while getting "dinner on the table in no time" (Marley Spoon). Time spent cooking is presented as a good use of time:

> Spend your time where it matters. *Peak-season produce, meats raised without antibiotics, and artisanal specialty items will be waiting with fun-to-follow recipes so you can get cooking sooner. Roll up your sleeves and* have fun *in the kitchen −* Real, delicious cooking, without all the extra work *(Plated).*

By encouraging consumers to see cooking as the preferred use of their time, with planning and shopping as problematic, meal-kit providers can position themselves as providing a solution by taking over those "negative" parts of food provisioning.

16.5 The professionalization of food provisioning and the transference of responsibility

So far, we have seen how consumers are situated by promotional text as being short of time, resenting certain aspects of food provisioning as a nuisance, yet desiring good food and enjoying cooking. This enjoyment is reflected in Abel & Cole customer testimonials which enthuse "this has made mealtimes fun again—everyone wants to cook!" But forms of cooking facilitated by meal-kit providers need to be sufficiently differentiated from the routine meals that the consumer may have previously prepared and also from other convenient food options. So, consumers are promised that they will cook "real, delicious" meals, indeed "culinary creations" yet with "simple and speedy prep." In this way, meal-kits avoid the negative stigma of fast-foods ("At this rate, I may never need to order a take-away again" enthuses a Simply Cook customer). By focusing on a simulacrum of the cooking process, cooking has essentially become the combining of ingredients; in other words, just the final part of the process. In order for consumers to see the value of meal-kits meals, we show below how they are associated with professionalism, while consumers' own cooking skills are devalued. From here, meal-kit providers assume a greater role in the food provisioning process and attempt to create change in the food system.

16.5.1 Deskilling consumers through expert discourse: cook like a pro

Promotional text positions the meal-kit provider as "expert," enabling the consumer to "Cook *Restaurant Quality* Meals in 20 Mins" (Simply Cook). Meal-kit chefs are "renowned" (Chef'd), "professional," and "expert" (Simply Cook), choosing ingredients so that consumers can "Cook *Like a Pro*—Cook *masterpiece* meals right at home. Make a *gourmet* meal any night of the week" (Chef'd). At the same time as emphasizing the expertize of chefs and the quality of meals, promotional texts construct the consumer as relatively unskilled, lacking food knowledge and in need of some educational advice:

> *Cooking needn't be a* cataclysmic *event resulting in the* ruin *of your kitchen. With the right kit at your disposal it's actually great fun! (Gousto FAQs).*

> *Techniques and recipes you'll love − at every skill level. From* getting more comfortable *in the kitchen to* mastering new shortcuts *and cuisines, you'll find yourself* learning, advancing*, and looking forward to cooking more and* living better *(Plated).*

Recipes are always "easy to follow" (Simply Cook), interpreted by chefs who "are the best when it comes to breaking down complex recipes into just a few simple steps" (Plated). Consumers need help not only because "there are lots of recipes out there" (Plated) but because, as Blue Apron state, our food system is "*complicated*, and making *good choices* for your family can be *difficult*." It is a small step then to proposing the transference of more responsibility for food provisioning from the consumer to meal-kit providers.

16.5.2 Apportioning responsibility and taking control: leave the hard work to us

Meal-kit promotional text positions the provider as assuming much of the "hard work" around food provisioning, while the consumer becomes more fully involved only at the "fun" end of the process, that is, cooking. "Leave the hard work to us" suggests Mindful Chef, while Plated tells consumers to "Enjoy your day—tonight is taken care of." These assertions suggest a transfer of responsibility for food provisioning to the provider, while consumers become less active participants. Paradoxically, meal-kit providers explicitly stress that the consumer remains in control, exerting agency by managing how they use their time and by having command of their order:

> *You decide when you want your Hello Fresh. When we say "You have control", we mean it! Order when you want, skip or pause when you don't (Hello Fresh).*

This control, however, is limited, when compared to alternative forms of food provisioning, as choice remains constrained by the meal (ingredients, recipe, and portion size) that the provider intends to deliver. In also assuming some responsibility for food education, as we illustrated earlier, providers potentially replace the learning traditionally acquired in the home with their own alternative food-related practices within a consumption relationship. Many meal-kit providers go further still, identifying themselves as having a social conscience, aiming to change food provisioning for the better. Hello Fresh, for example, "set out to change the way the nation eats", Blue Apron promises that "together we can build a better food system", while Plated are "reconnecting people to their food, one plate at a time." Marley Spoon acknowledges the role of technology in changing food provisioning, proposing that it can facilitate a nostalgic return to a lost relationship between consumers, food, and food suppliers:

> *Technology allows us to become* your local greengrocer, butcher and fishmonger. *Or at least it gives us the opportunity* to reconnect you with the core values of knowing where your food comes from. *Building relationships with sustainable suppliers ensures that we are a lot more than just an online shop (Marley Spoon).*

Such examples illustrate how meal-kit providers propose taking responsibility for planning meals, sourcing and measuring ingredients, developing "step by step" recipes, and delivering full meal-kits to the doorstep. Their promotional messages offer an alternative "food script" to consumers, encompassing different sequences of actions or routines (Block et al., 2011). This script changes the process of consumer food provisioning and, as we discuss in the next section, potentially discourages or even excludes consumers from participating in parts of the food system.

16.6   The new rules of food provisioning and some potential consequences

Promotional texts circulate ideas about contemporary values (Hadlaw, 2011). In meal-kit promotion, time is a valuable, scarce commodity which must be selectively expended in the process of food provisioning. By examining these promotional messages, a number of implicit rules (Benwell & Stokoe, 2006) for making the best use of food-related time can be discerned:

1. Cooking is a good use of time; it is fun and can be enjoyed with the family. Home cooking must be simultaneously gourmet yet easy, so as not to detract too much from that "fun" and from other pursuits.
2. Any aspect of the process that is hard work, requiring effort on the part of the consumer, can be transferred to meal-kit providers—that is, thinking, planning, and grocery shopping. These activities are time and energy consuming obstacles that get in the way of a busy lifestyle.
3. Time for oneself is prized and prioritized and this can be bought by relinquishing some control over food provisioning. Simultaneously, retaining some semblance of control remains important.
4. Supermarkets are certainly to be avoided by the time-poor consumer. These are the food suppliers against which meal-kit providers position themselves, with no mention of alternative sources such as local markets, farmers' markets, smaller grocery stores, grow-your-

own schemes—indeed any of those movements which foster a closer relationship with food.

5. Finally, there is no conflict in the meal-kit vision between technology and human values, relationships and historic food cultures; indeed food technology can bring a consumer closer to the origins of their food.

Accepting these "rules" prompts a change in the food provisioning process and in the relationship between consumers and providers and it is worth considering in whose favor such change works. Meal-kit promotion encourages a reallocation of time across the various activities which constitute the food provisioning process. Removing the consumer from important aspects of food preparation (planning, shopping, and the physical and mental effort of meal preparation) *reduces* their investment in the process while increasing that of the food provider. This change fosters a further shift in the power relationship between consumers and food providers who increasingly control the process, and it creates a greater disconnect between consumers and food (Kneafsey et al., 2008). This supports Walker's (2016) contention that we seem "intent on putting as much distance as possible between food and ourselves. As a result, we're quickly losing our ability to discern where this reshaped connection to food is carrying us." (p. 296). Such distancing obscures the real labor and involvement required to produce food, potentially leading to consumer complacency, deskilling, and wastefulness, and ultimately reducing the role and resilience of the consumer as the arbitrator of what is eaten at home.

The decline in food knowledge and the contribution of convenience foods in facilitating "a society of individuals lacking in food preparation skills" (Mc Dowell, McMahon-Beattie, & Burns, 2015, p. 629) is well rehearsed. Any further disconnection between consumers and food exacerbates these problems. Meal-kits, despite their advantages, contribute to this disconnection by excluding consumers from the procurement process so that we become simply recipients rather than active participants. Consumers neither smell nor touch the produce before it arrives on the doorstep. Instead, choice is based on the visual appeal of completed meals in an online image gallery. Essential food knowledge, such as the ability to visually identify ingredients in their unadulterated form (not prepared, portioned, or packaged), is potentially compromised by meal-kits. Consumers reliant on this process may lack confidence in identifying substitute products when natural or weather-related shortages occur, may be unable to evaluate fresh or good value produce, and may be unable to take advantage of discounts on produce in times of glut as they lack knowledge of what to do with them. Meal-kit services provide a model of cooking which consists of minimum preparation and the combining of premeasured ingredients, so that food skills are no more than superficial. Yet cooking can be creative without being gourmet; a good knowledge of food enables consumers to "make do" with whatever is available, recognizing that cooking creatively does not require either homogeneity or professionalism. Finally, meal-kit options remove aspects of food traditions and it is talk of such traditions, such as family recipes, that unites families (Walker, 2016).

16.7 Conclusion

It has not been the intention of this chapter to disparage meal-kit services; customer reviews are testament to how much they are appreciated and industry observers commend their nutritional and waste-reducing benefits. It is also recognized that the cost of such services limits their accessibility to wider markets. While many of the concerns expressed above apply equally to other convenient foods, meal-kits, as recent developments in the online grocery market, serve as a microcosm which exemplifies how technological innovation in food distribution creates change in consumption behaviors around food provisioning. We have attempted to show how, through the persistent appeal of convenience, meal-kit promotion encourages time-poor consumers to modify their food provisioning behaviors in ways which may not always be advantageous in the longer term. In this way, we contribute to consumer food science studies by illustrating how food consumption is influenced by powerful institutions (the marketing/promotional industry and technology providers), which shape the consumer's relationship with food and food providers. As Innes reminds us, "Food is *never* a simple matter of sustenance. How we eat, what we eat, and who prepares and serves our meals are all issues that shape society" (2001, p. 5). It is worth being more alert to the longer term consequences of that societal shaping.

16.8 Food for thought

1. The promotion of meal-kits relies on the concept of time as a commodity. What does this suggest about how time is understood? Is this a particularly Western view of time? Are there alternative ways to think about time in relation to food provisioning?
2. The case outlines how meal-kit promotion provides an alternative food script which has the potential to change consumer food-related routines and behaviors. What is implied here regarding the nature of "the consumer," the different roles adopted by consumers and the relative power of the individual and the food industry?
3. Marley Spoon claims that technology facilitates a closer relationship between consumers, food, and food suppliers. Consider whether this claim can be justified in the light of concerns regarding food security and ease of access to food.

References

Alm, S., & Olsen, S. (2017). Coping with time pressure and stress: Consequences for families' food consumption. *Journal of Consumer Policy, 40,* 105−123.

Benwell, B., & Stokoe, E. (2006). *Discourse and identity*. Edinburgh: Edinburgh University Press.

Block, L., Grier, S., Childers, T., Davis, B., Ebert, J., Kumanyika, S., ... van Ginkel Bieshaar, M. (2011). From nutrients to nurturance: A conceptual introduction to food well-being. *Journal of Public Policy & Marketing, 30*(Spring (1)), 5−13.

BMI Research. (2016). *Key themes for food & drink in 2017*. Business Monitor Online. December 6, 2016.

Brewis, J., & Jack, G. (2005). Pushing speed? The marketing of fast and convenience food. *Consumption, Markets & Culture, 8*(1), 49–67.

Brunner, T. A., van der Horst, K., & Siegrist, M. (2010). Convenience food products: Drivers for consumption. *Appetite, 55*(3), 498–506.

Cheng, S.-L., Olsen, W., Southerton, D., & Warde, A. (2007). The changing practice of eating: Evidence from UK time diaries, 1975 and 2000. *The British Journal of Sociology, 58*(1), 39–61.

European Commission. (2007). *Food consumer science: Lessons learnt from FP projects in the field of food and consumer science*. Brussels: European Commission. Available from http://cordis.europa.eu/pub/food/docs/booklet-consummer.pdf.

Hadlaw, J. (2011). Saving time and annihilating space: Discourses of speed in AT&T advertising, 1909–1929. *Space and Culture, 14*(1), 85–113.

Halkier, B. (2017). Normalising convenience food? *Food, Culture & Society, 20*(1), 133–151.

Hall, S. (2000). Who needs identity? In P. du Gay, J. Evans, & P. Redman (Eds.), *Identity: A reader*. London: Sage.

Henkes, D. (2017). The food market in 2017—The big foodservice trends to watch. *Just-food Research News*.

Herrington, J., & Capella, L. (1995). Shopper reactions to perceived time pressure. *International Journal of Retail & Distribution Management, 23*(12), 13–20.

Hertz, F., & Halkier, B. (2017). Meal box schemes a convenient way to avoid convenience food? Uses and understandings of meal box schemes among Danish consumers. *Appetite, 114*, 232–239.

Innes, S. (2001). *Kitchen culture in America*. Pennsylvania: University of Pennsylvania Press.

Jabs, J., & Devine, C. (2006). Time scarcity and food choices: An overview. *Appetite, 47*, 196–204.

Kalenkoski, C., & Hamrick, K. (2013). How does time poverty affect behavior? A look at eating and physical activity. *Applied Economic Perspectives and Policy, 35*(1), 89–105.

Kneafsey, M., Cox, R., Holloway, L., Dowler, E., Venn, L., & Tuomainen, H. (2008). Reconnecting consumers, producers and food: Exploring alternatives. Oxford and New York: Berg.

Mandemakers, J., & Roeters, A. (2015). Fast or slow food? Explaining trends in food-related time in the Netherlands, 1975–2005. *Acta Sociologica, 58*(2), 121–137.

Martino, V. (2017). Meal kits in the US—Don't believe the hype. *Just-Food Research News*.

Mc Dowell, D., McMahon-Beattie, U., & Burns, A. (2015). Schoolinary art: Practical cooking skills issues for the future. *British Food Journal, 117*(2), 629–650.

Mintel. (2016a). *Attitudes towards cooking in the home—UK—May (online)*. Mintel Oxygen. Last accessed 15 June 2017 https://store.mintel.com/uk-attitudes-towards-cooking-in-the-home-market-report.

Mintel. (2016b). *Attitudes towards home-delivery and takeaway food—UK—April (online)*. Mintel Oxygen. Last accessed 15 June 2017 https://store.mintel.com/attitudes-towards-home-delivery-and-takeaway-food-uk-april.

Serazio, M., & Szarek, W. (2012). The art of producing consumers: A critical textual analysis of post-communist Polish advertising. *European Journal of Cultural Studies, 15*(6), 753–768.

Shove, E. (2009). Everyday practice and the production and consumption of time. In E. Shove, F. Trentmann, & R. Wilk (Eds.), *Time, consumption and everyday life: Practice, materiality and culture* (2009, pp. 17−33). London: Bloomsbury.

Shove, E., & Southerton, D. (2000). Defrosting the freezer: From novelty to convenience. A narrative of normalization. *Journal of Material Culture*, *5*(3), 301−319.

Smith, C. H. (2001). Freeze frames: Frozen foods and memories of the postwar American family. In S. Innes (Ed.), *Kitchen culture in America* (2001, pp. 175−209). Pennsylvania: University of Pennsylvania Press.

Tashiro, S., & Lo, C.-P. (2012). Gender difference in the allocation of time. *Food, Culture & Society*, *15*(3), 455−471.

Walker, K. (2016). No time for food. *American Journal of Health Promotion*, *30*(4), 296−299.

Warde, A. (1999). Convenience food: Space and timing. *British Food Journal*, *101*(7), 518−527.

Warde, A., Cheng, S.-L., Olsen, W., & Southerton, D. (2007). Changes in the practice of eating: A comparative analysis of time-use. *Acta Sociologica*, *50*(4), 363−385.

Weinbren, E. (2016). Do recipe box schemes stack up? *The Grocer*, *30*, 22−27. Available at: https://www.thegrocer.co.uk/channels/online/do-recipe-box-schemes-stack-up/545674.article.

Welch, N., McNaughton, S., Hunter, W., Hume, C., & Crawford, D. (2008). Is the perception of time pressure a barrier to healthy eating and physical activity among women? *Public Health Nutrition*, *12*(7), 888−895.

Supermarkets, television cooking shows, and integrated advertising: New approaches to strategic marketing and consumer engagement

17

Michelle Phillipov
Department of Media, University of Adelaide, Adelaide, SA, Australia

17.1 Introduction

As supermarkets across the West have been the subject of mounting concern about their market control, impacts on the food system, and the ethics and healthfulness of their foods, they have increasingly sought to adopt approaches to strategic marketing and consumer engagement that counter media and public criticism of supermarkets and supermarket practices. The rise of popular food media—television cooking shows, in particular—has offered supermarket retailers a range of new communications and marketing resources with which to present consumers with more positive messages about supermarket food. Central to this have been the opportunities food television offers for supermarket sponsorship and "integrated" (Spurgeon, 2013) advertising, as part of which advertising messages are seamlessly integrated into television program content. This has provided supermarkets with more subtle and sophisticated methods for countering criticisms than more traditional media strategies; however, the fact that supermarkets' messages are not always clearly demarcated as paid promotional content places additional demands on consumers, who now need to develop increasingly sophisticated media literacies as part of their engagement with food media texts and supermarket brand messages.

This chapter considers these issues in relation to Australian food media and retail industries. With one of the world's most concentrated food retail sectors, criticisms of supermarkets have been especially intense, and new brand management strategies particularly urgent, in the Australian context. The chapter will look at three television cooking shows that are sponsored by Australia's two major supermarkets: *MasterChef Australia* and *My Kitchen Rules* (*MKR*), which are both sponsored by Coles; and *Recipe to Riches*, which is sponsored by Woolworths. It will show how integrated advertising has been used as a strategy to counter a range of criticisms directed at supermarkets and their practices—from their treatment of farmers to the

Case Studies in Food Retailing and Distribution. DOI: https://doi.org/10.1016/B978-0-08-102037-1.00017-7

ascendency of supermarket "own brands" (Burch & Lawrence, 2007)—by associating major supermarkets with the "culinary cultural capital" (de Solier, 2005) of the television programs and the artisan practices of ordinary home cooks. With many of these strategies an attempt to align with (some may say "appropriate") the discourses through which media and consumers have attempted to critique, and offer an alternative to, supermarket food, integrated advertising has implications for how consumers now engage with food media and food retailers, as well as for how the meanings associated with supermarket food are being progressively reconstructed and redefined in line with media and food industry interests.

17.2 Supermarkets under attack

Before considering the television case studies in more detail, it is important to first locate supermarket strategies of integrated advertising within the context of an intensified media focus on food. That is, supermarkets now operate in a time in which there is both a proliferation of food media texts that reflect (and help produce) growing consumer interest in the pleasures of food and cooking, and increasing media and consumer concern about food risk. In the case of the latter, there are now large media and food industries devoted to critiquing the negative health, environmental, animal welfare, and broader food systems impacts of industrialized foodways—of which the supermarket frequently serves as the public "face." In popular international media accounts, supermarkets are the key beneficiaries of anonymous, standardized, globalized, and unsustainable food systems. They are "nonplaces" (Augé, 1995) governed by an ethic of "permanent global summertime" in which fresh produce is available all year regardless of seasonality (Blythmann, 2007, p. 76), and filled with "foodish products" and "edible food-like substances" containing ingredients that "your ancestors simply wouldn't recognize as food" (Pollan, 2008, pp. 1, 149).

Such critiques have generated a range of popular texts—television cooking shows, cookbooks, and online and mobile media—that seek to reengage consumers with the pleasures of food and "re-enchant" (Lewis, 2008, p. 232; see also Versteegen, 2010) their experiences of the food system. From *River Cottage* to *Gourmet Farmer*, these texts frequently do this by adopting the language of alternative food networks (see Goodman, Dupuis, & Goodman, 2012) and an increasingly coherent set of representations as to what constitutes "good" food. "Good" food is frequently small-scale, local, seasonal, and unprocessed—it is "real" food that stands in stark contrast to the apparently "unreal" food of the supermarket.

While these critiques of supermarkets and supermarket food are likely recognizable to readers across the Anglophone West (and possibly beyond), they have been felt especially intensely in Australia. Australia's unusually concentrated food retail sector means that just two supermarkets, Coles and Woolworths, control around 70% of the grocery retail market (Roy Morgan Research, 2016). This substantial market power has been a cause for concern. Both Coles and Woolworths have been subject to increased regulatory oversight (including Senate Inquiries), as well as investigations by the consumer watchdog Australian Competition and Consumer Commission. They have also experienced ongoing media and public criticism. For example, when Coles and Woolworths slashed the price of their own-brand milk to a considerably discounted AU\$1/litre in 2011, this generated significant media backlash against both retailers, and galvanized public support for farmers hurt by supermarket discounting. In 2014–2015, a series of exposés by journalist Malcolm Knox revealed how supermarkets' asymmetrical contracts with growers and suppliers, along with their duopolistic control of the retail market, has worked to systemically disempower farmers and food manufacturers across a range of food industry sectors.

As they have for supermarkets elsewhere in the world, these kinds of criticisms have posed a brand management problem for both Coles and Woolworths. They have necessitated the adoption of a range of strategies to reengage consumer trust in supermarkets and supermarket food. Drawing on Lindgreen's (2003) work on trust as a "strategic variable" for the food industry, Richards, Lawrence, and Burch (2011, p. 29) call this process the "strategic manufacturing of consumer trust," and point to a range of strategies supermarkets typically adopt to do this, including reputational enhancement and direct quality claims through private standards certification, and practices of discursive claims-making that invoke notions of "authenticity" and "tradition." But in addition to these other strategies, supermarkets have also sought to forge closer relationships with the food media industry and with popular food television productions in particular.

17.3 Food media and integrated advertising

In addition to an unusually concentrated food retail sector, Australia has a thriving food media industry. This reflects a global rise of food and lifestyle media in the 1990s, which has typically been understood as a response to the growing centrality of "life politics" in contemporary cultural life and as a reflection of the increasing global influence of neoliberal, consumer-oriented modes of citizenship, in which consumption and lifestyle "choices" are posed as methods for investing in ethical, social, and civic concerns and for constructing ourselves as "good," self-governing subjects (Lewis, 2012; Ouellette & Hay, 2008). The rise of food and lifestyle media is also a response to structural changes in the global media industry. Deregulation of media markets, fragmentation of audiences, declining advertising revenues, and increased competition from online and user-generated media have each posed challenges for "traditional" media industries and necessitated the development of new media products, formats, and strategies.

Food television has been especially successful at navigating this climate of media industry change. It has proven to be particularly adaptable to the logic of formatting, which has enabled the production of predictably successful, low-cost programs that can be sold to global markets and/or customized for local contexts (see Oren, 2013). By featuring "ordinary" people (amateur home cooks, in particular), food television has stimulated new forms of audience engagement, including what one report called the "social sofa," where audience participation in online and social media is now a key dimension of their television experience (Ashton & Houston, 2011).

While television cooking shows are now a staple of broadcast, cable, online and streaming television services across the Anglophone West, they have proven especially popular in Australia. For example, *MasterChef* was only moderately successful in its original UK incarnation, but became a ratings hit when adapted and remade for Australian audiences. The 2010 series 2 finale of *MasterChef Australia* broke ratings records, making it the most watching nonsporting event since OzTAM television ratings began in 2001 (ABC News, 2010). It has also been credited with having a so-called *"MasterChef* effect" on the eating, shopping, and cooking habits of Australians: supermarket sales of ingredients surge by as much as 480% after appearing on the show (Sinclair, 2010). Similarly, *MKR*, a format that originated in Australia, has been so successful in its home country that it is now a global sensation franchised around the world. Both programs have proven highly lucrative for the television production companies involved. While the details of sponsorship deals are closely guarded secrets, *MasterChef Australia* and *MKR* have each been estimated to generate between $15 and $20 million in sponsorship revenue per series from their suite of sponsors (Mitchell, 2014); this is a significant amount of money by Australian television industry standards.

As these figures suggest, food television's success has made it attractive to both media and other industries. Its focus on food has made it especially appealing to the food industry, and particularly to the supermarkets that are the major sponsors of all of Australia's primetime competition cooking formats. Coles has been the major sponsor of both *MasterChef Australia* and *MKR* for each of the programs' ten and nine seasons, respectively. Woolworths has been the major sponsor of *Recipe to Riches*. While *MasterChef Australia* and *MKR* are more conventional programs in which amateur cooks are progressively eliminated in weekly cooking challenges, *Recipe to Riches* is a slightly more unusual program adapted from the Canadian format of the same name. *Recipe to Riches* brings together amateur home cooks who have an idea for a product they hope to commercialize, and a new group of contestants competes each week to have their product selected to be sold in Woolworths supermarkets the following day. *Recipe to Riches* performs much more poorly in the ratings than either *MasterChef Australia* or *MKR* (around 600,000 viewers an episode compared to *MKR's* 1.9 million, for example), but its success with key advertiser demographics has been enough to sustain two series of the program (Bodey, 2013; Jackson, 2013), with applications open for contestants for a third series at the time of writing.

These two different types of programs (*MasterChef Australia* and *MKR*, on the one hand, *Recipe to Riches*, on the other) reflect two different models of sponsorship. In the case of *MasterChef Australia* and *MKR*, Coles offers a "top tier" sponsorship contribution estimated to be between $3 and $5 million (Canning, 2010; Lee, 2010) in exchange for a range of brand placement and embedded marketing opportunities, which sit alongside those offered to range of other sponsors—from Holden cars to Finish dishwasher detergents. With programs beaming into Australian homes up to five or six nights per week, combined with significant online and social media presences, these sponsorship deals provide numerous opportunities for brand messages, and have been credited with helping Coles to "close the gap" with long-term market leader Woolworths (Janda, 2010). Woolworths' relationship with *Recipe to Riches* is more like the relationships negotiated between record labels and reality TV singing contests, such as *Idol*: in this case, a Woolworths representative sits on the show's judging panel, and the supermarket chain offers a production contract on the Woolworths' Select private label to the show's winners.

Supermarket sponsorship of television cooking shows is not particularly new— for example, product placement has long been a staple of daytime instructional cooking formats (de Solier, 2005)—but the mainstream "reach" of primetime sponsorship deals is substantial, and brand messages are presented in increasingly sophisticated forms. As well as traditional forms of advertising and product placement, they also use the techniques of integrated advertising to integrate brands in storylines, as well as offering branded tie-ins and other kinds of branded program content (Spurgeon, 2013).

Only industry insiders operating under the secrecy of commercial-in-confidence know exactly what is negotiated as part of these deals—for example, we do not know how much of what we see in a food television program is explicitly requested by, or offered to, sponsors and how much is the result of serendipitous alignments between the conventions of food television and sponsors' core brand messages. But through analysis of the television texts, we can deduce something of their effects as brand management tools, as well as some of their implications for consumers.

17.4 Strategic marketing and integrating brands

As the discussion so far indicates, food television has risen to popularity during a period of structural change for the media industries (which has necessitated the development of new sources of revenue and new forms of audience engagement) and at a time of mounting media and public scrutiny of, and concern about, the activities of major supermarkets. The confluence of these two factors has made integrated advertising and sponsorship of television cooking shows attractive to supermarkets not just as a lucrative marketing opportunity, but as a vehicle through which to strategically reengage consumer trust in supermarkets and supermarket food. The effectiveness of integrated advertising and sponsorship lies in its capacity to exploit multiple points of contact between brands and consumers (see Jenkins, 2006). In the context of television cooking shows, successful examples of supermarket integrated advertising and sponsorship are those that tend to draw upon the connotations already associated with the television program, and connect them to those of the supermarket brand. In what follows, I will offer three examples of branded content and the integration of brands in storylines across *MasterChef Australia*, *MKR*, and *Recipe to Riches* to highlight the different ways that supermarkets are using integrated advertising as a resource for offering alternative (and more positive) media messages about supermarkets as means of combatting ongoing criticism. Each of these examples reflects attempts to respond to key brand management issues facing supermarkets today: the nature of "supermarket food"; supermarkets' treatment of farmers; and their treatment of suppliers.

17.4.1   Cooking like a MasterChef

Of the three programs, *MasterChef Australia* was the first to embrace the full potential of integrated advertising. Although the show features branded tie-ins across its suite of sponsors, its central focus on food and cooking with fresh produce enables a deeper level of integration for Coles than is possible for many of *MasterChef Australia*'s other sponsors. Coles is the major supplier of ingredients for all of *MasterChef Australia*'s on- and off-site challenges. Off-site challenges are typically preceded by contestants shopping for ingredients at a nearby Coles supermarket. With Coles-branded shopping bags in hand, contestants are shown purchasing unusual or exotic ingredients, such as quail or octopus, that viewers may not normally associate with the supermarket chain, along with abundant fresh fruit and vegetables. Among this bounty of fresh produce, the processed foods and "edible food-like substances" (Pollan, 2008, p. 1) typically associated with the supermarket are nowhere to be seen. What we see instead is an emphasis on whole, fresh foods, prepared from scratch.

These scenes, in combination with the on-site challenges within the MasterChef kitchen, work to perform powerful ideological work for Coles. On-site, the MasterChef pantry is the first port of call to collect ingredients for a number of the regular weekly challenges. The pantry is a space filled with shelves and baskets piled high with beautiful, pristine-looking produce—all emblazoned with the Coles logo. Although the pantry includes some packaged ingredients (such as sauces, spices, and dairy products), the fresh produce is positioned in the center of the room, drawing the viewers' eye to the colorful variety of food on offer. Again, there are no "foodish products" (Pollan, 2008, p. 149) here, just real food, and in forms far more exotic than the typical suburban Coles supermarket—from exotic finger limes and tropical fruits to a dazzling array of multicolored potatoes.

These ingredients are the building blocks of the gourmet, high-end meals that are the goal of most contestant challenges, with the judges' highest praise reserved for those that produce dishes of "restaurant quality." If viewers are at all unclear that contestants' success is dependent on the quality of Coles' fresh produce, they are reminded during the commercial breaks and through a range of other in-store, online, and print advertisements that (in the words of a successful Coles advertising campaign) if they want to "cook like a MasterChef cooks, shop where a MasterChef shops." These advertisements were fronted by Curtis Stone, a celebrity chef well-known for his commitments to sustainable and seasonal produce (Lewis & Huber, 2015), and who periodically appears as a guest chef on *MasterChef Australia*. In these ways, messages about the quality and desirability of Coles' fresh produce are imbued throughout the televisual landscape of the program. The effect of this is to align the meanings associated with the supermarket brand with those associated with *MasterChef Australia*'s "restaurant quality" gourmet credentials and the "culinary cultural capital" (de Solier, 2005) these represent. This assists in positively reframing the meanings of supermarket food as no longer a source of concern or critique, but as the food of choice for "foodies" and aspiring (master)chefs.

17.4.2 Farmer's Choice

Integrated advertising offers opportunities not just to reframe the meanings associated with the types of foods supermarkets sell, but to also manage public perceptions of the conditions under which this food is produced. *MKR* shares many of the same integrated advertising approaches as *MasterChef Australia*, especially with respect to the off-site challenges. In the "instant restaurants" rounds, for example, contestants are shown shopping for ingredients at Coles, and *MKR*'s producers ensure that Coles-branded products always have their labels prominently facing the cameras. But *MKR* also offers a more subtle integration of brands in storylines. A key example of this is the now-annual event in which contestant cook for the farmers that supply to Coles. These challenges begin with farmers entering the *MKR* kitchen carrying colorful fruit, vegetables, meat, and fish; they then stay on to watch the contestants as they cook and to assist the judges in evaluating the final dishes.

The "farmer's choice" challenges were introduced at a time of especially intense negative media coverage about the major supermarkets' (allegedly poor) treatment of farmers and suppliers. Indeed, the first "Farmer's Choice" episode in 2014 was an activation of Coles' "Helping Australia Grow" campaign, itself an attempt to "neutralize the noise" of farmer and media criticisms in wake of intense supermarket discounting of own-brand milk (see The Guardian, 2013). In contrast to the angry and anxious farmers that appeared in a lot of the news and current affairs coverage of the issue at the time (see Phillipov, 2017), *MKR* depicts Coles' farmers as content, hardworking, and highly satisfied with their relationship with the supermarket chain: this is conveyed through their smiling demeanors, effusive praise of the contestants' dishes, and their protectiveness toward their produce (expressed through regular warnings to contestants to not "stuff it up").

For their part, the judges and contestants respond to the farmers with discourses of gratitude and respect. In the contestant interviews that intersperse the actions on-screen, contestants frequently describe it as an "honor" and a "privilege" to have the opportunity to cook for the farmers. Contestant interviews are an important device used by reality television to provide commentary on, and narrative structure to, the on-screen action. While contestants are not "forced" to say anything they do not want to, the fact that the seconds of commentary that appear in television episodes are compiled from hours of interview footage enables the show's producers to select a small number of comments, focused around particular messages, that otherwise appear as genuine and unsolicited expressions of contestants' views (Phillipov, 2017). These contestant interviews have been shown to be persuasive with *MKR*'s audience, with much of the social media commentary related to these episodes adopting the same discourses of respect and gratitude for farmers—even down to word choices identical to those used by judges and contestants (Phillipov, 2016). With contestants and audiences (wittingly or unwittingly) drawn into apparently unsolicited endorsements of key messages about how much Coles' farmers are valued and supported, the integration of the Coles brand into the storyline of "Farmer's Choice" offers an effective counter-narrative to other media representations of farmers suffering under supermarket power.

17.4.3   Turning recipes into riches

Perhaps in a reflection of the show's unusual model of sponsorship, Woolworths' relationship with *Recipe to Riches* results in some quite different approaches to integrated advertising than either *MasterChef Australia* or *MKR*. In contrast to these other two programs, where the integration of brands in storylines ensures lots of opportunities to feature Coles supermarkets, its products and brand logos, Woolworths supermarket appears less frequently on *Recipe to Riches* than one might perhaps expect. Sometimes contestants' product launches are held inside or in front of the supermarket, but apart from the short promotion at the end of episodes encouraging viewers to go and buy the winning product from Woolworths, the supermarket itself is often strangely absent.

This is in part a by-product of the conventions of the original Canadian format, but its effect is to present the products that appear on *Recipe to Riches* as the antithesis of "typical" mass produced supermarket food. For example, much of what is prepared during the episodes' "batch up" round bears little resemblance to what is required to produce products on the scale required for a national supermarket chain. In fact, much of the preparation looks like what contestants would normally do in their own home kitchens, batched up perhaps to catering or restaurant quantities, rather than those for mass manufacture. We see, for instance, contestants chopping chilies and dates by hand, individually zesting and juicing lemons, and preparing apple purée in a regular home blender—practices quite far removed from industrial-scale production (Lewis & Phillipov, 2016). At the point in the series where the products are batched up on a genuinely industrial scale, the judges return to the contestants seeking affirmations of the "lengths" that Woolworths went to in order to ensure that the products were identical to their original home-cooked recipes.

Recipe to Riches first aired at a time of ongoing media coverage of supermarket bullying of suppliers of branded products. Much of this coverage suggested that both Coles and Woolworths were targeting branded products through practices such as "cliffing" (where suppliers are given an ultimatum to agree to tougher terms or have their products pushed off the shelf) in order to replace them with supermarket own brands (Phillipov, 2017). Although less successful than *MKR* in enlisting contestants and audiences in the replication of core brand messages (see Phillipov, 2017), at the level of the television text at least, *Recipe to Riches* works to recast supermarket own brands within a much friendlier space of innovation connected to the artisan cooking practices of ordinary home cooks, and in which suppliers (in the form of contestants) are offered support and assistance to bring their products to the supermarket shelves.

17.5   Implications for consumers

As the above examples show, supermarket brands are no longer managed simply through traditional media management techniques, but also through more subtle brand messaging. In a climate where there has been considerable media and public criticism of supermarkets and supermarket practices, such approaches are examples of how these media and public concerns are not just difficulties that supermarkets need to overcome—they are also opportunities to discursively reconstruct and redefine supermarket food in new, more positive, ways. In the case of *MasterChef Australia*, strategies of integrated advertising assist in reframing the meanings of supermarket food as strongly linked to freshness, quality, and gourmet cachet. In the case of *MKR*, supermarkets are reimagined as places characterized by the gratitude and respect they show for the (always happy) farmers that produce this food. In the case of *Recipe to Riches*, supermarket own brands are reenvisioned as a space of innovation and community participation rather than as a threat to traditional brand manufacturers.

Such strategies to "strategically manufacture" (Richards et al., 2011) consumer trust and engagement with supermarkets have significant implications for consumers, who are now subject to increasingly sophisticated advertising messages that may not be immediately recognizable as paid promotional content. Viewers of food television programs are now required to be much more critical media consumers than may have been necessary in the past. While decades of media research shows us that audiences have always been active in their engagement with media texts, there is also some evidence to suggest that, in some cases at least, the blurred lines between advertising and program content on television cooking shows creates the potential that some brand messages will be uncritically accepted (and, indeed, perhaps not even recognized as the advertising that it actually is). The media literacies now required of food television audiences include understanding not just of the broader media landscape in which debates about supermarkets are played out, they also demand better understanding of the industry contexts in which food television is produced, and how these contexts work to create particular types of media text. For instance, it is increasingly necessary for viewers to maintain an awareness of what aspects of their shows might be shaped by the international formats from which local programs are derived, which aspects are determined by the program's producers with ideas of audience engagement and "good television" in mind, and which aspects are the results of deals with major sponsors. By remaining attentive to these competing forces, it becomes possible to enjoy pleasurable and entertaining experiences of television cooking shows while at the same time keeping an eye toward the other messages for which such shows also seek to enlist our support.

References

ABC News. (26 July 2010). *MasterChef smashes ratings record.* Available from ⟨http://
www.abc.net.au/news/2010-07-26/masterchef-smashes-ratings-record/918950⟩ Accessed
23 February 2015.

Ashton, E., & Houston, J. (2011). *Reality rules—2011 reality TV insights survey.* Available
at ⟨https://www.realityravings.com/2011/11/23/reality-rules-social-sofa-is-driving-real-
ity-tv-engagement/⟩ Accessed 22 March 2017.

Augé, M. (1995). In J. Howe (Trans.), *Non-places: Introduction to an anthropology of super-
modernity.* London and New York: Verso.

Blythmann, J. (2007). *Shopped: The shocking power of Britain's supermarkets.* London:
Harper Perennial.

Bodey, M. (2 April 2013). My kitchen rules rises again for seven. *The Australian.* Available
from ⟨http://www.theaustralian.com.au/business/media/my-kitchen-rules-rises-again-for-
seven/news-story/92f15fe52124caffb9cd48f3f106af71⟩ Accessed January 2017.

Burch, D., & Lawrence, G. (2007). Supermarket own brands, new foods and the reconfigura-
tion of agri-food supply chains. In D. Burch, & G. Lawrence (Eds.), *Supermarkets and
agri-food supply chains: Transformations in the production and consumption of foods.*
Cheltenham and Northampton: Edward Elgar.

Canning, S. (July 2010). Coles masters recipe for lucrative partnership. *The Australian,* 3.

de Solier, I. (2005). TV dinners: Culinary television, education and distinction. *Continuum:
Journal of Media & Cultural Studies, 19*(4), 465−481.

Goodman, D., Dupuis, M., & Goodman, M. K. (2012). *Alternative food networks:
Knowledge, practice, politics.* London and New York: Routledge.

Jackson, S. (November 2013). Ratings undercooked but Woolies loves it. *The Australian,* 25.

Janda, M. (26 July 2010). Coles sales climb on MasterChef bandwagon. ABC News.
Available from ⟨http://www.abc.net.au/news/2010-07-26/coles-sales-climb-on-masterch-
ef-bandwagon/920014⟩ Accessed 1 February 2017.

Jenkins, H. (2006). *Convergence culture: Where old and new media collide.* New York: New
York University Press.

Lee, J. (July 2010). Supermarkets continue to wage war. *The Sydney Morning Herald,* 10.

Lewis, T. (2008). Transforming citizens? Green politics and ethical consumption on lifestyle
television. *Continuum: Journal of Media & Cultural Studies, 22*(2), 227−240.

Lewis, T. (2012). 'There grows the neighbourhood': Green citizenship, creativity and life
politics on eco-TV. *International Journal of Cultural Studies, 15*(3), 315−326.

Lewis, T., & Huber, A. (2015). A revolution in an eggcup? Supermarket wars, celebrity chefs
and ethical consumption. *Food, Culture & Society, 18*(2), 289−307.

Lewis, T., & Phillipov, M. (2016). A pinch of ethics and a soupçon of home cooking: Soft-
selling supermarkets on food television. In P. Bradley (Ed.), *Food, media and contemporary culture: The edible image.* Hampshire: Palgrave Macmillan.

Lindgreen, A. (2003). Trust as a valuable strategic variable in the food industry: Different
types of trust and their implementation. *British Food Journal, 105*(6), 310−327.

Mitchell, J. (21 July 2014). Sponsors flock to support TV shows about cooking and home
renovation. *The Australian Financial Review,* 22.

Oren, T. (2013). On the line: Format, cooking and competition as television values. *Critical
Studies in Television, 8*(2), 20−35.

Ouellette, L., & Hay, J. (2008). *Better living through reality TV: Television and post-welfare
citizenship.* Malden, MA: Blackwell Publishing.

Phillipov, M. (2016). 'Helping Australia grow': Supermarkets, television cooking shows and the strategic manufacture of consumer trust. *Agriculture and Human Values, 33*(3), 587–596.

Phillipov, M. (2017). *Media and food industries: The new politics of food.* Cham: Palgrave Macmillan.

Pollan, M. (2008). *In defence of food.* London: Penguin Books.

Richards, C., Lawrence, G., & Burch, D. (2011). Supermarkets and agro-industrial foods: The strategic manufacturing of consumer trust. *Food, Culture & Society, 14*(1), 29–47.

Roy Morgan Research. (2016). *Supermarket weep: Woolies' share continues to fall and Coles and Aldi split the proceeds.* Available from ⟨http://www.roymorgan.com/findings/7021-woolworths-coles-aldi-iga-supermarket-market-shares-australia-september-2016--201610241542⟩ Accessed 1 April 2017.

Sinclair, L. (June 2010). MasterChef sparks Coles sales surge. *The Australian,* 32.

Spurgeon, C. (2013). Regulated integrated advertising. In M. P. Mcallister, & E. West (Eds.), *The Routledge companion to advertising and promotional culture.* New York: Routledge.

The Guardian. (30 September 2013). *Coles presentation boasts about silencing 'milk war' critics—in full.* Available at ⟨https://www.theguardian.com/business/interactive/2013/sep/30/coles-presentation-silencing-critics-in-full⟩ Accessed 1 February 2017.

Versteegen, H. (2010). Armchair epicures: The proliferation of food programmes on British TV. In M. Gymnich, & N. Lennartz (Eds.), *The pleasures and horrors of eating: The cultural history of eating in Anglophone literature.* Goettingen: V&R Unipress.

Premium private labels (PPLs): From food products to concept stores

18

Elisa Martinelli, Francesca De Canio and Gianluca Marchi
Department of Economics Marco Biagi, University of Modena and Reggio Emilia, Modena, Italy

18.1 Introduction

Private labels (PLs) or store brands are defined as brands owned, controlled, and sold exclusively by one retailer under its own brand name (Sethuraman & Cole, 1999). Their importance in the food sector has increased enormously over the past two decades, especially with Europe's shoppers (PLMA, 2016).

Traditionally, these products have been generally positioned as low price/good value for money offerings with a perceived quality differential with national brands (NBs) (Baltas, Doyle, & Dyson, 1997; De Wulf, Odekerken-Schroëder, Goedertier, & Van Ossel, 2005). But the role of PLs has evolved greatly over time (Martinelli, Belli, & Marchi, 2015): the width of PL offerings has enlarged to nonfood categories (e.g., clothes, appliances, etc.) and services (travel booking, broadband communications, etc.), while retailers have invested heavily in enhancing the control and quality level of these products. At the same time, the depth of PL offerings has increased, and retailers have introduced different lines of store brands in order to satisfy consumers' demands in different market segments (Sayman & Raju, 2004). This has brought a much more positive consumer attitude toward PLs in general, thanks to an increase in perceptions of their quality (Steenkamp, Van Heerde, & Geyskens, 2010).

On the basis of positioning differentiation, store brands can now be divided into three tiers: economy, standard, and premium PLs (PPLs) (Geyskens, Gielens, & Gijsbrechts, 2010; Lamey, Deleersnyder, Steenkamp, & Dekimpe, 2012). Under this strategy, PPLs are gaining increased interest, as the fastest-growing (IRI, 2016) and most profitable tier (Ter Braak, Geyskens, & Dekimpe, 2014). In Italy, their market share accounts for 7% of PLs, gaining the best increase year-on-year (+14% in the first nine months of 2016—Netti, 2016), and showing strong market potential. PPLs are defined as "consumer products, produced by or on behalf of retailers with high quality and priced close to NBs, that contribute to differentiating the retailer from its competitors" (Huang & Huddleston, 2009, p. 978). Examples are Tesco's "Finest" in the United Kingdom, Loblaw's "President's Choice" in Canada, and Conad's "Sapori & Dintorni (S&D)" in Italy (Lincoln & Thomassen,

Case Studies in Food Retailing and Distribution. DOI: https://doi.org/10.1016/B978-0-08-102037-1.00018-9

2008). Actually, the growing importance that the PPL is acquiring in stores' assortments is leading some retailers to develop food concept stores branded with the PPL brand name. In so doing, PPLs are becoming an important tool that grocery retailers are now using to upgrade chain image and strengthen customer loyalty to the retailer as a brand.

Within this context, this chapter describes a specific organizational case study of PPL extension from food products to concept stores, analyzing, with a qualitative approach, the PPL strategies of the Italian-based retailer Conad Scrl, with particular reference to the launch and development of its PPL flagship stores, which are branded S&D.

The chapter proceeds as follows: first, Conad's company profile is briefly presented, followed by a discussion of the company's PPL strategies and the specific case of an S&D ice-cream concept store called "Cremerie S&D." The chapter then moves on to examine the organizational implications deriving from the case study, as well as the possible implications for consumers.

18.2   Conad Scrl: company profile

Conad Scrl, an acronym of Consorzio Nazionale Dettaglianti, is an Italian-based retailer founded in 1962 as a retail buying group. Retail buying groups are quite important in the Italian retailing sector. During the 1970s, a defensive reaction of small retail entrepreneurs against multiples gave birth to an increasing number of retail buying groups and voluntary chains. This trend was also supported by a law on the retailing sector (Law 426/71) that created an entry barrier for multiples. The result was a slowdown in retail evolution that explains the present specificities of the Italian retailing system, which is still fragmented and less developed compared to other EU countries such as France, the United Kingdom, and Germany.

Conad is the biggest Italian retail buying group with a corporate structure organized as a consortium of grocery store owners and arranged on the following three levels (Conad, 2017):

1. At the baseline, 2713 independent retailers; owners of the stores are associated with local cooperatives.
2. At the intermediate level, seven cooperatives operating on a regional/multiregional basis. Cooperatives maintain self-determination in logistics decisions, partly in buying and contract renewals for local products, in assigning category's roles, and in defining promotions.
3. At the top, there is the national buying center, which coordinates the system and to which the cooperatives delegate negotiation and buying terms with the main brand suppliers, as well as some marketing policies (communication, PLs, information systems, etc.).

This results in a complex network model, which is not easy to manage in an efficient way, but with the advantage of being locally rooted and very close to consumers. Conad's corporate values are based on social responsibility, listening, being professional and coherent, respect, participation, affinity, and orientation to change and innovation. The aim to be close to customers is well supported by the advertising campaign of the retailer, which is based on the claim *"People beyond things"* (Conad, 2017). This aims to convey Conad's core values, which are based on the daily work of the associated store owners who are physically present inside their stores and personally know and interact with each customer.

The store structure responds to this positioning. There were 3169 grocery stores in 2016 in various formats (Conad, 2017): convenience stores—Margherita (493 stores; 140 square meters and 1500 SKUs on average) and Conad City (970 stores; 400 square meters and 4500–5500 SKUs on average); supermarkets—Conad (1077 stores; 800 square meters and 7000–8500 SKUs on average) and S&D (17 stores; 500 square meters and 3500–4000 SKUs on average); superstores—Conad Superstore (205 stores; 1800 square meters and 10,500–12,000 SKUs on average); hypermarkets—Conad Ipermercato (25 stores; 4800 square meters and 16,000–19,000 SKUs on average); discounters—Todis (215 stores; 530 square meters and 5000–7000 SKUs on average); and other stores (169). Conad is strongly focused on small-medium sized stores, confirming itself as a market leader in the supermarket format in 2016, with a market share of 20.7%.

In the last decade, Conad also started to offer stores focused on specific product categories/targets in order to better respond to market trends and consumer needs. These include: pharmacies (108), petrol stations (36), pet stores (14), and opticians (19). All are branded with the retailer's brand name, and are located in malls where a Conad hypermarket/superstore is also present. In 2016, Conad reached annual sales of €12.4 bn, reporting an increase of 1.5% on the previous year, and a market share equal to 11.9% (Conad, 2017). It is the second largest grocery retailer operating in Italy.

18.3 Conad's PPL strategy: Sapori & Dintorni

The case study analysis was performed by examining company reports and documents and conducting semistructured interviews (in April 2017) with Conad's brand manager and other important key informants involved in the PPL concept store planning, design, launch, and management—namely, Conad's Store Marketing Manager and the Store Proximity Development and Innovation Manager—in order to understand goals and objectives, with particular reference to the role of PPL food products.

18.3.1 Conad's PL architecture

Different PL types play different roles in Conad's assortment strategy (Fig. 18.1).

On one hand, to support everyday low price (EDLP) positioning, the standard PL (Conad red logo) is positioned as a category leader in quality with a differential price which is 25%–30% lower than the NB category leader. On the other hand, the company's ability to extend its control on the supply chain, to offer highly

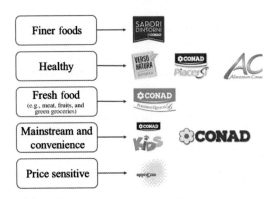

Figure 18.1 Conad's PL architecture.

selected quality and local products (S&D; Percorso Qualità), and to interpret successfully healthy (Piacersi), organic (Verso Natura), and "free from" trends, has rewarded Conad with a leading position amongst grocery retailers operating in Italy. Nowadays Conad's PLs account for €3 bn sales in 2016, with a market share of 29%: around one PL product out of three sold in the Italian grocery retailing system is branded Conad. In 2016, the Italian PL market accounted for €10 bn sales, with a market share of 19.2%; an increase compared to the 18.8% of the previous year (IRI, 2016). Within this context, Conad's PLs are particularly successful, as they performed 8.2% higher than the Italian PL average market share. Furthermore, in some product categories, Conad proposes an EDLP strategy named "bassi & fissi" ("low and fixed"), with the aim to communicate to its customers that they can find a valuable shopping option but at a low and fixed price on everyday shopping items.

18.3.2   Sapori & Dintorni

In 2001, the company launched its PPL, S&D, which in Italian literally means "flavors & surroundings," with a brand image inspired by some well-established values traditionally connected with the retailer's brand. From a strategic point of view, the company extended the traditional assortment in specific product categories (typically based on NBs with a branded alternative offering) with high-quality standards and good value for money.

In the launch stage, S&D appeared with attractive packaging, with colors supporting a "luxury" image (dark blue with gold writing). At this point, S&D covered 80 SKUs with a product strategy that was pretty much focused on fresh food and on a restricted number of grocery items such as cheeses, preserves, pasta, honey, salami, oils, and bread substitutes. These were selected among a basket of nearby 3000 products classified at that time as typical Italian food by the Italian Ministry of Agricultural Food and Forestry Policies (MIPAAF). However, in the early years (2001–04), S&D products experienced low rates of penetration in the PL market. The retailer identified problems both at the product range level, with a limited and not always coherent number of assorted products, and at the communication level, since S&D products were not sufficiently supported by advertising and were not adequately promoted in-store and on flyers. Therefore, in 2005, Conad took the decision to undertake a "deep restructuring" of the PPL strategy, by revising the product range and the communication approach. To better accomplish the original brand mission, "to offer the flavors (Sapori) of the surroundings (Dintorni)," Conad revised its PPL strategy in order to favor greater involvement in assortment decisions from the local cooperatives and associated retailers.

The S&D assortment became increasingly and substantially based on products with quality logos, namely products with protected designation of origin (PDO), protected geographical indication (PGI), and traditional specialty guaranteed (TSG), as well as food products related to the various regional cuisines of Italy. The European Commission (2017) defines these quality logos as brands that "attest to the specific traditions and qualities of food, agricultural products and wines, aromatized wines and spirit drinks, produced in the European Union [...]. Through the logos, consumers can easily recognize these traditional quality products and can rely on their authenticity in terms of regional origin or traditional production."

To date, S&D counts 250 SKUs, of which approximately 60% are PDO and PGI products. These quality logos generate almost 80% of the overall PPL turnover of €300 million. Conad offers its PPLs in some strategic product categories, and sometimes, if necessary, the same product category (for instance extra-virgin olive oils) can include two or more premium branded alternatives (e.g., Riviera Ligure DOP, Terra Di Bari DOP, Aprutino Pescarese DOP, etc.). The premium assortment strategy is not category driven, but it is based on product feasibility and on consumer demand. *"We consider the potential that each product that we want to include in the PPL assortment can have and its coherence with the S&D brand"* (Conad's Brand Manager). Italy is the EU country with the highest number of food quality logos (293) (MIPAAF, 2017)—and more than 4900 Italian agrofood traditional specialties are listed by MIPAAF (MIPAAF, 2016). Ninety-two percent of the 149 S&D suppliers are SMEs representing Italian food excellence.

The new brand strategy was successful and paved the way to double-digit sales growth for S&D products since the mid-2000s. In 2016, PPL made an increase of 10.1% compared to the previous year. The ability to reference and resell Italian regional food excellence has become the core competence for S&D's strategy. *"We exchange food culture, as our associate entrepreneurs are widespread all over Italy and know perfectly the best local products"* (Conad's Brand Manager). In fact, a distinctive trait of Conad as a retail service provider, which is strongly recognized by Italian consumers, relies on its ability to select high-quality fresh food such as delicatessen goods (e.g., salami and cheese) and match this with the specific needs of local markets. The S&D PPL enhances and builds on this strength.

Two main pillars have, thus, supported the growing penetration in the market of the S&D brand:

- *The retailer's expertise.* The organizational structure of the retail company, based on local cooperatives and store owners with a strong knowledge of the local supply market and a strong relationship with their customers, facilitated the selection of food products in the PPL assortment, as well as scouting the best PPL suppliers. In this way, Conad's pyramid structure guarantees professionalism in retailing and a strong ability to select local products that enrich the overall retailing offer.
- *The offer of a PPL strongly connected with the company's values.* In fact, the retailer has been able to establish a competitive advantage offering a PPL that reflects the origins and the historical identity of the company as a player "embedded in the territory." Therefore, the coherence between the company and the PPL values led consumers to a quick and strong acceptance of the premium offer proposed by Conad's S&D products. Conad provides products of high emotional value and a strong connection to the local territory. Correspondingly, consumers have consolidated over the time their willingness to pay a premium price for products with high-quality standards and produced by local manufacturers, which preserve the geographical origins of food production and culture and the heritage of local food recipes.

The contribution of the PPL brand to the overall strategy of the retailer has, therefore, significantly evolved over the time: from a traditional role of tactical support to category profitability, to a more strategic role as an autonomous value driver for the retailer.

18.3.3 From food products to concept stores: the Sapori & Dintorni store

The first two stores branded S&D were launched in June 2010 in Florence, on the basis of a corporate project aimed at:

1. Upgrading from a qualitative point of view the presence of the retailer in town centers: "*the proximity of our store is a must for us as we are the national market leader in the supermarket format. Thus, we started to think that we had to propose this specific quality format to our customers inside the town centers*" (Store Marketing Manager);
2. Addressing consumers' increasing demands for local products as synonymous with quality, and enhancing the presence of PDO and PGI products using a successful brand such as S&D.

Based on these considerations, Conad chose two strategic locations in Florence aimed at offering a concept store designed to enhance the idea of "quality" in assortment, consumption, and the consumer's shopping experience (Fig. 18.2).

> At the beginning, we found two locations that suit the development of a high-standard, quality supermarket, able to provide a great shopping experience and where there was an area in which customers can 'eat in' the products bought. Thus, we needed to find a brand name capable of characterizing a store with these features. We took a brand well known by our consumers, Sapori & Dintorni, traditionally associated with product quality, and we made a store brand out of it. The reputation of the Sapori & Dintorni brand is very high among our consumers; accordingly our flagship S&D stores reflect the high quality standards of our PPL (Store Marketing Manager).

Figure 18.2 Florence's Sapori & Dintorni store.

Positioned in city centers with high footfall, and in the train stations of the main tourist cities such as Florence, Naples, or Rome; the retailer's flagship stores aim to bond the Italian passion for food with culture and art, offering a unique food shopping experience to Conad's customers and tourists. In so doing, Conad wanted to enhance its local roots and involvement and to grow the culture of quality, authenticity, preservation, and respect for traditional food and Italian cuisine, strengthening Conad's policies in supporting local products and suppliers. The identity of the S&D brand has been strongly built on its recognized ability to select quality products and to give cognitive support to consumers by keeping them constantly informed of products and the territories from which these originate. When extending the PPL from a product category to a new store format, the positive effect on consumers of the brand's ability to tell the story of a place by telling the stories of its products is amplified, since the physical attributes of the store (e.g., location, layout, etc.), in contrast to a traditional supermarket or hypermarket format, can be more freely used to create a vivid brand experience (see Figs. 18.3 and 18.4).

To date, Conad has opened 17 S&D stores, with a new store due to open this year. S&D offers a narrower assortment (3500−4000 SKUs) compared to that present in the average Conad supermarket (6000 SKUs). These S&D stores are greatly focused on Italian food excellence, especially fresh food (60%, compared to the average of 50% in other Conad stores), and with a particular importance given to PLs (7%−8% more than those present in a standard Conad supermarket) and local and regional products. Promotion intensity is lower but in-store communication is still the key: the concept must transfer the importance of the origins of food, stressing its authenticity in accordance with the Italian tradition. The concept should communicate an emotional journey of Italian flavors and culinary traditions to novices as well as food lovers (see Table 18.1).

Figure 18.3 The Sapori & Dintorni store in Milan.

Figure 18.4 The Sapori & Dintorni store in Naples.

   Albeit with a consistent look and layout, every store has its own characteristics and architectural peculiarities that reflect the city and the place in where it is located, with the logic being to enhance the distinctiveness that each territory expresses. Some store owners have also started to develop a restaurant offer when space allows: *"The more space you have, the higher the possibility to invite important chefs and to arrange events with dedicated menus [...] Our Sapori & Dintorni store in Naples has become the city reference point for aperitifs [...] it is a "must" now among Naples's upper class people"* (Store Marketing Manager).

Table 18.1 Sapori & Dintorni store concept guidelines

Assortment	• 4500 SKUs: 3000 groceries; 1100 fresh food; 250 frozen food; and 150 nonfood • Focus on food quality, local and regional products, and store experience • Strong presence of Premium and Standard PL products: around 300 SKUs of PPL food products Sapori & Dintorni; 1300 SKUs of PL Conad
Layout	• Warm and traditional atmosphere; display made with wood or cardboard; special end caps and displays; store areas specifically dedicated to: PPL products; local products, wine, and food tasting
In-store communication	• Videos on the counters narrating food and its origins • High presence of the logos of the brand • End caps panels focusing on local foods, food authenticity, food origin, and Italian traditions
Customer service	• High level: no queues; no stock-outs; and clean and neat stores • Expert personnel in fresh food, able to give suggestions on how to cook and eat it • Long opening hours • Home delivery • Food product tasting

18.4   Cremerie Sapori & Dintorni

Subsequently, Conad launched "Cremerie S&D," an ice-cream shop branded with Conad's PPL. The first one opened in Rimini in 2012 on the basis of two considerations:

1. The possibility of utilizing free space in the mall gallery in which one Conad hypermarket was already present;
2. The tradition and reputation of the local town for ice-cream production: *"the ice-cream raw-material producers are all based in the Romagna area; the most important exhibition for ice-cream (SIGEL) takes place in Rimini; Italy is the ice-cream world leader"* (Store Proximity Development and Innovation Manager).

Currently, Conad has 12 Cremerie outlets, most of them located inside a mall gallery outside the hypermarket, in order to create synergy with the store and to deseasonalize ice-cream consumption. The concept—25−50 square meters of retail space offering 12−24 ice-cream flavors—was developed through an external agency with the purpose of evoking an image of tradition, as shown in Fig. 18.5.

Using S&D in the naming of these stores works well because ice-cream features are pretty much consistent with the S&D brand: local origins and Italian tradition (the pistachio flavor is made with Bronte's pistachios, a PDO product typical of the Sicilian village of Bronte, the lemon flavor with Sicilian lemons, etc.). At the beginning, Conad carried out the overall production process internally. The local

Figure 18.5 Cremerie Sapori & Dintorni.

supplier helped a lot in the launch phase as he coached one store employee on making the ice-creams and supported the local cooperative in devising the ice-cream recipes; *"today, this is no longer necessary, but it was necessary at the beginning in order to create a standard for all the other stores"* (Store Proximity Development and Innovation Manager). In fact, due to the high manufacturing costs, space and equipment requirements, and possible problems of quality standardization, the ice-cream bases are now sourced from a local manufacturer and whipped in-store. The concept is performing very well, assuring good profitability.

18.4.1 Implications for organizational strategy

Analysis of the S&D case offers some insightful implications for retailers engaged in extracting more value from their PPL strategy. First, we observe that the S&D brand extension strategy can be represented as an evolutionary process whose driving forces reside mainly in two aspects:

1. The creation of specific brand values that could be recognized by customers as distinctly different from those typically associated with the economy or standard PLs;
2. The retailer's ability to establish strong relationships with consumers by leveraging on the distinctive traits of the PPL.

In the early years after the S&D launch, Conad worked hard to identify the core values of the new PPL brand and to find better fine-tuning with the changing needs of consumers and with their growing interest in food excellence, regional cuisine, and ancestral ties with territories. Once the relationship with consumers had consolidated and reached a critical mass in sales volume and number of

SKUs, the brand extension strategy was progressively implemented over time: from an initial product-based extension to a new and malleable store format (S&D stores), then the creation of a new ice-cream concept (Cremerie S&D), and finally moving toward an institutional high-end brand used for all the events in the Conad world. Thus, the S&D experience can suggest to retail managers that a brand extension strategy cannot only address finding a new avenue for sales growth and higher profitability, but can also be used to strengthen the positioning of the overall identity of the retailer.

Second, implementation of the S&D strategy has also highlighted some specific, highly valuable marketing learning processes concerning consumer behavior and changing attitudes toward a PPL. Knowledge has been acquired about how, in the late 2000s, supermarket consumers behaved in relation to S&D products and other PL items and has been transferred and applied by the retailer to the formal design phase of the S&D stores in 2010. Then, learnings from the first S&D stores have allowed Conad to gain a deeper knowledge of how consumers behave when S&D products are offered in a store environment specifically designed to stimulate more involving and experiential approaches to the brand, and more complex relationships between consumers and brand associations. Recently, starting from what was learned in the S&D stores, and capitalizing on the understanding of which store characteristics more directly explain positive consumer experiences, Conad has identified some new guidelines for designing layouts and assortments, and for the use of materials that the retailer is gradually transferring back to the traditional formats:

> Today, the Sapori & Dintorni concept is used in Conad at 360°. Also in our supermarkets, it is no more simply a product positioned on a shelf; it is rather a concept that drives us in formulating new marketing strategies, in designing in-store events and promotional campaigns, in innovating the visual merchandising. By exploiting brand values strictly associated with the PPL brand, we are spreading marketing innovation in our traditional formats (Store Marketing Manager).

Thus, lessons from the S&D case study suggest to retail managers that PPL formats can be conceived as learning platforms for more intensive and bidirectional cognitive relationships with customers and from which innovative practices for the overall format portfolio can be implemented. For example: the visual merchandising and layouts used in the S&D supermarkets are now adapted in Conad supermarkets and Conad City stores in order to create special areas devoted to the Conad PPL.

In general, the Conad case study can help retailers to understand that the PPL's brand extension strategy cannot be simply formulated and implemented in a unidirectional mode: from core business (e.g., supermarkets) to new businesses (new formats and new concepts) and to new businesses again, in line with the traditional value-capturing scheme of a diversification process. PPL's brand extension is a more complex strategy in which learning and innovation are also relevant outcomes for retailers, and are not secondary to economic and financial results.

18.5   Implications for consumers

In the development of the PPL S&D, Conad showed a great ability to interpret the search for tradition to modern consumers, highlighting the reciprocity between the growth of the retailer and the growth of the PPL brand itself. In fact, it was able, on the one hand, to enhance its company's brand values in order to transmit the image of a retailer strongly connected with its territory. On the other hand, it was successful in interpreting the spread of new consumption trends by offering to consumers traditional and local PDO and PGI products. Acquiring the right expertise needed to offer a sufficiently extensive PPL assortment, Conad developed a concept store where the company's values, its constant connection with the territory, and the search for quality products were prioritized. Thus, the S&D concept stores and, subsequently, the Cremerie ice-cream shops have been able to offer to Conad's customers a unique shopping experience. In fact, as usually happens in NB flagship stores, the consumer experience with a brand is amplified in a dedicated store (Jones, Comfort, Clarke-Hill, & Hillier, 2010).

Another important strength of S&D is in its marketing strategy. As stated by a Conad's manager:

> we develop promotional and communicational campaigns, with specific flyers aimed on the one hand to promote S&D products, and on the other hand to let consumers know the origin of products and the history and traditions connected with those territories and their products. Moreover, through visual merchandising strategies performed in-store, we emphasize not only the brand and the product, but also the values connected with the brand (Store Marketing Manager).

Once again, Conad's strategies are designed to strengthen its bonds with its members, with the territory in which its stores are present, with the territory from which its top products originate, and consequently with its customers.

Finally, the case of the S&D flagship stores confirms that the great experience that consumers feel during their grocery shopping reinforces the strength of the connection between consumers and the retailer, having a robust and durable effect on brand attachment. In fact, the main scope of this new format store is to enhance store image and brand attachment by increasing the "in-store brand experiences by appealing to consumers' emotions, senses, behavior and cognition" (Dolbec & Chebat, 2013, p. 460). As confirmed by both the rise of Conad's national market share and by the average PPL market share, the brand extension strategy has consolidated consumers' attachment to the S&D brand.

References

Baltas, G., Doyle, P., & Dyson, P. (1997). A model of consumer choice for national vs store brands. *Journal of the Operational Research Society.*, *48*(10), 988−995.

Conad. (2017). *Annual report 2016.* ⟨http://www.conad.it/conad/home/global/chi-siamo/info-e-contatti-per-i-giornalisti/editoriale-annual-report-2016.html⟩.

De Wulf, K., Odekerken-Schroëder, G., Goedertier, F., & Van Ossel, G. (2005). Consumer perceptions of store brands versus national brands. *Journal of Consumer Marketing, 22* (4), 223–232.

Dolbec, P. Y., & Chebat, J. C. (2013). The impact of a flagship vs. a brand store on brand attitude, brand attachment and brand equity. *Journal of Retailing, 89*(4), 460–466.

European Commission. (2017). *EU quality logos.* ⟨https://ec.europa.eu/agriculture/quality/schemes_it⟩ Accessed 14 June 2017.

Geyskens, I., Gielens, K., & Gijsbrechts, E. (2010). Proliferating private-label portfolios: How introducing economy and premium private labels influences brand choice. *Journal of Marketing Research, 47*(5), 791–807.

Huang, Y., & Huddleston, P. (2009). Retailer premium own-brands: Creating customer loyalty through own-brand products advantage. *International Journal of Retail & Distribution Management, 37*(11), 975–992.

IRI. (2016). *Private label in Western economies.* Special Report, June.

Jones, P., Comfort, D., Clarke-Hill, C., & Hillier, D. (2010). Retail experience stores: Experiencing the brand at first hand. *Marketing Intelligence & Planning, 28*(3), 241–248.

Lamey, L., Deleersnyder, B., Steenkamp, J.-B. E. M., & Dekimpe, M. J. (2012). The effect of business-cycle fluctuations on private-label share: What has marketing conduct got to do with it? *Journal of Marketing, 76*(January), 1–19.

Lincoln, K., & Thomassen, L. (2008). *Private label: Turning the retail threat into your biggest opportunity.* London: Kogan Page.

Martinelli, E., Belli, A., & Marchi, G. (2015). The role of customer loyalty as a brand extension purchase predictor. *The International Review of Retail, Distribution and Consumer Research, 25*(2), 105–119.

MIPAAF. (2016). *Elenco dei prodotti agroalimentari tradizionali [List of traditional agri-food products].* ⟨https://www.politicheagricole.it/flex/cm/pages/ServeBLOB.php/L/IT/IDPagina/10241⟩.

MIPAAF. (2017). *Elenco dei prodotti DOP, IGP e STG [List of PDO, PGI and TSG products].* ⟨https://www.politicheagricole.it/flex/cm/pages/ServeBLOB.php/L/IT/IDPagina/2090⟩.

Netti, E. (2016). *Private label, è la corsa al premium [Private label: it's a premium run]. Il Sole 24 Ore,* Available at ⟨http://www.ilsole24ore.com/art/impresa-e-territori/2016-12-09/private-label-e-corsa-premium-110740.shtml?uuid = ADL8peAC⟩.

PLMA. (2016). *International private label yearbook.* ⟨http://www.plmainternational.com/industry-news/private-label-today⟩.

Sayman, S., & Raju, J. S. (2004). Investigating the cross-category effects of store brands. *Review of Industrial Organization, 24*(2), 129–141.

Sethuraman, R., & Cole, C. (1999). Factors influencing the price premiums that consumers pay for national brands over store brands. *Journal of Product & Brand Management, 8* (4), 340–351.

Steenkamp, J. B. E. M., Van Heerde, H., & Geyskens, I. (2010). What makes consumers willing to pay a price premium for national brands over private labels? *Journal of Marketing Research, 47*(6), 1011–1024.

Ter Braak, A., Geyskens, I., & Dekimpe, M. G. (2014). Why premium private label presence varies by category. *Journal of Retailing, 90*(2), 125–140.

Index

Printed in the United States
By Bookmasters